Triangles

Right Triangles

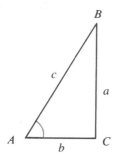

$$\sin \theta = \frac{a}{c} \qquad \cot \theta = \frac{b}{a}$$

$$\cos \theta = \frac{b}{c} \qquad \sec \theta = \frac{c}{b}$$

$$\tan \theta = \frac{a}{b} \qquad \csc \theta = \frac{c}{a}$$

Oblique Triangles

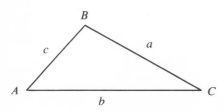

Law of Sines

$$\frac{\sin A}{a} = \frac{\sin B}{b} = \frac{\sin C}{c}$$

Law of Cosines

$$a^2 = b^2 + c^2 - 2bc \cos A$$
$$b^2 = a^2 + c^2 - 2ac \cos B$$
$$c^2 = a^2 + b^2 - 2ab \cos C$$

Plane Trigonometry Fourth Edition

Bernard J. Rice   Jerry D. Strange

University of Dayton

Prindle, Weber & Schmidt   Boston

PWS PUBLISHERS

Prindle, Weber & Schmidt • ♣ • Duxbury Press • ♠ • PWS Engineering • ◭ • Breton Publishers • ⬡
Statler Office Building • 20 Park Plaza • Boston, Massachusetts 02116

PWS Publishers is a division of Wadsworth, Inc.

Library of Congress Cataloging-in-Publication Data

Rice, Bernard J.
    Plane trigonometry.

    Includes index.
    1. Trigonometry, Plane.   I. Strange, Jerry D.
II. Title
QA533.R5 1986   516.2'4   85–12154

ISBN 0-87150-913-X

Printed in the United States of America

86 87 88 89 90 — 10 9 8 7 6 5 4 3 2 1

Editor: Dave Pallai
Production Coordinator and Designer: S. London
Production: Lifland Bookmakers
Cover Photo: Evans & Sutherland
Typesetting: Composition House Limited
Cover Printing: New England Book Components
Printing and Binding: R.R. Donnelley & Sons Company

Preface

This fourth edition of *Plane Trigonometry* retains the essential features that made the first three editions successful. They are (1) an early emphasis on the trigonometric definitions for acute angles imposed on the coordinate plane with the applications to the right triangle as a special focus, (2) an abundance of examples and exercises, and (3) a large variety of applications.

To simplify the early discussion of angles and the evaluation of the trigonometric ratios, we have delayed the introduction of radian measure until needed for arc length measurements, angular velocity and transition to the trigonometry of real numbers. Other changes to this edition include:

- an expanded discussion of vectors in Chapter 3 and a new section on sinusoidal modeling in Chapter 5.
- the use of the calculator as the primary tool for evaluating the trigonometric functions. (Tables and related topics such as interpolation are included as optional material.) We believe the change from the use of tables to the use of the calculator is both efficient and realistic since students are now using calculators at all levels of mathematics education.
- the addition of new applications to almost every exercise set. These range in emphasis from the natural sciences, to aerospace, to psychology.
- the addition of a geometry appendix. This appendix provides a handy reference for the elementary definitions and formulas of plane geometry.
- the inclusion of *WARNING* and *COMMENT* labels to highlight common errors or some pertinent explanation.

As with most books, there are more topics than even the most ambitious instructor could comfortably cover in a one-semester course. Not all of the applications in Chapter 3 need be covered and certain formulas (such as the formula for the Area) may be omitted without loss of continuity. Chapter 9 on complex numbers and polar coordinates is dependent on earlier chapters but is not used elsewhere in the book. Chapter 10 on logarithms may be covered at any time.

A special thanks in this edition goes to Professor Lois Miller of El Camino College for her suggestions for improving the text. The section in Chapter 5 on sinusoidal modeling was significantly influenced by her suggestions on including more and different applications.

This edition of *Plane Trigonometry* has benefited from critical reviews and comments from the following: Professor Arthur P. Dull, Diablo Valley College; Professor Ralph Esparza, Richland College; Professor Allen E. Hansen, Riverside City College; Professor Ferdinand Haring, North Dakota State University; Professor Don. L. Merrill, Utah Technical College; Professor John Spellman, Southwest Texas State University; Professor Donna M. Szott, Community College of Allegheny County–South Campus. We wish to thank all of these people for their valuable comments.

It is also a pleasure to acknowledge the fine cooperation of the staff of Prindle, Weber & Schmidt—particularly our editor Dave Pallai and production editor Susan London. Finally, we want to express our thanks to the production staff of Lifland et al.. Bookmakers who handled much of the work in actually producing the text.

Bernard J. Rice
Jerry D. Strange

Contents

1

Some Fundamental Concepts, 1

1.1. The Rectangular Coordinate System, 2
1.2. Functions, 6
1.3. Approximate Numbers, 11
1.4. Angles, 15
1.5. Some Facts About Triangles, 22

2

The Trigonometric Functions, 31

2.1. Definitions of the Trigonometric Functions, 31
2.2. Fundamental Relations, 37
2.3. The Values of the Trigonometric Functions for Special Angles, 42
2.4. Finding Values of Trigonometric Functions with a Calculator, 49
2.5. Trigonometric Tables and Interpolations (Optional), 52

3

The Solution of Triangles, 61

3.1. Solving Right Triangles, 61
3.2. Vectors, 72
3.3. Applications of Vectors, 78
3.4. Oblique Triangles: The Law of Cosines, 87
3.5. The Law of Sines, 93
3.6. The Ambiguous Case, 100
3.7. Analysis of the General Triangle, 104
3.8. Area Formulas, 105

4

Radian Measure, 111

4.1. The Radian, 111
4.2. Arc Length and Area of a Sector, 117
4.3. Angular Velocity, 121

5

Analytic Trigonometry, 129

5.1. Trigonometric Functions of Real Numbers, 129
5.2. Analytic Properties of the Sine and Cosine Functions, 135
5.3. Expansion and Contraction of Sine and Cosine Graphs, 144
5.4. Vertical and Horizontal Translation, 149
5.5. Sinusoidal Modeling, 154
5.6. Addition of Ordinates, 164
5.7. Graphs of the Tangent and Cotangent Functions, 166
5.8. Graphs of the Secant and Cosecant Functions, 168
5.9. A Fundamental Inequality, 170

6

Identities, Equations, and Inequalities, 175

6.1. Fundamental Trigonometric Relations, 175
6.2. Trigonometric Identities, 180
6.3. Trigonometric Equations, 189
6.4. Parametric Equations, 194
6.5. Graphical Solutions of Trigonometric Equations, 198
6.6. Trigonometric Inequalities, 198

7

Composite Angle Identities, 203

7.1. The Cosine of the Difference or Sum of Two Angles, 203
7.2. Other Addition Formulas, 209
7.3. Double- and Half-Angle Formulas, **214**
7.4. Sum and Product Formulas, 221

8

The Inverse Trigonometric Functions, 229

8.1. Relations, Functions, and Inverses, 229
8.2. The Inverse Trigonometric Functions, 235
8.3. Graphs of the Inverse Trigonometric Functions, 243

9

Complex Numbers and Polar Equations, 249

9.1. Complex Numbers, 249
9.2. Polar Representation of Complex Numbers, 254
9.3. DeMoivre's Theorem, 259
9.4. Polar Equations and Their Graphs, 262

10

Logarithms, 269

10.1. Definition of the Logarithm, 269
10.2. Basic Properties of the Logarithm, 271
10.3. Exponential and Logarithmic Equations, 274

Appendix A

Review of Elementary Geometry, 281

Angles, 282
Intersection of Lines, 283
Polygons, 283
Triangles, 285
Quadrilaterals, 288
Circles, 288
Formulas from Geometry, 289

Appendix B

Tables, 291

Table A   Values of the Trigonometric Functions for Degrees, 292
Table B   Values of Trigonometric Functions—Decimal Subdivisions , 300
Table C   Values of the Trigonometric Functions for Radians and
Real Numbers, 312

Answers to Odd-Numbered Exercises, 317

Index, 353

1

Some Fundamental Concepts

Historical Background

Trigonometry is one of the oldest branches of mathematics. An ancient scroll called the Ahmes Papyrus, written about 1550 B.C., contains problems that are solved by using similar triangles, the heart of the trigonometric idea. There is historical verification that, in about 1100 B.C., the Chinese made measurements of distance and height using what is essentially right-triangle trigonometry. The subject eventually became intertwined with the study of astronomy. In fact, the Greek astronomer Hipparchus (180–125 B.C.) is credited with compiling the first trigonometric tables and thus has earned the right to be known as "the father of trigonometry." The trigonometry of Hipparchus and the other astronomers was strictly a tool of measurement, and it is, therefore, difficult to classify the early uses of the subject as either mathematics or astronomy.

In the fifteenth century, trigonometry was developed as a discipline within mathematics by Johann Muller (1436–1476). This development created an interest in trigonometry throughout Europe and thus placed Europe in a position of prominence with respect to astronomy and trigonometry.

In the eighteenth century, trigonometry was systematically developed in a completely different direction, highlighted by the publication in 1748 of the now-famous "Introduction to Infinite Analysis" by Leonhard Euler (1707–1783). From this new viewpoint, trigonometry did not necessarily have to be considered in relation to a right triangle. Rather, the analytic or functional properties became paramount. As this wider outlook of the subject evolved, many new applications arose, especially for describing physical phenomena that are "periodic."

To read this book profitably, you should have some ability with elementary algebra, particularly manipulative skills. Some of the specific background knowledge you will need is presented in this chapter.

1.1 The Rectangular Coordinate System

We often wish to make an association between points on a line (or in a plane) and numbers, a process called coordinatization. The number (or numbers) assigned to a point is called the **coordinate** (or coordinates) of the point.

To associate points with real numbers, choose any straight line, and then choose any point on the line to be the starting point, or origin. Take any unit distance and measure that distance to the right of the origin. Then the number 0 is associated with the origin, the number 1 with the point a unit distance to the right of the origin, the number 2 with the point two units to the right of the origin, etc. In this way, the so-called **integral points** are determined. Points in between are coordinated by noninteger real numbers. The line, illustrated in Figure 1.1, is called a **real number line**.

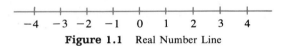

Figure 1.1   Real Number Line

For purposes of elementary trigonometry, the most important type of coordinatization is the association of each point in the plane with a pair of numbers. In this case, we choose two mutually perpendicular intersecting lines, as shown in Figure 1.2. Normally, the horizontal line is called the **x-axis**, the vertical line is called the **y-axis**, and their intersection is called the **origin**. When considered together, the two axes form a rectangular coordinate system.* As you can see, the coordinate axes divide the plane into four zones, or **quadrants**. The upper right quadrant is called *the first quadrant*, and the others are numbered consecutively in a counterclockwise direction from this one, as in Figure 1.2. The coordinate axes are not considered to be in any quadrant.

To locate points in the plane, use the origin as a reference point and lay off a suitable scale on each of the coordinate axes. The displacement of a point in the plane to the right or left of the y-axis is called the **x-coordinate**, or **abscissa**, of the point, and is denoted by x. Values of x measured to the right of the y-axis are *positive* and to the left are *negative*. The displacement of a point in the plane above or below the x-axis is called the **y-coordinate**, or **ordinate**, of the point, and is denoted by y. Values of y above the x-axis are *positive* and below the x-axis are *negative*. Together, the abscissa and the ordinate of a point are called the **coordinates** of the point. The coordinates of a point are conventionally written in parentheses, with the abscissa written first and separated from the ordinate by a comma—that is, (x, y).

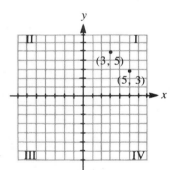

Figure 1.2   Cartesian Coordinate System

* This system is also called the Cartesian coordinate system in honor of René Descartes, who invented it.

We see that a point (x, y) lies

- in quadrant I if both coordinates are positive,
- in quadrant II if the x-coordinate is negative and the y-coordinate is positive,
- in quadrant III if both coordinates are negative,
- in quadrant IV if the x-coordinate is positive and the y-coordinate is negative.

Since the first number represents the horizontal displacement and the second the vertical displacement, order is significant. For example, the ordered pair $(3, 5)$ represents a point that is displaced 3 units to the right of the origin and 5 units up, whereas the ordered pair $(5, 3)$ represents a point that is 5 units to the right and 3 units up. The association of points in the plane with ordered pairs of real numbers in an obvious extension of the concept of the real line.

To be precise, we should always distinguish between the point and the ordered pair; however, it is common practice to blur the distinction and say "the point (x, y)" instead of "the point whose coordinates are (x, y)."

Each point in the plane can be described by a unique ordered pair of numbers (x, y), and each ordered pair of numbers (x, y) can be represented by a unique point in the plane called the **graph** of the ordered pair.

Example 1. Locate the points (a) $P(-1, 2)$, (b) $Q(2, 3)$, (c) $R(-3, -4)$, (d) $S(3, -5)$, and (e) $T(\pi, 0)$ in the plane.

Solution
(a) $P(-1, 2)$ is in quadrant II because the x-coordinate is negative and the y-coordinate is positive.
(b) $Q(2, 3)$ is in quadrant I because both coordinates are positive.
(c) $R(-3, -4)$ is in quadrant III because both coordinates are negative.
(d) $S(3, -5)$ is in quadrant IV because the x-coordinate is positive and the y-coordinate is negative.
(e) $T(\pi, 0)$ is not in any quadrant, but lies on the positive x-axis.
The points are plotted in Figure 1.3.

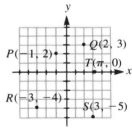

Figure 1.3 Locating Points in the Plane

When an entire set of ordered pairs is plotted, the corresponding set of points in the plane is called the **graph** of the set.

Example 2. Graph the set of points whose abscissas are greater than -1 and whose ordinates are less than or equal to 4.

Solution. This set is described by the two inequalities

$$x > -1$$

$$y \leq 4$$

The shaded region in Figure 1.4 is the graph of the set. The solid line is part of the region, whereas the broken line is not.

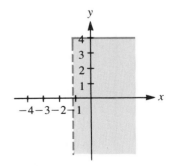

Figure 1.4

4

1 Some Fundamental Concepts

Sometimes we want to find the distance between two points in the plane. Consider two points P_1 and P_2 on the x-axis, as shown in Figure 1.5. The distance between these two points can be found by counting the

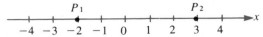

Figure 1.5 Distance between Two Points on the x-axis

units between them or, algebraically, by subtracting their coordinates. To ensure that the distance will be positive, we define it in terms of the absolute value of the difference between the coordinates of P_1 and P_2. Thus

$$d(P_1, P_2) = |P_2 - P_1| \tag{1.1}$$

Computing the distance in Figure 1.5, we have

$$d(-2, 3) = |3 - (-2)| = |5| = 5$$

A similar scheme is followed if the points lie on the y-axis.

Now consider two points $P_1(x_1, y_1)$ and $P_2(x_2, y_2)$ that determine a slanted line segment, as shown in Figure 1.6. Draw a line through P_1 parallel to the x-axis and a line through P_2 parallel to the y-axis. These two lines intersect at the point $M(x_2, y_1)$. Hence, by the Pythagorean theorem,*

$$[d(P_1, P_2)]^2 = [d(P_1, M)]^2 + [d(M, P_2)]^2 \tag{1.2}$$

We see from Figure 1.6 that $d(P_1, M)$ is the horizontal distance between P_1 and P_2. Therefore, the distance $d(P_1, M)$ is given by

$$d(P_1, M) = |x_2 - x_1|$$

Likewise, the vertical distance $d(M, P_2)$ is given by

$$d(M, P_2) = |y_2 - y_1|$$

Recalling that $|A|^2 = A^2$ and denoting $d(P_1, P_2)$ by d, we make these substitutions into equation (1.2):

$$d^2 = (x_2 - x_1)^2 + (y_2 - y_1)^2$$

Taking the square root of both sides yields a formula for the distance d.

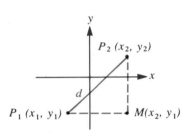

Figure 1.6 Distance between Two Points

---

* The Pythagorean theorem gives the relationship between the lengths of the sides of a right triangle. Specifically, in the triangle shown, $a^2 + b^2 = c^2$.

> **The Distance Formula:** The length of the line segment connecting $P_1(x_1, y_1)$ and $P_2(x_2, y_2)$ is
>
> $$d = \sqrt{(x_2 - x_1)^2 + (y_2 - y_1)^2} \qquad (1.3)$$

Equation (1.3) is used to find the distance between two points in the plane directly from the coordinates of the points. The order in which the two points are labeled is immaterial since

$$(x_2 - x_1)^2 = (x_1 - x_2)^2 \qquad \text{and} \qquad (y_2 - y_1)^2 = (y_1 - y_2)^2$$

Example 3. Find the distance between $(-3, -6)$ and $(5, -2)$. (See Figure 1.7.)

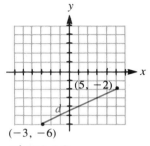

Figure 1.7

Solution. Let $(x_1, y_1) = (-3, -6)$ and $(x_2, y_2) = (5, -2)$. Substituting these values into the distance formula, we have

$$d = \sqrt{(x_2 - x_1)^2 + (y_2 - y_1)^2}$$
$$= \sqrt{[5 - (-3)]^2 + [-2 - (-6)]^2}$$
$$= \sqrt{64 + 16} = \sqrt{80} = 4\sqrt{5} \approx 8.9$$

Notice the inclusion of the numerical sign of each number in the substitution of values into the distance formula.

Example 4. Find the distance between $(2, 5)$ and $(2, -1)$.

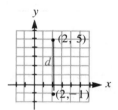

Figure 1.8

Solution. In this case the two given points lie on a vertical line since they have the same abscissa. (See Figure 1.8.) The distance between the two points, therefore, can be found directly.

$$d = |5 - (-1)| = |5 + 1| = 6 \text{ units}$$

The distance can also be found from the distance formula [Equation (1.3)]. Letting $(x_1, y_1) = (2, 5)$ and $(x_2, y_2) = (2, -1)$, we have

$$d = \sqrt{(2 - 2)^2 + (-1 - 5)^2} = \sqrt{36} = 6 \text{ units}$$

Example 5. Find the distance from the origin to any point (x, y).

Solution. From the distance formula, the length of r is given by

$$r = \sqrt{(x - 0)^2 + (y - 0)^2} = \sqrt{x^2 + y^2}$$

where r is always positive. (See Figure 1.9.)

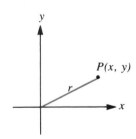

Figure 1.9

Exercises for Section 1.1

In Exercises 1–6, plot the ordered pairs.

1. $(3, 2)$
2. $(4, 6)$
3. $(-2, \frac{1}{2})$
4. $(-6, -5)$
5. $(\frac{1}{4}, -\frac{1}{2})$
6. $(-2.5, 1.7)$

7. In what two quadrants do the points have positive abscissas?

8. In what two quadrants do the points have negative ordinates?

9. In what quadrant are the abscissa and the ordinate both negative?

10. In what quadrant is the ratio y/x negative?

11. What is the ordinate of a point on the x-axis?

In Exercises 12–19, plot the pairs of points and find the distance between the points.

12. $(1, 2)$, $(5, 4)$

13. $(0, 4)$, $(-1, 3)$

14. $(-1, 5)$, $(-1, -6)$

15. $(\frac{1}{2}, \frac{1}{2})$, $(\frac{1}{2}, -\frac{3}{4})$

16. $(-5, 3)$, $(2, -1)$

17. $(0.5, 1.6)$, $(6.2, 7.5)$

18. $(-3, 4)$, $(0, 4)$

19. $(2, -6)$, $(-\sqrt{3}, -3)$

In Exercises 20–27, graph the set of points for which the coordinates satisfy the given condition(s).

20. $y > 0$

21. $x = 0$

22. $y = 2$

23. $x > 0$

24. $x > -1$ and $y > 0$

25. $x = y$

26. $x > 0$ and $y > 0$

27. $x > -1$ and $y < -1$

28. The point $(x, 3)$ is 4 units from $(5, 1)$. Find x.

29. Find the distance between the points $(\sqrt{x}, \sqrt{y})$ and $(-\sqrt{x}, -\sqrt{y})$.

30. Find the distance between the points (x, y) and $(-x, y)$.

31. Find the distance between the points (x, y) and $(x, -y)$.

32. Find the point on the x-axis that is equidistant from $(0, -1)$ and $(3, 2)$.

In Exercises 33 and 34, find the distance from the origin to the given point.

33. $(1, 2)$

34. $(-1, -5)$

1.2 Functions

There are basically two kinds of number symbols used in algebraic and trigonometric discussions: constants and variables. **Constants** have fixed values throughout a discussion, whereas **variables** may take on different values, called the set of permissible values of the variable, or, more technically, the **domain** of the variable. If the domain is not specifically mentioned, we usually allow the variable to take on all real numbers permissible in a discussion. For example, if the expression $1/(x - 3)$ is a part of the discussion, the variable x may not take on the value 3 since division by zero is prohibited. Hence, the domain consists of all real numbers except 3.

Sometimes the values of one variable determine the values of some second variable. For example, the set $\{2, 3, 4\}$ is said to determine the set $\{4, 6, 8\}$ by the rule of doubling each of the elements of the first set. The notion that two variables may be related to each other by some rule of correspondence is used extensively in mathematics and is basic to our understanding of the physical world.

> **Definition 1.1:** If the rule of correspondence between two variables x and y is such that there is exactly one value of y for each value of x, then we say **y is a function of x**.

The x variable is called the **independent variable**, y is called the **dependent variable**, and the defining relation is called a **functional relationship**. The set of values that the independent variable can assume is called the **domain**; the corresponding set of values for the dependent variable is called the **range**. The rule of correspondence along with the domain and range make up the **function**.

Example 1. The equation $y = x^2$ essentially defines a function with x as the independent variable and y as the dependent variable. The function has an understood domain of all real numbers and a range of nonnegative numbers.

Functions are frequently defined by some formula or expression involving x and y. The following remarks should help to clarify the definition of a function as it relates to the use of formulas.

- A formula such as $y = \pm\sqrt{x}$ does not give a functional relationship since two values of y correspond to each value of x. However, each of the formulas $y = \sqrt{x}$ and $y = -\sqrt{x}$ taken separately *does* define a function.

- The definition does not require that the value of the dependent variable change when the independent variable changes, but only that the dependent variable have a unique value corresponding to each value of the independent variable. Thus, $y = 5$ is a function, since y has the unique value 5 for any value of the independent variable. However, $x = 5$ is not a function because many (in fact, infinitely many) values of y correspond to the value $x = 5$.

- The fact that expressions such as $y = \pm\sqrt{x}$ and $x = 5$ do not describe functions does not mean that they are unimportant. Instead, they describe a broader concept called a **relation**, which is any pairing of two sets of numbers.

In some discussions, we want to indicate that y is a function of x without specifying the relationship. The notation commonly used to indicate that such a functional relationship exists between x and y is

$y = f(x)$. This is read "y equals f of x." The letter f is the name of the function; it is not a variable. The letter x represents the domain value of the function and is sometimes called the **argument** of the function.

> **Warning:** The notation $f(x)$ does NOT mean the multiplication of f times x.

Example 2. Find the value of the function $f(x) = x^2 + 3x$ at $x = 2$.

Solution. We denote the value of $f(x)$ at $x = 2$ by $f(2)$. To find $f(2)$, we substitute 2 for x in $x^2 + 3x$; that is,

$$f(2) = (2)^2 + 3(2) = 10$$

> **Comment:** In general, if $f(x)$ is the value of the function f at x, then $f(a)$ is the value of f at $x = a$.

You can think of x as representing a blank and the functional notation as telling you what to put in the blank. The function in Example 2 could be written

$$f(\ \ ) = (\ \ )^2 + 3(\ \ )$$

Then, for example,

$$f(a^2) = (a^2)^2 + 3(a^2) = a^4 + 3a^2$$

Calculators are indispensable in evaluating functions for specified values of the independent variable. With a calculator, it is just as easy to approximate $f(x) = \pi x^3 + \sqrt{x + 2}$ *for $x = 3.105$ as it is for $x = 2$.*

A function f determines a set of ordered pairs

$$\{(x, y)\} \quad \text{where} \quad y = f(x)$$

and, conversely, *some* sets of ordered pairs determine a function. Can you tell which sets of ordered pairs determine a function? The key is that for any given x there must be *only one* value of y. Thus, if the set of ordered pairs has two pairs with the same first element and different second elements, the set does not represent a function. In this case the set represents a relation.

Example 3. The set of ordered pairs $\{(2, 5), (3, -1), (-1, 0), (0, 2)\}$ represents a function. The domain elements are $2, 3, -1, 0$ and the range elements are $5, -1, 0, 2$. In functional notation,

$$f(2) = 5 \qquad f(3) = -1 \qquad f(-1) = 0 \qquad f(0) = 2$$

Example 4. The set of ordered pairs $\{(2, 3), (3, -1), (2, 5), (0, 2)\}$ does not represent a function because the first and third pairs have the same first element. This set represents a relation.

The **graph** of a function is the geometric picture of the equation defining the function. It consists of the set of points corresponding to the set of ordered pairs given by the function. Usually to graph a function we plot a "reasonable" number of the points and connect them with a smooth curve.

Example 5. Sketch the graph of $y = \sqrt{4 - x}$.

Solution. A table is constructed by selecting some reasonable values of x and determining the corresponding values of y. Both the table and graph are shown in Figure 1.10. Notice that the domain is all $x \leq 4$ and the range is all $y \geq 0$.

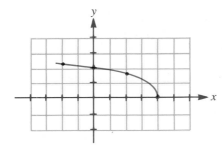

x	y
4	0
2	$\sqrt{2}$
0	2
-2	$\sqrt{6}$

Figure 1.10

Example 6. Sketch the graph of $y = 32x - 16x^2$.

Solution. See Figure 1.11.

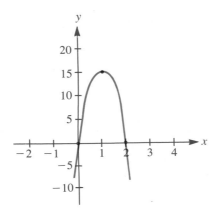

x	y
-1	-48
0	0
1	16
2	0

Figure 1.11

Just as certain sets of ordered pairs determine functions, some sets of points of the Cartesian coordinate system determine functions. Usually the sets of points will be in the form of continuous curves, as in Figure 1.12. Not all of the graphs of Figure 1.12 determine functions; the first two represent functions and the other two do not. Again, as in the case of sets of ordered pairs, the key lies in the fact that for any one value of x there

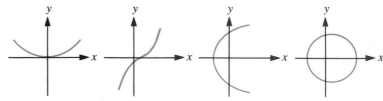

(a) These are graphs of functions. (b) These graphs are not functions.

Figure 1.12

must be *only one* value of y. Geometrically, this means that if a line is drawn parallel to the y-axis, there should be, at most, one point of intersection with the graph.

Exercises for Section 1.2

1. Given $f(x) = x^2 - 5x$, find $f(2)$ and $f(-1)$.

2. Given $g(x) = x^2 - 4$, find $g(2)$ and $g(-4)$.

3. Given $f(t) = (t - 2)(t + 3)$, find $f(-4)$ and $f(0)$.

4. Given $\phi(y) = y(y + 1)$, find $\phi(3)$ and $\phi(-3)$.

5. Given $h(x) = 2(x^2 - 3)$, find $h(5)$ and $h(1)$.

6. Given $g(z) = (1 - z^2)/(1 + z^2)$, find $g(-1)$ and $g(4)$.

In Exercises 7–9, symbolize the expressions using functional notation.

7. The area A of a circle as a function of its radius r.

8. The circumference C of a circle as a function of its diameter d.

9. The perimeter P of a square as a function of its side length s.

In Exercises 10–15, give the domain and range of the functions.

10. $f(x) = x^2$ 11. $f(t) = t - 2$

12. $f(x) = x^3$ 13. $f(x) = \sqrt{x - 25}$

14. $f(x) = \sqrt{x}$ 15. $f(x) = 1/x$

In Exercises 16–24, graph each of the functions.

16. $y = x^3$ 17. $f(x) = -x^2$ 18. $z = t^2 + 4$

19. $i = r - r^2$ 20. $y(x) = \sqrt{x}$ 21. $\phi = w^2/2$

22. $p = z^2 - z - 6$ 23. $v = 10 + 2t$ 24. $y = \sqrt{16 - 4x^2}$

In Exercises 25–30, which of the sets of ordered pairs determine a function?

25. $\{(1, 1), (2, 1)\}$ 26. $\{(1, 1), (1, 2)\}$

27. $\{(2, 0), (3, 0), (-1, 1)\}$

28. $\{(4, 1), (3, 2), (5, 4)\}$

29. $\{(-1, -2), (0, 0), (1, 3), (0, 5)\}$

30. $\{(0, 1), (1, -5), (\sqrt{2}, 3)\}$

In Exercises 31–34, which of the graphs represent a function?

31.

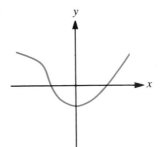

32.

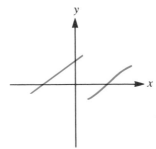

33.

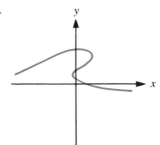

34.

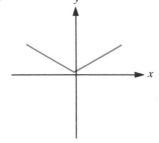

1.3 Approximate Numbers

Significant Digits

If you count each card in a deck of 52 playing cards, you can say there are exactly 52 cards in the deck. However, if you use a meter stick to measure the distance between two points, you realize that a measurement such as 3.5 cm is only an approximation to the precise answer and that the accuracy of the measurement depends on the accuracy of the measuring device being used. Most calculations in the physical sciences involve measurements that are accurate to some specified number of digits. The digits that are known to be accurate are called **significant digits**. Zeros that are required to locate the decimal point are not considered to be significant digits.

Example 1
(a) 5.793 has four significant digits.
(b) 20.781 has five significant digits.
(c) 0.000059 has two significant digits.
(d) 0.08300 has four significant digits.

(e) $\sqrt{2}$, π, and 3 are examples of *exact numbers*. When used in computations, an exact number is considered to have as many significant digits as any other number in the same computation.

Example 2. How many significant digits does 9480 have?

Solution. A number like 9480 is difficult to categorize unless we know something about the number. It could represent the exact number of cards in a computer program, or it could represent a measurement accurate to either three or four significant digits. Sometimes we use scientific notation for numbers like this to avoid confusion. Thus,

$$9.48 \times 10^3$$

is used to indicate three significant digits, and

$$9.480 \times 10^3$$

indicates four significant digits.

Comment: Note that the last significant digit of a measured number is not completely accurate. For instance, if you measure a length of wire to be 2.56 cm, you realize that the length could be anywhere between 2.555 and 2.565 cm since it has been obtained by estimation.

The process of reducing a given number to a specified number of significant digits is called **rounding off**. There are several popular schemes for rounding off numbers. The following method is used in many calculators and computers.

(1) If the last digit is less than 5, drop the digit and use the remaining digits.

Example: 8.134 rounds off as 8.13 to three significant digits.

(2) If the last digit is 5 or greater, drop the digit and increase the last remaining digit by 1.

Example: 0.0225 rounds off as 0.023 to two significant digits.

Another method used to reduce the digits in an approximate number is **truncation**. *In this scheme, the unwanted digits are simply dropped or truncated. Thus, the numbers 23.157 and 23.154 both truncate to the four-digit number 23.15. Likewise, under the truncation scheme, $\frac{2}{3}$ to seven decimal places is carried as 0.6666666, not 0.6666667. Check your calculator to see if it rounds off or truncates.*

Example 3. Round off each of the following numbers to three significant digits: (a) 18.89, (b) 0.0003725, (c) 99430, and (d) 4.996.

Solution

(a) 18.89 is rounded off to 18.9.
(b) 0.0003725 is rounded off to 0.000373.
(c) 99430 is rounded off to 99400.
(d) 4.996 is rounded off to 5.00.

Precision

The two numbers 28,500 and 0.285 are both accurate to three significant digits. However, 0.285 is a more precise measurement than 28,500. The decimal position of the last significant digit of a number determines its precision. Both accuracy and precision are important when computations are made with approximate numbers.

Example 4

(a) The measurements 395 cm and 0.0712 cm are both accurate to three significant digits, but 395 cm is precise to the nearest 1 cm and 0.0712 cm is precise to the nearest 0.0001 cm.
(b) 497.3 is more accurate than 0.025 but less precise.

Operations with Approximate Numbers

There is a great temptation, especially when you are using a calculator, to write the answer to arithmetic calculations to as many digits as the calculator will display. In doing so, you make the answer seem more accurate or precise than it really is. For instance, writing

$$8.4 \times 12.137 = 101.9508$$

implies that the product is accurate to seven significant digits, when the numbers being multiplied are only accurate to two and five significant digits, respectively. To avoid this problem, we adopt the following convention when performing arithmetic operations on approximate numbers.

- When adding or subtracting approximate numbers, express the result with the precision of the least precise number.

  Example: $0.74 + 0.0515 - 0.3329 = 0.4586$ is rounded off to 0.46.

- When multiplying or dividing approximate numbers, express the result with the accuracy of the least accurate number.

  Example: $1.93(13.77) = 26.5761$ is rounded off to 26.6.

■ When finding the root of an approximate number, express the root with the accuracy of the number.

Example: $\sqrt{29.14} = 5.398$

Example 5
(a) The answer to $R = 1.90(63.21) + 4.9072$ should have three significant digits. Thus, $R = 125$.
(b) The answer to $Q = 3.005\sqrt{2}$ should have four significant digits. Thus, $Q = 4.250$.
(c) The answer to $y = 3 - \sqrt{29}$ where 3 is an exact number can be written with as many significant digits as desired since there are no approximate numbers being used.

Example 6. The length of a rectangle is measured with a meter stick to be 95.7 cm, and the width is measured with a vernier caliper to be 8.433 cm. What is the area of the rectangle?

Solution. The area is given by

$$A = 8.433 \times 95.7 = 807.0381 \text{ cm}^2$$

Since the length has only three significant digits, the area must be rounded off to three significant digits. Therefore, $A = 807 \text{ cm}^2$ is the proper answer to the question.

> **Comment 1:** In the previous example, we should write $A \approx 807$ to indicate that the area is approximately 807 cm². However, we shall adopt the convention of using an equal sign for quantities expressed to a specified number of significant digits.

> **Comment 2:** The answers appearing in the answer section have been rounded off to the appropriate number of significant digits.

Exercises for Section 1.3

In Exercises 1–10, indicate the number of significant digits in the given numbers.

1. 3.37	2. 2.002	3. 812.0	4. 6161
5. 0.03	6. 0.000215	7. 0.40	8. 57.001
9. 500.0	10. 0.06180		

In Exercises 11–20, round off the given number to three significant digits.

11. 9818	12. 72267	13. 54.745	14. 1.002

15. 0.06583   16. 2435   17. 39.75   18. 0.4896

19. 0.9997   20. 900,498

In Exercises 21–32, perform the indicated operations and round off the answer to the appropriate number of significant digits.

21. 23.45(0.91669)

22. 4.7(54.75)

23. 0.5782 + 1.34 + 0.0057

24. 50.68 + 9.666 − 24.059

25. 2.9(3.57 + 10.28) + 25.0

26. 0.20 + 3.86(0.127 − 0.097)

27. $\sqrt{2.4^2 + 1.93^2}$

28. 2.176$\sqrt{3}$

29. $\dfrac{25(0.9297)}{0.0102}$

30. $\dfrac{5.0887(2.20)}{8813}$

31. $\dfrac{0.9917(771.33)}{\sqrt{30.04}}$

32. $\sqrt{2.14^2 + 3.9^2}$

1.4 Angles

When two line segments meet, they form an **angle**. We ordinarily think of an angle as formed by two rays OA and OB that extend from a common point O, called the **vertex**. The rays are called the **sides** of the angle. (See Figure 1.13.)

We refer to an angle by mentioning a point on each of its sides and the vertex. Thus the angle in Figure 1.13 is called "the angle AOB" and is written $\angle AOB$. If there is only one angle under discussion whose vertex is at O, we sometimes simply say, "the angle at O," or, more simply, "angle O." It is also customary to use Greek letters to designate angles. For example, $\angle AOB$ might also be called the angle θ (read "theta").

Sometimes it is useful to think of an angle as being "formed" by rotating one of the sides about its vertex while keeping the other side fixed, as shown in Figure 1.14. If we think of OA as being fixed and OB as

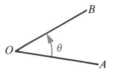

Figure 1.13 An Angle

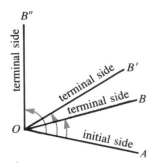

Figure 1.14 Generated Angles

rotating about the vertex, OA is called the **initial side** and OB the **terminal side** of the generated angle. Other terminal sides such as OB' and OB'' result in different angles. The *size* of the angle depends on the amount of rotation of the terminal side. Thus, $\angle AOB$ is smaller than $\angle AOB'$ which, in turn, is smaller than $\angle AOB''$. Two angles are equal in size if they are formed by the same amount of rotation of the terminal side.

The Degree

The most common unit of angular measure is the **degree**. We will define a degree as $\frac{1}{360}$ of the measure of an angle formed by one complete revolution of the terminal side about its vertex. The measure of an angle formed by the complete revolution is then 360 degrees, written 360°. One half of this angle, 180°, is called a **straight angle**, and one fourth of it, 90°, is called a **right angle**. (See Figure 1.15.)

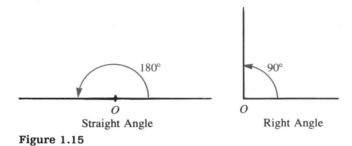

Straight Angle   Right Angle

Figure 1.15

An angle is **acute** if it is smaller in size than a right angle and is **obtuse** if it is larger than a right angle but smaller than a straight angle. (See Figure 1.16.)

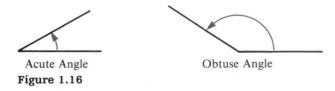

Acute Angle   Obtuse Angle

Figure 1.16

Figure 1.17 shows two angles that are larger than a straight angle.

Figure 1.17   Angles Larger Than a Straight Angle

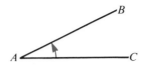

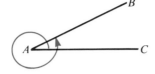

Figure 1.18   Coterminal Angles

Angles with the same initial and terminal sides are said to be **coterminal**. The two angles shown in Figure 1.18 are coterminal, but they are obviously not equal. There are many important considerations, both practical and theoretical, that make it necessary to distinguish between coterminal angles formed in different ways.

Sometimes we make note of the direction in which the terminal side was rotated to form the angle: a counterclockwise rotation of the terminal side is called **positive**, and a clockwise rotation is called **negative**, as shown in Figure 1.19. The measure of an angle has no numerical limit since a terminal side may be rotated as much as desired.

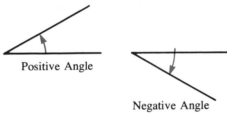

Figure 1.19

Example 1.   Draw the following angles: (a) θ (theta) of measurement $42°$, (b) ϕ (phi) of $-450°$, (c) β (beta) of $1470°$, and (d) α (alpha) of $-675°$. Indicate an acute coterminal angle for each generated angle.

Solution.   See Figure 1.20.

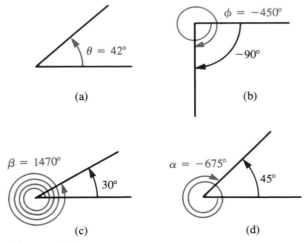

Figure 1.20

Example 2. Determine an angle between 0° and 360° that is coterminal with an angle of 1650°.

Solution. To find a coterminal angle between 0° and 360° of an angle greater than 360°, we subtract the number of times 360° is contained in the given angle. In this example we note that $4(360) = 1440$ and $5(360) = 1800$, which means we should subtract 1440° from 1650°. Thus,

$$1650° - 1440° = 210°$$

is the desired coterminal angle.

Example 3. A fan blade rotates 150 times per minute. Through how many degrees does a point on the tip of one of the blades move in 7 sec?

Solution. Since the blade rotates 150 times per minute, it will rotate

$$\frac{150}{60} = 2.5 \text{ times in one second}$$

Therefore, in 7 sec the blade will rotate

$$2.5 \times 7 = 17.5 \text{ times}$$

Since each rotation is 360°, the angle generated by the fan blade is

$$17.5 \times 360° = 6300°$$

We note that if the sum of the measures of two angles is 90°, the two angles are **complementary**. If the sum of the measures is 180°, they are **supplementary**. (See Figure 1.21.)

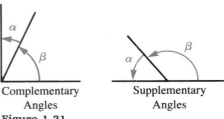

Complementary
Angles

Supplementary
Angles

Figure 1.21

The basic angular unit of the degree is subdivided into 60 parts, each of which is called a **minute** and is denoted by the symbol ′. The minute is further subdivided into 60 parts, each of which is called a **second** and is denoted by the symbol ″. As the next example shows, arithmetic calculations are sometimes a bit more cumbersome with these subdivisions than with the decimal system.

Example 4. Find the sum and difference of the two angles whose measurements are 45°41′09″ and 32°52′12″.

Solution. The sum of the two angles is found by adding the corresponding units; that is, degrees to degrees, minutes to minutes, and seconds to seconds. Thus,

$$45°41'09'' + 32°52'12'' = (45 + 32)°(41 + 52)'(09 + 12)'' = 77°93'21''$$

Here we see that $93' = 1°33'$, so we write our answer as $78°33'21''$.

To find the difference in the two angles, we must write $45°41'09''$ in the following form:

$$45°41'09'' = 45°40'69'' = 44°100'69''$$

Thus,

$$45°41'09'' - 32°52'12'' = (44 - 32)°(100 - 52)'(69 - 12)''$$
$$= 12°48'57''$$

The use of minutes and seconds for angular subdivisions (and, indeed, the degree measurement itself) is based on an ancient Babylonian numeral system. Although the decimalization of angular measurement has been accelerated with the widespread use of the calculator, both systems will continue to be used for the foreseeable future. Therefore, you should know how to make conversions between the two systems.

To convert an angle measured in degrees, minutes, and seconds to a decimal representation in degrees, simply divide the minutes by 60 (since $60' = 1°$) and the seconds by 3600 (since $3600'' = 1°$) and then add the results. A comment on conversion accuracy is appropriate at this point. So that the converted decimal does not suggest more accuracy than the given angle, we adopt the following convention:

■ An angle measured to the nearest minute should contain two decimal places in the converted form.

■ An angle measured to the nearest second should contain four decimal places in the converted decimal form.

Example 5. Convert $15°35'$ to decimal degrees.

Solution.
$$15°35' = 15° + \left(\frac{35}{60}\right)°$$
$$= 15° + (0.583\ldots)°$$
$$= 15.58°$$

Example 6. Convert $37°47'23''$ to decimal degrees.

Solution. $37°47'23'' = 37° + \left(\frac{47}{60}\right)° + \left(\frac{23}{3600}\right)°$
$$= 37° + (0.78333\ldots)° + (0.006388\ldots)°$$
$$= 37.7897°$$

To convert decimal notation to degrees, minutes, and seconds, multiply the fractional part of a degree by 60 to obtain minutes and multiply the fractional part of this result by 60 to obtain seconds.

Example 7. Convert 67.8235° to degrees, minutes, and seconds.

Solution.
$$67.8235° = 67° + (0.8235 × 60)'$$
$$= 67° + (49.41)'$$
$$= 67°49' + (0.41 × 60)''$$
$$= 67°49'25'' \qquad \text{To the nearest second}$$

You may be lucky enough to have a calculator with a built-in conversion button, usually designated by **DMS**. *On such a calculator, enter the angle in the order degrees/minutes/seconds and then press the* **DMS** *key to obtain the angle in decimal degrees. If the angle is in decimal degrees, push* **inv** **DMS** *to convert to degrees/minutes/seconds. Check your manual for specific instructions.*

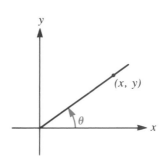

Figure 1.22 An Angle in Standard Position

In trigonometry we often locate angles in the Cartesian plane. An angle is said to be in **standard position** in the plane if its vertex is at the origin and its initial side is along the positive half of the x-axis, as shown in Figure 1.22. The magnitude of an angle in standard position is measured from the positive x-axis to the terminal side.

Example 8. Draw an angle in standard position whose terminal side passes through $(-2, -2)$. What is the measure of this angle in degrees?

Solution. The terminal side of θ obviously bisects the third quadrant, and therefore $\theta = 180° + 45° = 225°$. (See Figure 1.23.) You should also observe that the measure of this angle is not unique, since there are many angles with the indicated side as the terminal side. Each of these angles differs by 360°. Thus, $\theta = 225° + m \cdot 360°$ where m is any integer.

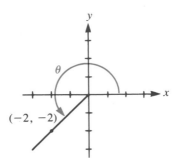

Figure 1.23

Any angle in standard position is coterminal with one whose measure is between 0 and 360°. For example, $-45°$ is coterminal with an angle of 315°.

An angle is called a **first-quadrant angle** if the terminal side is in the first quadrant, a **second-quadrant angle** if the terminal side is in the second quadrant, and so on for the other quadrants. Angle θ in Figure 1.23 is a third-quadrant angle. If the terminal side of an angle is on one of the axes, the angle is called a **quadrantal angle**; that is, the angle does not lie in any quadrant.

Measuring Instruments

Figure 1.24 shows a simple form of a **protractor**, the simplest instrument used to measure angles. It is marked in degrees around its rim.

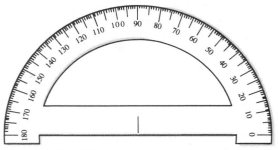

Figure 1.24   A Protractor

A more accurate device used by engineers and surveyors is called a **transit**. A transit measures an angle by locating two different line-of-sight objects. (See Figure 1.25.)

Figure 1.25   A Transit

Exercises for Section 1.4

1. What is the sum of two complementary angles?

2. What is the difference in degree measure of two coterminal angles?

In Exercises 3–12, find the sum $A + B$ and the difference $A - B$ of the two given angles.

3. $A = 45°10'$, $B = 30°5'$

4. $A = 72°12'$, $B = 30°38'$

5. $A = 58°35'40''$, $B = 50°34'20''$

6. $A = 42°40'10''$, $B = 65°50'50''$

7. $A = 60°10'15''$, $B = 70°45'$

8. $A = 138°40'20''$, $B = 23°52'30''$

9. $A = 240°45'40''$, $B = 333°25'14''$

10. $A = 320°50'20''$, $B = -30°55'10''$

11. $A = -40°42'57''$, $B = -80°18'13''$

12. $A = -90°0'49''$, $B = 269°57'1''$

In Exercises 13–20, convert the given angle to degree decimal representation. Express angles to four decimal places.

13. 18°25′36″   **14.** 54°50′16″   **15.** 94°17′08″   **16.** −90°5′48″

17. 283°36′30″   **18.** 480°45′45″   **19.** 183°14′40″   **20.** 71°12′20″

In Exercises 21–28, convert the given angle to degree/minute/second representation.

21. 48.2572°   **22.** −34.5618°   **23.** −235.4500°   **24.** 30.5052°

25. 45.7575°   **26.** 234.5831°   **27.** 15.2575°   **28.** 68.3040°

In Exercises 29–44, draw the angle and name the initial and terminal sides. If the given angle is greater than 360°, indicate an angle between 0° and 360° that is coterminal with the given one.

29. 420°   **30.** −300°   **31.** −317.5°   **32.** 500°

33. −225°   **34.** −270°   **35.** 590°   **36.** 489.1°

37. 720°   **38.** 780°   **39.** 840°   **40.** 765°

41. 1485°   **42.** 2000°   **43.** −1290°   **44.** −1205°

In Exercises 45–50, draw an angle in standard position whose terminal side passes through the given point. Give a degree measure of the angle.

45. (−1, 1)   **46.** (5, 0)   **47.** (0, −3)

48. (4, −4)   **49.** (−4, 0)   **50.** (1000, −1000)

51. A contractor surveying a building site records the angle 79.473° in the log. Convert this angle to degrees, minutes, and seconds.

52. A road that makes an angle of 30°40′ north of east intersects a road that makes an angle of 76°45′ south of east. What is the angle between the two roads?

53. During a lab experiment a student measures an angle as 16°50′. Another member of the group measures the same angle as 16.75°. What is the difference to the nearest one hundredth of a degree between the two measurements?

54. A pine tree grows vertically on a hillside that makes an angle of 25.7° with the horizontal. (See Figure 1.26.) What angle θ does the tree make with the hillside above it?

55. A bicycle wheel rotates 50 times in 1 min. (See Figure 1.27.) Through how many degrees does a point on the tip of the wheel move in 15 sec?

56. A searchlight at an airport makes 5 revolutions in 1 min. (See Figure 1.28.) Through how many degrees does the searchlight rotate in 15 sec?

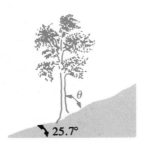

Figure 1.26

Figure 1.27

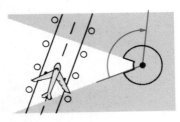

Figure 1.28

1.5 Some Facts About Triangles

Much of Chapters 2 and 3 is devoted to a discussion of triangles and how trigonometry is used to compute unknown parts of a triangle. Therefore, in this section some geometrical facts about triangles are summarized.

A triangle is said to be **equiangular** if the measures of each of its three angles are exactly the same; it is said to be **equilateral** if all three sides have the same length. A theorem of geometry tells us that a triangle is equiangular if and only if it is equilateral. A triangle is said to be **isosceles** if two of its sides are equal; in such a triangle, the angles opposite the two equal sides are also equal. (See Figure 1.29.)

A **right** triangle is one in which one of the angles is a right angle. An **oblique** triangle is one without a right angle. In any triangle, the sum of the measures of the angles is 180°. Thus, in an equilateral triangle, each of the angles measures 60°. In a right triangle, each of the two non-right angles is acute, and the sum of their measures is 90°.

There is a relatively standard method of referencing the sides and the angles of any triangle. For example, in a triangle with vertices A, B, and C, the sides AB and AC are called the sides **adjacent** to the angle at vertex A. The side BC is called the side **opposite** angle A. There are similar statements concerning the sides opposite and adjacent to angle B and those opposite and adjacent to angle C. In the special case of a right triangle, the side opposite the right angle is called the **hypotenuse**.

Referring to the right triangle in Figure 1.30, we see that side AC is called the adjacent side to angle A, side BC is called the side opposite angle A, and side AB is called the hypotenuse.

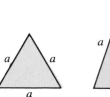

Equilateral
Triangle

Isosceles
Triangle

Figure 1.29

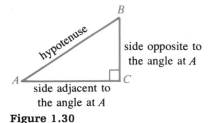

side opposite to
the angle at A

side adjacent to
the angle at A

Figure 1.30

In trigonometry, as in practically every other branch of mathematics, we use the famous theorem of Pythagoras relating the squares of the lengths of the sides of a right triangle.

Theorem 1.1:

Pythagorean Theorem   In a right triangle, the square of the hypotenuse is equal to the sum of the squares of the other two sides.

$$c^2 = a^2 + b^2$$

Through use of the Pythagorean theorem, the third side of a right triangle may be found if any two of the sides are known. For instance, if a and b are given, the hypotenuse c can be computed from the relationship

$$c = \sqrt{a^2 + b^2}$$

Example 1. The line-of-sight distance to the top of an antenna attached to the chimney of a house is known to be 15 m. If the sighting is taken 6.0 m from the house, how high is the top of the antenna above the ground?

Solution. A diagram of the situation is shown in Figure 1.31. As you can see, the unknown measurement is the third side of a right triangle in which two of the sides are known. Hence, from the Pythagorean theorem,

$$h^2 + 6^2 = 15^2$$

so

$$h^2 = 225 - 36$$

and

$$h = \sqrt{189} = 13.75 = 14 \text{ m} \qquad \text{To two significant digits}$$

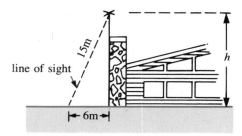

Figure 1.31

Another basic concept of trigonometry is that of similar triangles. Generally, two triangles are **similar** if they have the same shape (not necessarily the same size). Thus, similar triangles have equal angles but not necessarily equal sides. The relationship between the sides of similar triangles has been known for centuries; it is given in the following theorem of Euclid.

Theorem 1.2: If two triangles are similar, their sides are proportional.

Comment: The fact that the two triangles are similar is written

$$\triangle ABC \sim \triangle A'B'C'$$

Note that it is incorrect to denote the similarity by

$$\triangle ABC \sim \triangle B'A'C'$$

for this would incorrectly suggest that angles A and B' and B and A' were congruent.

In terms of the triangles in Figure 1.32, the content of Theorem 1.2 can be written

$$\frac{a}{a'} = \frac{b}{b'} = \frac{c}{c'}$$

Combinations of any two of the three ratios will yield an equation with four parts. If we know three of these parts, we can find the fourth. Similar

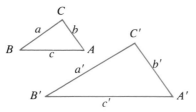

Figure 1.32 Two Similar Triangles

triangles are commonly used to compute distances that are difficult to measure by direct means. The next two examples show how this is done.

Example 2. A spectator is watching high divers dive from some unusually large heights. The spectator, who knows he is 200 yd from the diving site, notes that his pencil of length 6.0 in. is just large enough to cover the diving height when he holds the pencil about 3.00 ft from his eye. How high is the dive? (See Figure 1.33.)

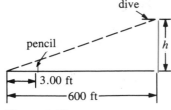

Figure 1.33

Solution. Because of the relatively large distances involved, we may, for the sake of approximation, ignore the fact that the sighting is taken 5 ft or so above ground level. Then, since the small triangle and the larger triangle are obviously similar, we have that

$$\frac{3 \text{ ft}}{600 \text{ ft}} = \frac{\frac{1}{2} \text{ ft}}{h \text{ ft}}$$

and thus

$$h = \frac{1}{6}(600) = 100 \text{ ft}$$

Example 3. A group of physics students was provided with a pencil and a 12-in. ruler and told to determine the height of the steeple on the campus chapel. Describe how they did it.

Solution. First the students measured the length of the shadow of the steeple on the ground, using the ruler, and found it to be 28 ft. Then, holding the ruler vertically with one end on the sidewalk, they marked the end of its shadow. (See Figure 1.34.) Finding the length of the shadow of the ruler to be 5.3 in., they used the pencil to write

$$\frac{h \text{ ft}}{28 \text{ ft}} = \frac{12 \text{ in.}}{5.3 \text{ in.}}$$

$$h = \frac{12}{5.3}(28) = 63 \text{ ft} \qquad \text{To two significant digits}$$

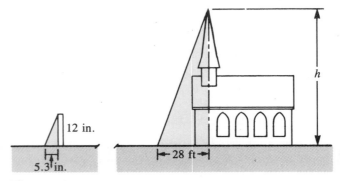

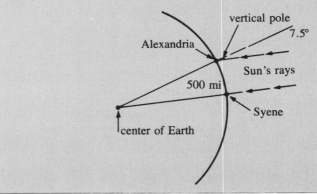

Figure 1.34

Historical Note: The early Greek mathematician Eratosthenes (240 B.C.) closely approximated the circumference of the Earth by simple arc length and shadow measurements. He noticed that at Syene (now the site of the Aswan dam) the rays of the noon sun on the summer solstice shone straight down a deep well. At Alexandria, approximately 500 miles due north, the sun rays were simultaneously measured to be 7.5° from the vertical. Thus he determined that the circumference of the Earth must be 360/7.5 times the distance between Syene and Alexandria, or approximately 24,000 miles. (The actual value is 24,875 miles.)

Exercises for Section 1.5

1. If the angle between the equal sides of an isosceles triangle is 32°, how large is each of the other angles?

2. How many degrees are there in each angle of an equilateral triangle?

3. A baseball diamond is a square 90.0 ft on a side. What is the distance across the diamond from first to third base?

4. What is the line-of-sight distance to an airplane known to be directly over the center of a city 3 mi away, if the plane is flying at 5000 ft?

5. The solar collector of a water heating system is held in a right-angle bracket, as shown in Figure 1.35. Calculate the length of the solar collector.

6. One side of a rectangle is half as long as the diagonal. The diagonal is 4 m long. How long is the other side of the rectangle?

7. What is the length of the diagonal of a cube that is 5.0 cm on an edge?

8. If a room is 21 ft long, 15 ft wide, and 10 ft high, what is the length of the diagonal of (a) the floor, (b) an end wall, and (c) a side wall of the room?

In Exercises 9–15, show that the triangles with sides having the given measures are right triangles.

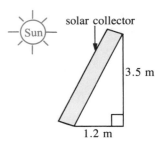

Figure 1.35

9. 6, 8, 10   10. 5, 12, 13   11. 7, 24, 25   12. 9, 40, 41

13. 11, 60, 61   14. 10, 24, 26   15. 28, 21, 35

The two triangles shown in Figure 1.36 are similar. Find the unknown sides for the following given conditions.

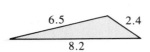

Figure 1.36

16. Given $a = 9.2$, find b and c.   17. Given $b = 3.0$, find a and c.

18. Given $b = 2.5$, find a and c.   19. Given $c = 12$, find a and b.

20. Given $c = 8.7$, find a and b.

21. If the sides of a triangle are 2, 4, and 5 cm, what is the perimeter of a similar triangle in which the longest side is 15 cm?

22. A snapshot is 7.60 cm wide and 10.1 cm long. It is enlarged so that it is 25.3 cm wide. How long is the enlarged picture? What is its area? its perimeter?

23. Is every equilateral triangle similar to every other equilateral triangle? Is every isosceles triangle similar to every other isosceles triangle? Give reasons.

24. At the same time that a yardstick held vertically casts a 5.0-ft shadow, a vertical flagpole casts a 30-ft shadow. How high is the flagpole?

25. At a certain time of day, a television relay tower casts a shadow 100 m long, and a nearby pole 12 m tall casts a shadow 15 m long. How tall is the tower?

26. Assume that the three triangles in Figure 1.37 are similar. Find the measure of the unknown sides.

 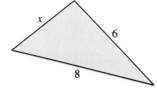

Figure 1.37

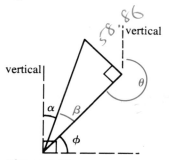

Figure 1.38

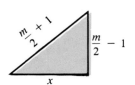

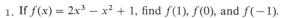vertical

vertical

α β

ϕ

Figure 1.39

5

31.14

27. If the measure of the hypotenuse of the right triangle in Figure 1.38 is $(m/2) + 1$ and one leg has measure $(m/2) - 1$, find the measure of the other leg.

28. Find angles α and ϕ in Figure 1.39 if $\theta = 215°$ and $\beta = 26.6°$.

29. Find angles α and ϕ in Figure 1.39 if $\theta = 211°14'$ and $\beta = 24°12'$.

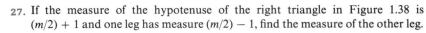

Define and/or discuss each of the following.

Cartesian Coordinate System	Degree Measure
Function	Complementary Angles
Significant Digits	Supplementary Angles
Rounding Off	Standard Position of an Angle
Angles	Pythagorean Theorem
Coterminal Angles	Similar Triangles

Review Exercises for Chapter 1

1. If $f(x) = 2x^3 - x^2 + 1$, find $f(1)$, $f(0)$, and $f(-1)$.

2. If $f(x) = 3x - 2$, find $f(r + s)$.

3. Graph $y = f(x)$ if $f(x) = 2x - 3$.

4. Graph $y = f(x)$ if $f(x) = x^2 - 5$.

5. Find the distance from $(1, f(1))$ to $(2, f(2))$ if $f(x) = x^2$.

6. Find the distance from $(0, 5)$ to $(3, 2)$.

7. Determine if the points $(0, 1)$, $(4, 2)$, and $(3, 5)$ are vertices of a right triangle.

8. Let $y = \dfrac{3 - x}{x + 1}$. Find the distance from $(1, f(1))$ to $(0, f(0))$.

9. Determine an angle between $0°$ and $360°$ that is coterminal with an angle whose measure is $3400°$.

10. Determine an angle between $0°$ and $360°$ that is coterminal with an angle whose measure is $800°$.

11. Change $38°43'23''$ to decimal degree form.

12. Change $67.5428°$ to degree/minute/second form.

13. Give the degree measure of an angle in standard position whose terminal side passes through $(-1, 1)$.

14. Give the degree measure of an angle in standard position whose terminal side passes through $(-\sqrt{2}, 0)$.

15. Subtract: $45°23'14'' - 35°35'54''$.

16. The Gateway Arch in St. Louis is known to be approximately 670 ft high. If you are 500 ft away from the foot of the arch, what is your line-of-sight distance to the top?

17. Determine an angle in standard position between 0° and 360° that is coterminal with an angle whose measure is 10,000°.

18. The hypotenuse and one leg of a right triangle are 54.6 ft and 34.9 ft, respectively. Find the other leg.

19. Find the length of arc subtended* by a 32° angle on a circle of radius 4.6 ft. (Hint: See the "Historical Note" in Section 1.5.)

20. Find the line-of-sight distance to the top of a 100-ft tower if you are standing 50 ft from its base.

In Exercises 21–26, round off to four significant digits.

21. 44.653   22. 131,450   23. 0.0021245

24. 0.871351   25. 352.97   26. 12999

In Exercises 27–30, indicate how many significant digits the answer should contain.

27. 22.6(0.9087)   28. 0.023(67.445)

29. $\dfrac{1.83\sqrt{15}}{2.005}$ 30. $3.7^2 + 0.936^2 + \pi^2$

31. A woman walks 1.8 km due east and then turns and walks 2.4 km due north. How far is she from her initial starting point? Round off the answer to the correct number of significant digits.

32. Find x and y if the two triangles in Figure 1.40 are similar triangles.

 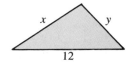

Figure 1.40

33. Under certain conditions the path of a thrown ball can be represented mathematically by the equation $h = a + bx + cx^2$, where x is the horizontal displacement of the ball, h is the corresponding height, and a, b, and c are constants. Draw the path of a ball for the interval $x = 0$ to $x = 7$, if $h = 7x - x^2$.

* A line segment or arc **subtends** an angle θ if the line segment or arc extends between the sides of the angle.

2

The Trigonometric Functions

2.1 Definitions of the Trigonometric Functions

Trigonometry was invented as a means of calculating distances that could not be measured. Although triangle applications are still an important part of the study of trigonometry, many modern applications have nothing to do with triangles. For this reason, we state the basic definitions of trigonometry in terms of a generated angle in standard position in the plane. [See Figure 2.1(a).]

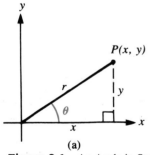

(a)   (b)

Figure 2.1   An Angle in Standard Position

Referring to Figure 2.1(a), we note that there are three important numbers relative to any point on the terminal side of angle θ; namely, the x- and y-coordinates of the point and the distance r from the origin to the point. For a given angle θ, we are interested in the ratios of the numbers x, y, and r. By inspection we can see that the following six ratios can be formed from the three numbers:

$$\frac{x}{r}, \frac{y}{r}, \frac{y}{x}, \frac{x}{y}, \frac{r}{x}, \frac{r}{y}$$

For a given angle θ, these six ratios are independent of the point chosen on the terminal side. The following argument proves that this is true. Choose two distinct points on the terminal side of angle θ, as shown in Figure 2.1(b), and then draw a line through each point perpendicular to the x-axis. In this way we visualize two similar triangles with a common vertex at the origin. From our knowledge of similar triangles, we know that corresponding sides of the two triangles are proportional. Therefore,

$$\frac{y}{y_1} = \frac{r}{r_1} \quad \text{and} \quad \frac{x}{x_1} = \frac{r}{r_1} \quad \text{and} \quad \frac{y}{y_1} = \frac{x}{x_1}$$

or, rearranging terms,

$$\frac{y}{r} = \frac{y_1}{r_1} \quad \text{and} \quad \frac{x}{r} = \frac{x_1}{r_1} \quad \text{and} \quad \frac{y}{x} = \frac{y_1}{x_1}$$

Each proportion says that the ratio of the two numbers associated with the smaller triangle is equal to the ratio of the corresponding numbers in the larger triangle. Recognizing that these are the first three ratios of the six ratios mentioned earlier and that the other three could be handled in the same manner, we conclude that the six ratios are independent of the point chosen on the terminal side of θ.

Although the six ratios are independent of the point selected on the terminal side, they are dependent on the generated angle θ. For instance, if in Figure 2.1(b) angle θ increased, the x-coordinate of P would decrease and the y-coordinate would increase, while r remained constant. Consequently, the ratio x/r would decrease as θ increased and the ratio y/r would increase as θ increased. The six ratios are functions of the angle θ and have come to be called the **trigonometric functions**. Definition 2.1 gives the names of the six ratios.

Definition 2.1: With reference to Figure 2.1, the six trigonometric functions of angle θ are as follows.

$$\text{sine } \theta = \frac{y}{r} \qquad \text{Abbreviated sin } \theta$$

$$\text{cosine } \theta = \frac{x}{r} \qquad \text{Abbreviated cos } \theta$$

$$\text{tangent } \theta = \frac{y}{x} \qquad \text{Abbreviated tan } \theta$$

$$\text{cotangent } \theta = \frac{x}{y} \qquad \text{Abbreviated cot } \theta$$

$$\text{secant } \theta = \frac{r}{x} \qquad \text{Abbreviated sec } \theta$$

$$\text{cosecant } \theta = \frac{r}{y} \qquad \text{Abbreviated csc } \theta$$

You should take time to learn the definition of each of the trigonometric functions since they are the building blocks of trigonometry. You should know them so well that when someone mentions $\sin \theta$ you automatically think "y to r."

Historical Note: The word *sine* has an interesting origin. The first trigonometric tables were of chords of a circle corresponding to an angle θ, as shown in the figure below. As you can see, if the radius of the circle is 1, then $\sin \theta$ is just one-half the chord length. The Hindus gave the name "jiva" to the half chord, and the Arabs used the word "jiba." In the Arabic language there is also a word "jaib," meaning "bay," whose Latin translation is "sinus." A medieval translator inadvertently confused the words "jiba" and "jaib" and thus the word *sine* is used instead of "half chord."

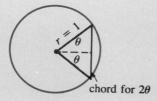

chord for 2θ

Example 1.   Determine the trigonometric functions of the angle θ in standard position whose terminal side passes through the point $(3, 5)$, as shown in Figure 2.2.

Solution.   Note from Figure 2.2 that r is given by $r = \sqrt{x^2 + y^2}$. Therefore, $r = \sqrt{3^2 + 5^2} = \sqrt{34}$. Using $x = 3$, $y = 5$, and $r = \sqrt{34}$ in Definition 2.1, we find that

$$\sin \theta = \frac{5}{\sqrt{34}} \qquad \csc \theta = \frac{\sqrt{34}}{5}$$

$$\cos \theta = \frac{3}{\sqrt{34}} \qquad \sec \theta = \frac{\sqrt{34}}{3}$$

$$\tan \theta = \frac{5}{3} \qquad \cot \theta = \frac{3}{5}$$

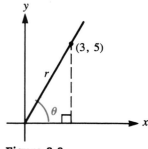

Figure 2.2

Of course, $\sin \theta$ and $\cos \theta$ can be written in rationalized form as $\sin \theta = 5\sqrt{34}/34$ and $\cos \theta = 3\sqrt{34}/34$, respectively.

Example 2.   A support line from the top of a 150-m antenna is anchored at a spot 50 m from the base of the antenna. Find the tangent of the angle of elevation of the cable. (The angle of elevation is the angle between the horizontal and one's line of sight when looking up at the object.)

Solution.   Here, we draw the support line and the antenna in the Cartesian plane, with the anchor point at the origin. The situation is shown in Figure 2.3.
From the figure we see that $x = 50$ and $y = 150$, so the tangent of the angle of elevation of θ is

$$\tan \theta = \frac{y}{x} = \frac{150}{50} = 3.0$$

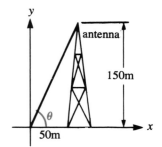

Figure 2.3

Example 3. An angle θ in standard position has the point $(-6, 3)$ on its terminal side. Find the values of the six trigonometric functions of θ. (See Figure 2.4.)

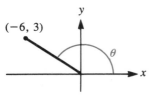

$(-6, 3)$

Figure 2.4

Solution. Using Definition 2.1 with $x = -6$, $y = 3$, and $r = \sqrt{(-6)^2 + 3^2} = \sqrt{45} = 3\sqrt{5}$, we get

$$\sin \theta = \frac{y}{r} = \frac{3}{3\sqrt{5}} = \frac{1}{\sqrt{5}} \qquad \csc \theta = \frac{r}{y} = \frac{3\sqrt{5}}{3} = \sqrt{5}$$

$$\cos \theta = \frac{x}{r} = \frac{-6}{3\sqrt{5}} = -\frac{2}{\sqrt{5}} \qquad \sec \theta = \frac{r}{x} = \frac{3\sqrt{5}}{-6} = -\frac{\sqrt{5}}{2}$$

$$\tan \theta = \frac{y}{x} = \frac{3}{-6} = -\frac{1}{2} \qquad \cot \theta = \frac{x}{y} = \frac{-6}{3} = -2$$

Example 4. An angle θ in standard position has its terminal side passing through the point $(3, -4)$. Find the six trigonometric functions of the angle. (See Figure 2.5.)

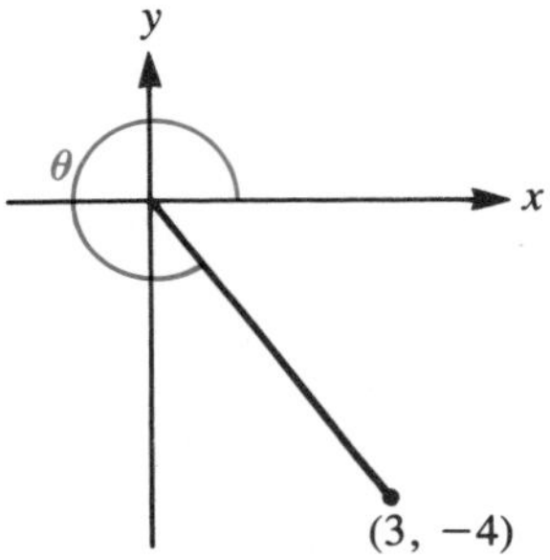

Solution. Using Definition 2.1 with $x = 3$, $y = -4$, and $r = \sqrt{3^2 + (-4)^2} = 5$, we get

$$\sin \theta = \frac{y}{r} = \frac{-4}{5} = -\frac{4}{5} \qquad \csc \theta = \frac{r}{y} = \frac{5}{-4} = -\frac{5}{4}$$

$$\cos \theta = \frac{x}{r} = \frac{3}{5} \qquad \sec \theta = \frac{r}{x} = \frac{5}{3}$$

$$\tan \theta = \frac{y}{x} = \frac{-4}{3} = -\frac{4}{3} \qquad \cot \theta = \frac{x}{y} = \frac{3}{-4} = -\frac{3}{4}$$

$(3, -4)$

Figure 2.5

In each of these examples, four of the six trigonometric values are negative. This follows from the fact that since r is positive, the signs of the functional values depend on the signs of x and y.

The sine function is the ratio of y to r, which means that it is positive for angles in the first and second quadrants and negative for angles in the third and fourth quadrants. This is because y is positive above the x-axis and negative below.

The cosine function, which is the ratio of x to r, is positive for angles in the first and fourth quadrants and negative for angles in the second and third quadrants. This is because x is positive to the right of the y-axis and negative to the left.

The tangent function, which is the ratio of y to x, is positive in the first and third quadrants because y and x have the same signs in these quadrants. The tangent is negative in the second and fourth quadrants. The signs of the remaining three functions can be analyzed in the same way. Table 2.1 summarizes the results for all six functions.

Table 2.1

Quadrant	sin θ	cos θ	tan θ	cot θ	sec θ	csc θ
I	+	+	+	+	+	+
II	+	−	−	−	−	+
III	−	−	+	+	−	−
IV	−	+	−	−	+	−

Example 5
(a) $\sin \theta > 0$ in quadrants I and II.
(b) $\tan \theta < 0$ in quadrants II and IV.
(c) $\sec \theta < 0$ in quadrants II and III.

Example 6.
Show that the terminal side of θ (in standard position) is in quadrant II if $\sin \theta > 0$ and $\cos \theta < 0$.

Solution. From Table 2.1, $\sin \theta > 0$ in quadrants I and II, and $\cos \theta < 0$ in quadrants II and III. Since both of the given conditions are satisfied in quadrant II, we conclude that the terminal side of θ must be in this quadrant.

Example 7
(a) If $\sec \phi > 0$ and $\sin \phi < 0$, the terminal side of angle ϕ is in quadrant IV.
(b) If $\csc \phi > 0$ and $\cot \phi < 0$, the terminal side of angle ϕ is in quadrant II.

Comment: If an angle is in standard position, the values of all six trigonometric functions can be determined if you know the value of one of them and the quadrant of the terminal side of the angle. If the quadrant is not given, two sets of values are possible.

Example 8.
Given that $\tan \theta = -\frac{4}{3}$ and that θ in standard position has its terminal side in quadrant II, find the values of the other five trigonometric functions.

Solution. We choose a convenient point on the terminal side—in this case $(-3, 4)$, as shown in Figure 2.6. (If we had been told to locate the point in quadrant IV, we would have chosen the point $(3, -4)$.) The desired trigonometric functions for the given angle are

$$\sin \theta = \frac{4}{5} \qquad \cos \theta = \frac{-3}{5} \qquad \cot \theta = \frac{-3}{4} \qquad \sec \theta = \frac{5}{-3} \qquad \csc \theta = \frac{5}{4}$$

Example 9.
Given that $\cos \theta = -\frac{5}{13}$, find the values of the other trigonometric functions. (See Figure 2.7.)

Solution. Since the quadrant is not specified, two angles between 0 and 360° will satisfy the given condition. One is in the second quadrant, and the other is in the third quadrant. For the second quadrant angle,

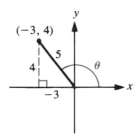

Figure 2.6

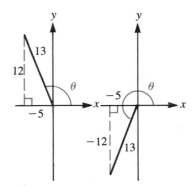

Figure 2.7

$$\sin\theta = \frac{12}{13} \quad \tan\theta = -\frac{12}{5} \quad \cot\theta = -\frac{5}{12} \quad \sec\theta = -\frac{13}{5} \quad \csc\theta = \frac{13}{12}$$

For the third quadrant angle,

$$\sin\theta = -\frac{12}{13} \quad \tan\theta = \frac{-12}{-5} = \frac{12}{5} \quad \cot\theta = \frac{-5}{-12} = \frac{5}{12}$$

$$\sec\theta = -\frac{13}{5} \quad \csc\theta = -\frac{13}{12}$$

Exercises for Section 2.1

In Exercises 1–9, find the values of the trigonometric functions of an angle in standard position whose terminal side passes through the points given.

1. $(2, 4)$
2. $(-1, 5)$
3. $(-9, 16)$
4. $(3, 1)$
5. $(2, -7)$
6. $(-1, -1)$
7. $(3, -1)$
8. $(1, -1)$
9. $(-\sqrt{3}, -1)$

10. Show that the values of the trigonometric functions of an angle θ are independent of the choice of the point P on its terminal side.

11. In which quadrants must the terminal side of θ lie for (a) $\sin\theta$ to be positive? (b) $\cos\theta$ to be positive? (c) $\tan\theta$ to be positive?

In Exercises 12–18, indicate the quadrant in which the terminal side of θ lies for the given conditions.

12. $\sin\theta > 0$ and $\tan\theta < 0$
13. $\sec\theta < 0$ and $\cot\theta < 0$
14. $\cos\theta > 0$ and $\sin\theta < 0$
15. $\tan\theta > 0$ and $\csc\theta < 0$
16. $\sin\theta < 0$ and $\cos\theta < 0$
17. $\sin\theta < 0$ and $\sec\theta > 0$
18. $\csc\theta < 0$ and $\sec\theta < 0$

In Exercises 19–38, find all of the trigonometric functions at an angle θ that satisfies the conditions.

19. $\tan\theta = \frac{3}{4}$ in Q I
20. $\sec\theta = -3$
21. $\tan\theta = \frac{3}{4}$ in Q III
22. $\tan\theta = \frac{1}{2}$ in Q III
23. $\cos\theta = \frac{\sqrt{3}}{2}$
24. $\cot\theta = -3$ in Q IV
25. $\sin\theta = \frac{2}{3}$ in Q II
26. $\sin\theta = \frac{\sqrt{3}}{2}$

27. $\sin \theta = -\dfrac{1}{2}$

28. $\sin \theta = \dfrac{1}{5}$

29. $\tan \theta = 10$ in Q I

30. $\csc \theta = 2$

31. $\cos \theta = \dfrac{12}{13}$

32. $\cos \theta = -\dfrac{\sqrt{3}}{2}$ in Q II

33. $\tan \theta = -\sqrt{3}$

34. $\cot \theta = -\sqrt{2}$

35. $\sin \theta = \dfrac{u}{v}$

36. $\tan \theta = \dfrac{u}{v}$

37. $\cos \theta = u$

38. $\sin \theta = \dfrac{1}{v}$

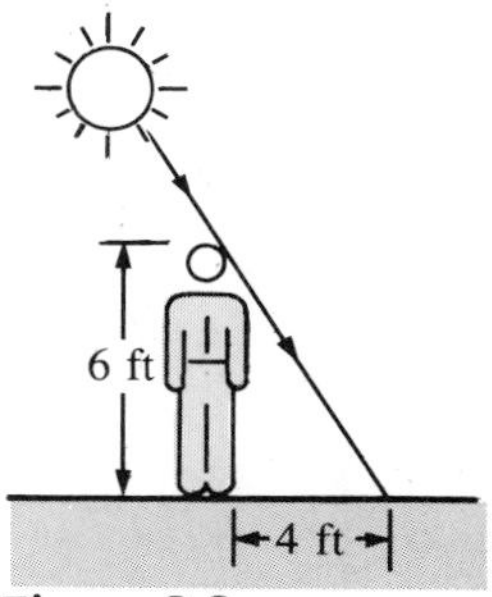

39. A 6.0-ft man casts a shadow of 4.0 ft. Find the tangent of the angle that the rays of the sun make with the horizontal. (See Figure 2.8.)

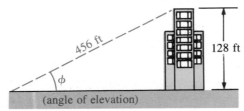

Figure 2.8

40. A wire 30 ft long is used to brace a flagpole. If the wire is attached to the pole 25 ft above the level ground, what is the cosine of the angle made by the wire with the ground?

41. The line-of-sight distance to the top of a 128-ft-high building is 456 ft. (See Figure 2.9.) What is the tangent of the angle of elevation?

456 ft

128 ft

ϕ

(angle of elevation)

Figure 2.9

θ (angle of depression)

255m

θ

75m

42. Suppose that a girl is flying a kite at the end of a 100-m string that makes an angle of 45° with the ground. Find the cosine of the angle that the string makes with the ground.

43. A man on a 255-m cliff looks down on a rowboat known to be 75.0 m from the base of the cliff. (See Figure 2.10.) What is the sine of the angle of depression? (The angle of depression is defined as the angle between the horizontal and one's line of sight when looking down on an object.)

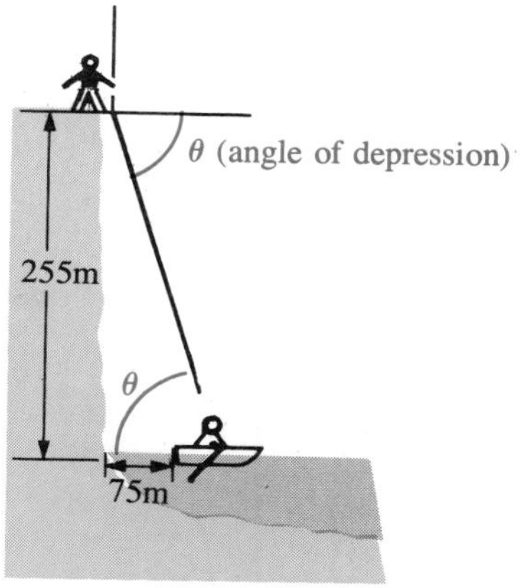

Figure 2.10

2.2 Fundamental Relations

The values of the trigonometric functions are interrelated by some interesting and useful formulas. Recall from Definition 2.1 that $\sin \theta = y/r$ and $\csc \theta = r/y$; consequently,

$$\sin \theta \csc \theta = \frac{y}{r} \cdot \frac{r}{y} = 1$$

Similarly,

$$\cos \theta \sec \theta = \frac{x}{r} \cdot \frac{r}{x} = 1$$

and

$$\tan \theta \cot \theta = \frac{y}{x} \cdot \frac{x}{y} = 1$$

Rearranging, we have

$$\sin \theta = \frac{1}{\csc \theta}$$

$$\cos \theta = \frac{1}{\sec \theta} \qquad (2.1)$$

$$\tan \theta = \frac{1}{\cot \theta}$$

The three relations in (2.1) are called the **reciprocal relations** for the trigonometric functions. Of course, they can also be written in the forms $\csc \theta = 1/\sin \theta$, $\sec \theta = 1/\cos \theta$, and $\cot \theta = 1/\tan \theta$.

Example 1. Given $\sin \theta = \frac{2}{3}$, use Relation (2.1) to find $\csc \theta$.

Solution

$$\csc \theta = \frac{1}{\sin \theta} = \frac{1}{2/3} = \frac{3}{2}$$

Example 2. Given $\sec \alpha = 2$, use Relation (2.1) to find $\cos \alpha$.

Solution

$$\cos \alpha = \frac{1}{\sec \alpha} = \frac{1}{2}$$

Two other relations that are of considerable importance are

$$\tan \theta = \frac{\sin \theta}{\cos \theta} \qquad \text{and} \qquad \cot \theta = \frac{\cos \theta}{\sin \theta} \qquad (2.2)$$

The first of these is verified by the following sequence of operations:

$$\tan \theta = \frac{y}{x} \qquad \text{Definition of } \tan \theta$$

$$= \frac{y/r}{x/r} \qquad \text{Divide numerator and denominator by } r$$

$$= \frac{\sin\theta}{\cos\theta} \qquad \text{Definition of } \sin\theta \text{ and } \cos\theta$$

The fact that $\cot\theta = \cos\theta/\sin\theta$ follows immediately from $\cot\theta = 1/\tan\theta$.

Example 3

(a) Given $\sin\phi = 1/\sqrt{5}$ and $\cos\phi = 2/\sqrt{5}$, use Relation (2.2) to find $\tan\phi$.
(b) Indicate the quadrant in which the terminal side of ϕ lies.

Solution

(a) $\tan\phi = \dfrac{\sin\phi}{\cos\phi} = \dfrac{1/\sqrt{5}}{2/\sqrt{5}} = \dfrac{1}{2}$

(b) The terminal side of ϕ lies in quadrant I since $\sin\phi > 0$ and $\cos\phi > 0$.

Finally, with the aid of the Pythagorean theorem we derive an important relation between the sine and cosine functions. For any angle θ whose terminal side passes through the point $P(x, y)$, the Pythagorean theorem requires that

$$y^2 + x^2 = r^2$$

Dividing both sides of this equation by r^2, we obtain

$$\left(\frac{y}{r}\right)^2 + \left(\frac{x}{r}\right)^2 = 1$$

or, in terms of the trigonometric functions,

$$(\sin\theta)^2 + (\cos\theta)^2 = 1$$

It is customary to write $(\sin\theta)^2$ as $\sin^2\theta$ and $(\cos\theta)^2$ as $\cos^2\theta$. (A similar convention holds for expressing powers of the other trigonometric functions.) Thus the equation reads

$$\sin^2\theta + \cos^2\theta = 1 \qquad (2.3)$$

which is often called the **Pythagorean relation** of trigonometry.

Two alternative forms of Relation (2.3) often prove useful. By dividing both sides by $\cos^2\theta$, we obtain

$$\frac{\sin^2\theta}{\cos^2\theta} + \frac{\cos^2\theta}{\cos^2\theta} = \frac{1}{\cos^2\theta}$$

Then, since $\sin\theta/\cos\theta = \tan\theta$ and $1/\cos\theta = \sec\theta$, we have

$$\tan^2\theta + 1 = \sec^2\theta \qquad (2.3a)$$

In a similar manner, we can show that

$$\cot^2 \theta + 1 = \csc^2 \theta \qquad\qquad (2.3b)$$

Example 4. Given that $\sin \theta = \frac{1}{4}$ and that θ is a second-quadrant angle, use the Pythagorean relation to find $\cos \theta$. Then find $\tan \theta$.

Solution. Solving the Pythagorean relation for $\cos \theta$, we have $\cos \theta = -\sqrt{1 - \sin^2 \theta}$. The negative square root is chosen because $\cos \theta < 0$ when θ is in quadrant II. Therefore,

$$\cos \theta = -\sqrt{1 - \left(\frac{1}{4}\right)^2} = -\sqrt{1 - \left(\frac{1}{16}\right)} = -\frac{\sqrt{15}}{4}$$

Finally, using $\sin \theta = \frac{1}{4}$ and $\cos \theta = -\sqrt{15}/4$ in Relation (2.2) we get

$$\tan \theta = \frac{\sin \theta}{\cos \theta} = -\frac{\frac{1}{4}}{\sqrt{15}/4} = -\frac{1}{\sqrt{15}}$$

Exercises for Section 2.2

In Exercises 1–25, use the Fundamental Relations (2.1)–(2.3) to find the exact value of the indicated trigonometric function.

1. $\sin \theta = \dfrac{1}{2}$, find $\csc \theta$

2. $\cos \phi = \dfrac{2}{3}$, find $\sec \phi$

3. $\sec \beta = 3$, find $\cos \beta$

4. $\tan \theta = \dfrac{10}{7}$, find $\cot \theta$

5. $\sin A = \dfrac{\sqrt{3}}{2}$, $\tan A < 0$, find $\cos A$

6. $\sin \alpha = \dfrac{\sqrt{2}}{2}$, $\sec \alpha > 0$, find $\cos \alpha$

7. $\cot \theta = \sqrt{2}$, find $\tan \theta$

8. $\csc \theta = \dfrac{2}{\sqrt{3}}$, find $\sin \theta$

9. $\csc \alpha = 2$, $\tan \alpha > 0$, find $\cos \alpha$

10. $\cos x = -\dfrac{1}{2}$, $\tan x > 0$, find $\sin x$

11. $\sin \phi = -\dfrac{5}{13}$, $\cos \phi = \dfrac{12}{13}$, find $\tan \phi$

12. $\sin \beta = -\dfrac{2}{\sqrt{7}}$, $\cos \beta = \sqrt{\dfrac{3}{7}}$, find $\tan \beta$

13. $\tan \theta = \dfrac{1}{2}$, $\cos \theta = -\dfrac{2}{\sqrt{5}}$, find $\sin \theta$

14. $\tan \theta = \dfrac{2}{3}$, $\sin \theta = \dfrac{2}{\sqrt{13}}$, find $\cos \theta$

15. $\sin x = -\dfrac{1}{\sqrt{10}}$, $\cos x = -\dfrac{3}{\sqrt{10}}$, find $\tan x$

16. $\sec \theta = \dfrac{13}{12}$, $\tan \theta = \dfrac{5}{12}$, find $\sin \theta$

17. $\csc B = \dfrac{\sqrt{5}}{2}$, $\sec B > 0$, find $\tan B$

18. $\sin \gamma = \dfrac{2}{3}$, $\sec \gamma < 0$, find $\cot \gamma$

19. $\cos \phi = -\dfrac{\sqrt{2}}{3}$, $\csc \phi > 0$, find $\cot \phi$

20. $\sec \alpha = -2$, $\sin \alpha < 0$, find the other five functions

21. $\csc \theta = 3$, $\cos \theta > 0$, find the other five functions

22. $\sin \theta = \dfrac{2}{3}$, $\sec \theta > 0$, find the other five functions

23. $\cos \theta = -\dfrac{2}{\sqrt{5}}$, θ in quadrant II, find the other five functions

24. $\tan \alpha = 1$, α in quadrant III, find the other five functions

25. $\tan \beta = \sqrt{2}$, β in quadrant III, find the other five functions

Use Relation (2.1) and a calculator to find the indicated trigonometric functions.

26. $\sin \theta = 0.4313$, find $\csc \theta$ 27. $\cos x = 0.1155$, find $\sec x$

28. $\tan \phi = 2.397$, find $\cot \phi$ 29. $\csc A = 1.902$, find $\sin A$

30. $\sec t = 2.030$, find $\cos t$

31. The slope of a line is equal to the tangent of its angle of elevation. Use the fundamental relations to find the slope of a line for which the cosecant of the angle of elevation is $\sqrt{2.6}$ in quadrant I.

32. In a problem involving the rotation of coordinate axes, a student calculates the tangent of the rotation angle to be 2. Use the fundamental relations to find the sine and cosine of the rotation angle. Assume the rotation angle is acute.

33. Do Exercises 19–34 in Section 2.1 again, but this time use the fundamental relations.

34. Prove Relation (2.3b).

2.3 The Values of the Trigonometric Functions for Special Angles

In the previous sections we discussed and computed values of the trigonometric functions from known points on the terminal side of an angle or by using the fundamental relations. No attempt was made to relate the measure of the angle to the values of its trigonometric functions. In practice it is important to know how to obtain the trigonometric functions for a specified angle. The values of the trigonometric functions of certain angles can be found geometrically, as illustrated in the next two examples.

Example 1. Find the values of the trigonometric functions for a 45° angle.

Solution. Drawing a 45° angle in standard position, we observe that the terminal side will bisect the first quadrant. Consequently, the x-coordinate of any point on the terminal side of a 45° angle will equal the y-coordinate. For convenience we choose the point $(1, 1)$, as shown in Figure 2.11. Then $r = \sqrt{1^2 + 1^2} = \sqrt{2}$. Using $x = 1$, $y = 1$, and $r = \sqrt{2}$ in the definitions, we get

$$\sin 45° = \frac{y}{r} = \frac{1}{\sqrt{2}} \qquad \csc 45° = \frac{r}{y} = \frac{\sqrt{2}}{1} = \sqrt{2}$$

$$\cos 45° = \frac{x}{r} = \frac{1}{\sqrt{2}} \qquad \sec 45° = \frac{r}{x} = \frac{\sqrt{2}}{1} = \sqrt{2}$$

$$\tan 45° = \frac{y}{x} = \frac{1}{1} = 1 \qquad \cot 45° = \frac{x}{y} = \frac{1}{1} = 1$$

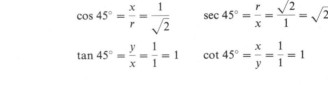

Figure 2.11

Example 2. Find the trigonometric functions of a 60° angle.

Solution. Consider the equilateral triangle shown in Figure 2.12. The bisector of any one of the angles divides the equilateral triangle into two congruent right triangles. Since a line that bisects an angle of an equilateral triangle also bisects the side opposite that angle, the length of the side opposite the 30° angle is

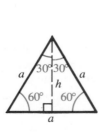

Figure 2.12

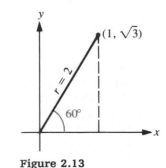

Figure 2.13

one-half the length of the hypotenuse. By the Pythagorean theorem, the length of h is $h = \sqrt{a^2 - (\frac{1}{2}a)^2} = a\sqrt{3}/2$. If $a = 2$, then $h = \sqrt{3}$ and $\frac{1}{2}a = 1$. From this we can conclude that the terminal side of a $60°$ angle in standard position will pass through the point $(1, \sqrt{3})$ if r is 2. (See Figure 2.13.) Hence, by definition,

$$\sin 60° = \frac{y}{r} = \frac{\sqrt{3}}{2} \qquad \csc 60° = \frac{r}{y} = \frac{2}{\sqrt{3}}$$

$$\cos 60° = \frac{x}{r} = \frac{1}{2} = 0.5 \qquad \sec 60° = \frac{r}{x} = \frac{2}{1} = 2$$

$$\tan 60° = \frac{y}{x} = \frac{\sqrt{3}}{1} = \sqrt{3} \qquad \cot 60° = \frac{x}{y} = \frac{1}{\sqrt{3}}$$

The values of the trigonometric functions for a $30°$ angle are found by the same right-triangle relationship used for a $60°$ angle. Table 2.2 summarizes the trigonometric functions for $30°$, $45°$, and $60°$. Study it carefully. You should know how to *derive* each of the values in the table. (Keep in mind Figures 2.11 and 2.13.)

Table 2.2 Values for Some Important Angles

θ (degrees)	sin θ	cos θ	tan θ	cot θ	sec θ	csc θ
30	$\frac{1}{2}$	$\frac{\sqrt{3}}{2}$	$\frac{\sqrt{3}}{3}$	$\sqrt{3}$	$\frac{2\sqrt{3}}{3}$	2
45	$\frac{\sqrt{2}}{2}$	$\frac{\sqrt{2}}{2}$	1	1	$\sqrt{2}$	$\sqrt{2}$
60	$\frac{\sqrt{3}}{2}$	$\frac{1}{2}$	$\sqrt{3}$	$\frac{\sqrt{3}}{3}$	2	$\frac{2\sqrt{3}}{3}$

An interesting and useful observation that can be made about the values in Table 2.2 is that for the complementary angles $30°$ and $60°$,

$$\sin 30° = \frac{1}{2} = \cos 60°$$

$$\tan 30° = \frac{\sqrt{3}}{3} = \cot 60°$$

$$\sec 30° = \frac{2\sqrt{3}}{3} = \csc 60°$$

Two trigonometric functions that have equal values for complementary angles are called **cofunctions**. The various cofunction relationships are apparent from the names of the trigonometric functions; for example, the names "sine" and "cosine" reflect the cofunction relationship. Similarly, the tangent and the cotangent are cofunctions, as are the secant and cosecant. Making use of the fact that θ and $90° - \theta$ are complementary angles, we have

$$\left.\begin{cases} \text{trigonometric function of} \\ \text{an acute angle } \theta \end{cases}\right\} = \{\text{cofunction of } (90° - \theta)\} \qquad (2.4)$$

Example 3

(a) $\sin 40° = \cos 50°$

(b) $\tan 5.6° = \cot 84.4°$

(c) $\cos 13°15' = \sin 76°45'$

Quadrantal Angles

An angle in standard position whose terminal side lies on a coordinate axis is called a **quadrantal angle**. Angles of $0°$, $\pm 90°$, and $\pm 180°$ are examples. For these angles, one of the coordinates of a point on the terminal side must be zero. Since division by zero is undefined, two of the six trigonometric functions will be undefined at each quadrantal angle.

Example 4. Find the trigonometric functions of an angle $\theta = 180°$ whose terminal side passes through the point $(-1, 0)$, as shown in Figure 2.14.

Solution. In this case, $x = -1$, $y = 0$, and $r = 1$. Since $\theta = 180°$, we write

$$\sin 180° = \frac{y}{r} = \frac{0}{1} = 0 \qquad \csc 180° = \frac{r}{y} = \frac{1}{0} \text{ (undefined)}$$

$$\cos 180° = \frac{x}{r} = \frac{-1}{1} = -1 \qquad \sec 180° = \frac{r}{x} = \frac{1}{-1} = -1$$

$$\tan 180° = \frac{y}{x} = \frac{0}{-1} = 0 \qquad \cot 180° = \frac{x}{y} = \frac{-1}{0} \text{ (undefined)}$$

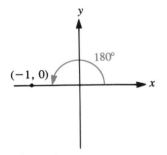

Figure 2.14

The preceding example exhibits the values for a quadrantal angle of $180°$, or one coterminal with it. The values of the other quadrantal angles can be found by a similar procedure; they are listed for your reference in Table 2.3.

Table 2.3

θ (degrees)	sin θ	cos θ	tan θ	cot θ	sec θ	csc θ
0	0	1	0	undefined	1	undefined
90	1	0	undefined	0	undefined	1
180	0	-1	0	undefined	-1	undefined
270	-1	0	undefined	0	undefined	-1

Comment: You should be able to verify all of the values in Table 2.2 and Table 2.3. Memorizing the values in these tables will help you in later chapters.

Reference Angles

The values of the trigonometric functions for 30°, 45°, and 60° can be used to find the values of the trigonometric functions for related angles in other quadrants. To accomplish this we introduce the concept of a reference angle.

Definition 2.2: The **reference angle** of a given angle θ ($0° \leq \theta < 360°$) is the positive acute angle α between the terminal side of θ and the x-axis.

The reference angle α of a given angle θ in standard position can be found by using the following rules. θ_1 represents an angle in quadrant I, θ_2 an angle in quadrant II, and so forth.

(1) First quadrant angle: $\alpha = \theta_1$
(2) Second quadrant angle: $\alpha = 180° - \theta_2$
(3) Third quadrant angle: $\alpha = \theta_3 - 180°$
(4) Fourth quadrant angle: $\alpha = 360° - \theta_4$

Example 5. Find the reference angle for (a) 196° and (b) 98°.

Solution
(a) $\alpha = 196° - 180° = 16°$ (See Figure 2.15(a).)
(b) $\alpha = 180° - 98° = 82°$ (See Figure 2.15(b).)

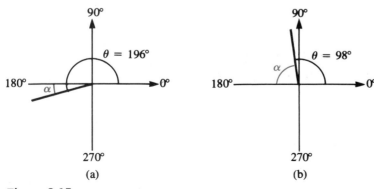

Figure 2.15

Comment: Note that the reference angle is always computed with respect to the *x*-axis.

Trigonometric Functions of Angles Greater Than 90°

The next example shows how the reference angle is utilized to evaluate tan 150°. The example illustrates the general procedure used to evaluate trigonometric functions of angles greater than 90° in terms of trigonometric functions of acute angles.

Example 6. Find tan 150°.

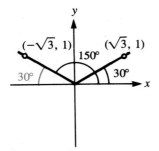

Figure 2.16

Solution. A generated angle of 150° is shown in Figure 2.16. The reference angle is 30° and is also shown as a generated angle in the first quadrant. Now, from our knowledge of the 30°-60° right triangle, we know that the terminal side of a 30° angle must pass through the point $(\sqrt{3}, 1)$. Thus $(-\sqrt{3}, 1)$ is a point on the terminal side of 150°. Therefore, by definition,

$$\tan 150° = \frac{1}{-\sqrt{3}} = -\frac{1}{\sqrt{3}} = -\tan 30°$$

We generalize the result in Example 6 in the following reference-angle rule.

Reference-Angle Rule: To find the value of a trigonometric function for a given angle θ ($0 \leq \theta < 360°$), determine the value of the trigonometric function of the reference angle and prefix the appropriate sign. The sign is determined from the location of the terminal side of θ. (See Table 2.1.)

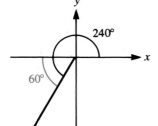

Figure 2.17

Example 7. Find sec 240°. (See Figure 2.17.)

Solution. We see that 240° is a third-quadrant angle, and, therefore, the reference angle is

$$\alpha = 240° - 180° = 60°$$

The reference angle is indicated in Figure 2.17. Since the secant function is negative in the third quadrant and sec 60° = 2, we have

$$\sec 240° = -\sec 60° = -2$$

Example 8. Find sin 330°. (See Figure 2.18.)

Solution. We see that 330° is a fourth-quadrant angle, and, therefore, the reference angle is

$$\alpha = 360° - 330° = 30°$$

The reference angle is indicated in Figure 2.18. Since the sine function is negative in the fourth quadrant and $\sin 30° = \frac{1}{2}$, we have

$$\sin 330° = -\sin 30° = -\frac{1}{2}$$

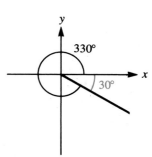

Figure 2.18

The definitions of the trigonometric functions of any angle show that the functional values are completely determined by the location of the terminal side when the angle is in standard position. Thus, **coterminal angles have equal functional values.** For instance, since 30° and 390° are coterminal, $\sin 30° = \sin 390°$, $\cos 30° = \cos 390°$, $\tan 30° = \tan 390°$, etc. Thus, in finding values of trigonometric functions, we need only consider angles between 0° and 360°.

Example 9. Find sin 945°.

Solution. Note that $945° = 2(360°) + 225°$, so 945° is coterminal with 225°. The reference angle for 225° is 45°. Since 225° is a third-quadrant angle and the sine function is negative in the third quadrant, we have that

$$\sin 945° = \sin 225° = -\sin 45° = -\frac{\sqrt{2}}{2}$$

By considering cases in which the angle θ has, in turn, its terminal side in each of the four quadrants, you can verify the following general relations.

$$\begin{aligned}
\sin(-\theta) &= -\sin\theta & \cos(-\theta) &= \cos\theta \\
\sin(180° - \theta) &= \sin\theta & \cos(180° - \theta) &= -\cos\theta \\
\sin(180° + \theta) &= -\sin\theta & \cos(180° + \theta) &= -\cos\theta \\
\sin(360° - \theta) &= -\sin\theta & \cos(360° - \theta) &= \cos\theta
\end{aligned} \tag{2.5}$$

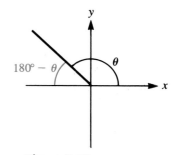

Figure 2.19

You should not attempt to memorize this type of relation but rather be ready to work out any of the results listed by the methods of this section. Figure 2.19 gives the idea for a second-quadrant angle. From Figure 2.19 we see that $180° - \theta$ is the reference acute angle for θ. Hence, since θ is in the second quadrant,

$$\sin\theta = \sin(180° - \theta) \quad \text{and} \quad \cos\theta = -\cos(180° - \theta)$$

Exercises for Section 2.3

1. Verify the six trigonometric functions of 30° given in Table 2.2.

2. Verify the six trigonometric functions of 90° given in Table 2.3. Note that (0, 1) lies on the terminal side of the angle of 90° when it is placed in standard position.

3. Verify the six trigonometric functions of 0° given in Table 2.3. Use (1, 0) as a point on the terminal side.

4. Verify the six trigonometric functions of 270° given in Table 2.3. Use (0, −1) as a point on the terminal side.

In Exercises 5–14, give the reference angle for each of the given angles.

5. 210°	6. 315°	7. 480°	8. 570°	9. −135°
10. −225°	11. 225°	12. 120°	13. 930°	14. 1020°

In Exercises 15–34, use the values of the trigonometric functions of the special angles (Table 2.2) to evaluate the indicated expression.

15. $\cos(-45°)$	16. $\tan 240°$
17. $\sin 150°$	18. $\sec 135°$
19. $\tan 300°$	20. $\cos 240°$
21. $\csc 210°$	22. $\sin 330°$
23. $\sec 315°$	24. $\cot 150°$
25. $\sin 210°$	26. $\tan 225°$
27. $\cot 510°$	28. $\cos 840°$
29. $\cos 1050°$	30. $\csc 600°$
31. $\tan 930°$	32. $\sin 1020°$
33. $\sin 945°$	34. $\cot 1140°$

35. Indicate all values of θ for which $\sin \theta = 1$

36. Indicate all values of θ for which $\cos \theta = 1$.

37. Indicate all values of θ for which $\cos \theta = 0$.

38. Indicate all values of θ for which $\tan \theta = 0$.

39. Indicate all values of θ for which $\sin \theta = \frac{1}{2}$.

40. Construct a table showing the exact values of $\sin \theta$, $\cos \theta$, and $\tan \theta$ for angles of 30°, 45°, 60°, 120°, 135°, 150°, 210°, 225°, 240°, 300°, 315°, and 330°.

41. The current in an a.c. circuit is given by $i = 3 \sin \theta$. Find the value of the current when $\theta = 0°, 30°, 60°, 120°, 150°$, and 180°.

42. The azimuth error in a radar measurement involves the term $\sec \theta$. Evaluate $\sec \theta$ when $\theta = 225°$.

Tables 2.2 and 2.3 of the previous section are obviously incomplete listings of values of the trigonometric functions; Table 2.2 only gives the functional values for 30°, 45°, and 60°, and Table 2.3 is restricted to the quadrantal angles 0°, 90°, 180°, 270°, and 360°. The trigonometric functions for the other angles can be found with a calculator or a table like Tables A and B in the appendix.

> **Comment:** Most scientific calculators operate in either a "degree" mode or a "radian" mode (another unit of angular measure). We will discuss the radian in Chapter 4, so for now just keep your calculator in the degree mode.

Scientific calculators have keys for sin θ, cos θ, and tan θ. The values of these three trigonometric functions for any angle θ, are found by entering the angle and then pushing the desired key. (Note: In some calculators the order of entry is reversed.)

Example 1. Use a calculator to find (a) sin 37° and (b) tan 122°.

Solution
(a) To find sin 37°, enter 37 and push the $\boxed{\text{sin}}$ key to obtain 0.60181. Thus,

$$\sin 37° = 0.60181$$

(b) To find tan 122°, enter 122 and push the $\boxed{\text{tan}}$ key to obtain -1.6003. Thus,

$$\tan 122° = -1.6003$$

> **Comment:** The number of digits displayed in the register will depend on the brand of calculator. The functional values are rounded off to five figures in this section. Keep in mind that calculator computations are often approximate.

Keys for csc θ, sec θ, and cot θ are not provided on calculators because these functions can be obtained from sin θ, cos θ, and tan θ by using the relations

$$\csc \theta = \frac{1}{\sin \theta} \qquad \sec \theta = \frac{1}{\cos \theta} \qquad \cot \theta = \frac{1}{\tan \theta}$$

These reciprocal relations, which were introduced in Section 2.2, make it possible to obtain csc θ, sec θ, and cot θ by using the keys provided for sin θ, cos θ, and tan θ and the key for the reciprocal of a number.

Example 2. Use a calculator to find (a) cot 17° and (b) csc 310°.

Solution

(a) To find cot 17°, enter 17 and push the [tan] key to obtain 0.30573.
Now push the [1/x] key to obtain 3.2708. Thus,

$$\cot 17° = 3.2708$$

(b) To find csc 310°, enter 310 and push the [sin] key to obtain −0.76604.
Now push the [1/x] key to obtain −1.3054. Thus,

$$\csc 310° = -1.3054$$

Angles are commonly expressed in units of degrees, minutes, and seconds ($1° = 60' = 3600''$). Some calculators allow angles to be entered directly in this format, but others accept only decimal equivalents of these units in degrees. If this is the case with your calculator, you must change degrees, minutes, and seconds into the equivalent decimal form before entering the angle in the register.

Example 3. Use a calculator to find cos 204°15′23″.

Solution. First express 204°15′23″ in degrees by dividing 15′ by 60 and 23″ by 3600. Thus,

$$204°15'23'' = 204° + \left(\frac{15}{60}\right)° + \left(\frac{23}{3600}\right)° = 204° + 0.25° + 0.0064° = 204.2564°$$

Now enter 204.2564 and push the [cos] key to obtain −0.91171. Thus,

$$\cos 204°15'23'' = -0.91171$$

Calculators are also used to find angles corresponding to a given trigonometric function. The procedure for finding an angle corresponding to a given functional value varies from brand to brand, but most calculators require that you enter the given number and then push an [inv] or [arc] button prior to pushing the trigonometric function button. Other models have single buttons for this purpose labeled [sin⁻¹], [cos⁻¹], or [tan⁻¹].

> **Comment:** There are two angles θ between 0° and 360° for which cos $\theta = 0.5$. However, if you enter 0.5 in a calculator and push [inv] [cos], only 60° is displayed in the register. To obtain the other angle, which is 300°, you must use the concept of a reference angle.

Example 4. Find the angles θ between 0° and 360° for which tan $\theta = 1.5$.

Solution. To find the first angle, enter 1.5 and push [inv] [tan] to obtain 56.31°. The other angle for which tan $\theta = 1.5$ is in the third quadrant. Specifically, it is the third-quadrant angle whose reference angle is 56.31°. Thus,

$$\theta_3 = 180° + 56.31° = 236.31°$$

The desired angles are 56.31° and 236.31°.

> **Warning:** The angle displayed in the register for a given trigonometric function is not always a positive, acute angle. For example, if $\sin \theta = -0.5$, your calculator will give only the one angle $\theta = -30°$. However, if $\cos \theta = -0.5$, your calculator will give the angle $\theta = 120°$. The reason we get a negative acute angle in the first case and a positive obtuse angle in the second is explained in Chapter 8. To avoid confusion at this time, we suggest that you use your calculator to find the reference angle for a given trigonometric function. **The reference angle will be obtained from your calculator if you enter the absolute value of the given function.**

Example 5. Find angles θ $(0 \le \theta < 360°)$ if $\sin \theta = -0.5664$.

Solution. Enter the absolute value of the given function and push `inv` `sin` . The resulting reference angle is

$$0.5664 \; \boxed{\text{inv}} \; \boxed{\text{sin}} \; = 34.5°$$

Now, since the sine is negative in both the third and fourth quadrants, we get

$$\theta_3 = 180° + 34.5° = 214.5°$$
$$\theta_4 = 360° - 34.5° = 325.5°$$

Example 6. Use a calculator to find θ $(0° \le \theta < 360°)$ if $\cot \theta = -2.573$ and $\cos \theta < 0$.

Solution. To obtain the reference angle for θ, enter 2.573, the absolute value of the given function. Next, push the `1/x` key to obtain 0.38865, which is $\tan \theta$. (Recall that $\tan \theta = 1/\cot \theta$.) Now, with 0.38865 in the register, push `inv` `tan` to obtain the reference angle 21.24°.

We require that both $\cot \theta$ and $\cos \theta$ be negative. Since $\cot \theta$ is negative in the second and fourth quadrants and $\cos \theta$ is negative in the second and third quadrants, the desired angle must be in the second quadrant. Thus,

$$\theta_2 = 180° - 21.24° = 158.76°$$

Exercises for Section 2.4

In Exercises 1–24, use a calculator to find the values of the trigonometric functions.

1. $\sin 13°$
2. $\cos 78°$
3. $\tan 17.3°$
4. $\cos 5.45°$
5. $\cos 100.2°$
6. $\cos 211.1°$
7. $\cos 399°$
8. $\sin 105.7°$
9. $\tan 1540°$

10. sec 50.8° 11. sec 142° 12. cot 305°

13. csc 111.9° 14. csc (−32.6°) 15. cot (−43.3°)

16. sin 54°32′ 17. cos 213°31′ 18. cos 335°56′

19. tan 950°52′ 20. csc 3°46′ 21. sec 59°38′

22. sec 720°58′ 23. cot 100°3′ 24. cot 470°24′

In Exercises 25–40, use a calculator to find θ where $0° \le \theta < 360°$. Express θ to the nearest minute.

25. $\sin \theta = 0.5567$ and $\cos \theta > 0$ 26. $\tan \theta = 1.802$ and $\sec \theta > 0$

27. $\tan \theta = 0.4414$ and $\cos \theta < 0$ 28. $\cos \theta = 0.9002$ and $\sin \theta < 0$

29. $\sin \theta = -0.4253$ and $\tan \theta < 0$ 30. $\sin \theta = 0.4331$ and $\cos \theta > 0$

31. $\cot \theta = 3.0326$ and $\csc \theta > 0$ 32. $\cos \theta = -0.8635$ and $\cot \theta < 0$

33. $\sec \theta = 1.345$ and $\tan \theta < 0$ 34. $\csc \theta = 2.026$ and $\cos \theta < 0$

35. $\tan \theta = -0.3378$ 36. $\sin \theta = 0.8279$

37. $\csc \theta = 1.505$ 38. $\cos \theta = -0.4642$

39. $\cot \theta = -0.5137$ 40. $\sec \theta = 1.104$

2.5 Trigonometric Tables and Interpolations (Optional)

Angles Less Than 90°

The values of the trigonometric functions for acute angles are computed and listed in tables, which may be consulted when a calculator is not available. Two such tables are included in the appendix. Table A is tabulated in degrees to the nearest 10′ increment; Table B is tabulated in degrees to the nearest 0.1° increment. The functional values in both tables are accurate to four decimal places.

Tables A and B are representative of most trigonometric tables tabulated in degrees in that they apparently include only those angles between 0° and 45°. This is because the values of the functions for angles between 45° and 90° are the same as the values of the cofunctions between 0° and 45°. Thus, sin 57° = cos (90° − 57°) = cos 33°. Most tables take advantage of this relation between the cofunctions of complementary angles by placing the complementary angle to the right of the table and the names of the function to be read for angles between 45° and 90° at the bottom. Thus, the table does double duty.

The procedure for using Tables A and B is summarized as follows:

(1) To find the values of the trigonometric functions for angles between 0 and 45 degrees, locate the angle at the left-hand side of the table and the name of the function at the top of the column.

(2) To find the values of the trigonometric functions for angles between 45 and 90 degrees, locate the angle at the right-hand side of the table and the name of the function at the bottom.

(3) The value of the trigonometric function is found in the appropriate column opposite the angle.

Example 1. Use Table A to verify that
(a) $\sin 34°10' = 0.5616$
(b) $\cos 54°30' = 0.5807$
(c) $\tan 65°40' = 2.2113$

and Table B to verify that

(a) $\cos 17.3° = 0.9548$
(b) $\tan 39.9° = 0.8361$
(c) $\csc 70.4° = 1.601$ (Find $\sin 70.4°$ and use $\csc 70.4° = 1/\sin 70.4°$)

Comment: Trigonometric tables are used in two ways—to find the value of a trigonometric function if an acute angle is given and to find the value of the angle when the value of the trigonometric function is known. For example, if you are given that $\cos \theta = \frac{1}{2}$ and θ is acute, then you know that $\theta = 60°$.

Example 2. Use Table A to find the acute angle θ if $\sin \theta = 0.3638$.

Solution. We examine Table A, running down the column headed "sin x" until we come to 0.3638. Then we read across and find that this value corresponds to an angle of 21°20′. This is written

$$\sin \theta = 0.3638$$
$$\theta = 21°20'$$

Example 3. Use Table B to find the acute angle β if $\tan \beta = 1.6066$.

Solution. Running down the column headed "tan θ" at the top of the page, we fail to reach 1.6066 before coming to 45°; therefore, we look in the column headed "tan θ" at the bottom of the page until we come to 1.6066. Reading "angle" from the right-hand column, we find that $\beta = 58.1°$; that is,

$$\tan \beta = 1.6066$$
$$\beta = 58.1°$$

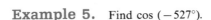

Angles Greater Than 90°

For angles that are not acute, the tables are used together with the concept of a reference angle. (See Section 2.3.)

Example 4. Find cos 145°20′. (See Figure 2.20.)

Solution. We see that 145°20′ is a second-quadrant angle, and, therefore, the reference angle is

$$\theta' = 180° - 145°20' = 34°40'$$

The reference angle is indicated in the figure. Since the cosine function is negative in the second quadrant, we have

$$\cos 145°20' = -\cos 34°40' = -0.8225 \qquad \text{Table A, Appendix}$$

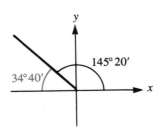

Figure 2.20

Example 5. Find cos (−527°).

Solution. This angle is coterminal with −167°, which, in turn, is coterminal with 193°. This 193° angle, being in the third quadrant, has a reference acute angle of 13°. Hence,

$$\cos (-527°) = \cos 193° = -\cos 13° = -0.9744$$

Example 6. Given tan $\beta = 0.3115$ and sin $\beta < 0$, find β on the interval $0° \leq \beta < 360°$.

Solution. Since tan β is positive and sin β is negative, β must be a third-quadrant angle. From Table B, tan 17.3° = 0.3115. Therefore,

$$\beta = 180° + 17.3° = 197.3°$$

Interpolation

Interpolation is a method of estimating a value between two given values. All mathematical tables, and in particular trigonometric tables, are of necessity tabulated in discrete steps. Thus, when using such tables you will often find it necessary to use interpolated values between tabulated values. For example, if Table A is being used for the trigonometric values, then the values for 41°15′ are not immediately available but must be estimated by using those in the table for 41°10′ and 41°20′.

Obviously, if you have a calculator with trigonometric function capability you will not need to interpolate, since you will not need to consult the trigonometric tables. However, since many other tabulated values in science and engineering are not available on calculators, an understanding of the technique of interpolation will often prove useful. We explain the interpolation process in the context of the trigonometric tables, since that is an immediate application.

The type of interpolation that is the easiest to use and understand is called **interpolation by proportional parts** or **linear interpolation**. Although it is only approximately true, we assume that a change in the angular measurement is proportional to a linear change in the value of the trigonometric function.

A graphical display of a typical error introduced by the assumption of linearity is shown in Figure 2.21 for a trigonometric function whose values are increasing from θ_1 to θ_2. The values for the given trigonometric ratio at θ_1 and θ_2 are assumed to be known from a table, and we assume no such tabulated value is known for the angle θ; hence the necessity to use an interpolated value.

The interpolated value of the function at any θ between θ_1 and θ_2 is the distance from the x-axis to the point P, and the actual value is the distance from the x-axis to the point Q. Thus, the error arising from linear interpolation is length of the line segment PQ. As long as the interval from θ_1 to θ_2 is relatively small and the difference between the known tabulated values is not too large, the error introduced will usually be acceptable.

Using Figure 2.21, we can derive an equation for determining the interpolated value in terms of tabulated values. Since triangles OPA and OCB are similar, the corresponding sides are proportional. Hence,

$$\frac{\overline{PA}}{\overline{CB}} = \frac{\theta - \theta_1}{\theta_2 - \theta_1}$$

where $\overline{PA}$ and $\overline{CB}$ are the distances between the points P and A and the points C and B, respectively. In words,

$$\frac{\text{interpolated value at } \theta - \text{tabulated value at } \theta_1}{\text{tabulated value at } \theta_2 - \text{tabulated value at } \theta_1} = \frac{\theta - \theta_1}{\theta_2 - \theta_1} \qquad (2.6)$$

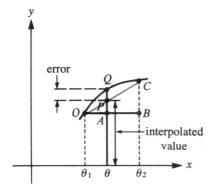

Figure 2.21

The next few examples will help you to learn the method of linear interpolation.

Example 7. Use Table A and the method of linear interpolation to approximate sin 31°14′.

Solution. From Table A, we have sin 31°10′ = 0.5175 and sin 31°20′ = 0.5200. Thus, for a 10′ increase in the angle, the sine function increases by 0.5200 − 0.5175 = 0.0025. For a 4′ increase (from 31°10′ to 31°14′) in the angle, the sine is assumed to increase proportionally by (0.4)(0.0025) = 0.001. Since the sine function is increasing, this correction is added to 0.5175 to obtain sin 31°14′ = 0.5185. The above discussion is summarized in the table below.

Angle	Sine	
31°10′	0.5175	
31°14′	....	0.0025
31°20′	0.5200	

$$\frac{c}{0.0025} = \frac{4}{10}$$

$$c = 0.001$$

Example 8. Using Table A, approximate cos 52°15′.

Solution. From Table A, cos 52°10′ = 0.6134 and cos 52°20′ = 0.6111.

Angle	Cosine	
52°10′	0.6134	
52°15′	....	0.0023
52°20′	0.6111	

$$\frac{c}{0.0023} = \frac{5}{10}$$

$$c = \frac{5}{10}(0.0023)$$

$$= 0.0012$$

Since cos 52°15′ < cos 52°10′, we subtract the correction from 0.6134. Thus,

$$\cos 52°15′ = 0.6134 - 0.0012 = 0.6122$$

Interpolation is also used to approximate the measure of an unknown angle for which a trigonometric functional value is given that does not correspond exactly to any of the tabulated values.

Example 9. Use Table A to find angle θ to the nearest minute if tan θ = 0.3.

Solution. From Table A, we find tan 16°40′ = 0.2994 and tan 16°50′ = 0.3026.

Angle		Tangent
$10\begin{bmatrix} c\begin{bmatrix} 16°40' \\ \cdots\cdots \\ 16°50' \end{bmatrix} \end{bmatrix}$	$\begin{bmatrix} 0.2994 \\ 0.3000 \\ 0.3026 \end{bmatrix}0.0006$	0.0032

$$\frac{c}{10} = \frac{0.0006}{0.0032}$$

$$c = \frac{0.0006}{0.0032}(10) = 2'$$

Adding this 2′ correction to 16°40′, we have $\theta = 16°42'$.

Exercises for Section 2.5

In Exercises 1–10, use Table A or B to find the value of the trigonometric function.

1. sin 13°
2. sin 46°10′
3. tan 17°30′
4. cos 61°40′
5. sec 5°10′
6. tan 44°
7. cot 17°52′
8. csc 38°27′
9. sin 75.5°
10. cot 56.5°

In Exercises 11–20, express the trigonometric function in terms of the same function of a positive acute angle.

11. cos 125°
12. tan 94°
13. sin 225°
14. csc 252°28′
15. tan 1243°
16. tan 100°
17. sec 1000°
18. sin 988°
19. cos 453°
20. cos 5000°

In Exercises 21–30, evaluate the trigonometric function using Table A.

21. sin 154°
22. sin 333°
23. tan 96°
24. cos 205°
25. sec 163°20′
26. tan 200°35′
27. cos 285°50′
28. csc 261°
29. sin 247°36′
30. cos 108°29′

In Exercises 31–40, use Table A to find θ where $0 \le \theta < 360°$. Express θ to the nearest minute.

31. $\sin \theta = -0.4253$, $\tan \theta < 0$
32. $\sin \theta = 0.4331$, $\cos \theta < 0$
33. $\cot \theta = 3.0326$, $\csc \theta < 0$
34. $\cos \theta = -0.8635$, $\cot \theta > 0$
35. $\tan \theta = -6.8269$, $\csc \theta > 0$
36. $\cos \theta = 0.9012$, $\sec \theta > 0$

37. $\cos \theta = 0.4832$

38. $\sec \theta = 1.5411$

39. $\csc \theta = -1.9835$

40. $\sin \theta = -0.5887$

Key Topics for Chapter 2

Define and/or discuss each of the following.

$\sin \theta$, $\cos \theta$, $\tan \theta$, $\cot \theta$, $\sec \theta$, $\csc \theta$ Quadrantal Angles
The Reciprocal Relations   Reference Angle
The Pythagorean Relation   Use of Trigonometric Tables
The Special Angles   Interpolation

Review Exercises for Chapter 2

1. Find the six trigonometric functions of the angle whose terminal side passes through $(-2, 5)$.

2. Determine the quadrant in which the terminal side of θ lies if $\cot \theta < 0$ and $\sec \theta < 0$.

3. Find the other five trigonometric functions of θ if $\cos \theta = -\frac{4}{5}$ and $\tan \theta > 0$.

4. Find the other five trigonometric functions of α if $\csc \alpha = \frac{13}{5}$ and $\cos \alpha < 0$.

In Exercises 5–10, indicate the reference angle for the given angle.

5. $315°18'$ 6. $109°41'$ 7. $241.9°$

8. $185.3°$ 9. $272.2°$ 10. $156.4°$

In Exercises 11–18, evaluate the function.

11. $\tan 203°$ 12. $\sin 310°20'$ 13. $\cos (-112.6°)$

14. $\csc (-9°15')$ 15. $\sec 98.3°$ 16. $\cot 189.6°$

17. $\sin 975°$ 18. $\tan (-1053°)$

19. Given $\sin x = -0.8102$, find x $(0° \le x < 360°)$ to the nearest minute.

20. Given $\tan \theta = 1.202$, find θ $(0° \le \theta < 360°)$ to the nearest minute.

21. Given $\sec \phi = 2.603$ and $\sin \phi < 0$, find ϕ $(0° \le \phi < 360°)$ to the nearest tenth degree.

22. Given $\csc \theta = -1.118$ and $\cot \theta > 0$, find θ $(0° \le \theta < 360°)$ to the nearest tenth degree.

23. Given $\tan \beta = -0.7761$ and $\cos \beta > 0$, find β $(0° \le \beta < 360°)$ to the nearest tenth degree.

24. For a new roadway being surveyed, the distance between two points must be calculated by

$$x = \frac{1500 \sin 25.7°}{\sin 105.3°} \text{ ft}$$

Find x.

25. In filing a flight plan, a pilot must calculate the ground speed of the plane by taking into account the speed and direction of the wind. The ground speed can be expressed as

$$v = \frac{210 \sin 135°}{\sin 40°} \text{ mph}$$

What is the numerical value?

26. In evaluating the resultant of two forces, a student encounters the expression

$$F = \sqrt{5^2 + 9^2 - 2(5)(9) \cos 133°18'}$$

Evaluate F.

3

The Solution of Triangles

A triangle is composed of six parts, the three sides and the three angles. A principal use of the trigonometric functions is for the solution of triangles. To *solve a triangle* means to find the value of each of the six parts. Since the six parts of a triangle are not independent, unknown values can be calculated if certain values are known. For example, if two of the angles are known, the other one is obtained by subtracting the sum of the two known ones from 180°. If the triangle is a right triangle and two of the sides are known, the third side may be obtained using the Pythagorean theorem.

Initially we will assume that one of the angles of the triangle is a right angle. We will look at a few of the numerous classical and modern applications associated with the solution of a right triangle. Later in the chapter we will generalize the discussion to oblique triangles.

3.1 Solving Right Triangles

In discussing right triangles it is customary to designate the vertices and corresponding angles by the capital letters A, B, and C, as shown in Figure 3.1(a). The right angle is usually denoted by the letter C, and the lowercase letters a, b, and c designate the sides opposite angles A, B, and C, respectively. Side c is then the hypotenuse.

To see how the trigonometric functions are related to the parts of a right triangle, consider the right triangle shown in Figure 3.1(b). With angle A in standard position, the coordinates of the vertex at B are (b, a), and the distance from A to B is c. Thus, the six trigonometric functions of angle A in terms of the sides of the standard right triangle are

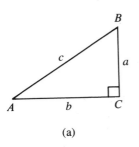

(a)

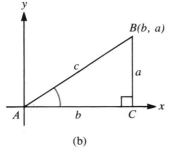

(b)

Figure 3.1

$$\sin A = \frac{a}{c} \qquad \csc A = \frac{c}{a}$$

$$\cos A = \frac{b}{c} \qquad \sec A = \frac{c}{b} \qquad (3.1)$$

$$\tan A = \frac{a}{b} \qquad \cot A = \frac{b}{a}$$

The sides of a right triangle are often referred to in terms of one of the two acute angles. For example, the side of length a is called the **side opposite** angle A, the side of length b is called the **side adjacent** to angle A, and the side of length c is called the **hypotenuse**. Using this terminology, the six trigonometric functions in Definition 3.1 become

$$\sin A = \frac{a}{c} = \frac{\text{opposite side}}{\text{hypotenuse}} \qquad \csc A = \frac{c}{a} = \frac{\text{hypotenuse}}{\text{opposite side}}$$

$$\cos A = \frac{b}{c} = \frac{\text{adjacent side}}{\text{hypotenuse}} \qquad \sec A = \frac{c}{b} = \frac{\text{hypotenuse}}{\text{adjacent side}} \qquad (3.2)$$

$$\tan A = \frac{a}{b} = \frac{\text{opposite side}}{\text{adjacent side}} \qquad \cot A = \frac{b}{a} = \frac{\text{adjacent side}}{\text{opposite side}}$$

The definitions in (3.2) are convenient for solving right triangles.

When a right triangle is given, the six parts may be completely determined if you know two parts other than the right angle, at least one of which is a side.

- If an angle and one of the sides is given, the third angle is simply the complement of the one given. The other two sides are obtained from the values of the known trigonometric functions.
- If two sides are given, the value of the third side is obtained from the Pythagorean theorem. The angles may then be determined by taking ratios of the sides. Each ratio will uniquely determine the value of some trigonometric function.

Thus, in solving right triangles, we make use of the trigonometric functions, the Pythagorean theorem, and the fact that the two acute angles are complementary. You will usually find it to your advantage to make a rough sketch of the triangle. This will help you to determine what is given and which trigonometric functions must be used to find the unknown parts.

The relationship between the accuracy of the sides and the angles is given in Table 3.1. We will use this convention in the examples and the answers to the exercises.

Table 3.1

Accuracy of Sides	Accuracy of Angles
2 significant digits	nearest degree
3 significant digits	nearest 0.1° or 10′
4 significant digits	nearest 0.01° or 1′

Example 1. Solve the triangle in Figure 3.2.

Solution. Since A and B are complementary angles, $B = 90° - 27° = 63°$. Also, $\tan A =$ opposite side/adjacent side, so

$$\tan 27° = \frac{a}{5.9}$$

Solving for a, we get

$$a = 5.9 \tan 27° = 5.9(0.5095) = 3.0$$

Similarly,

$$\cos 27° = \frac{5.9}{c}$$

so

$$c = \frac{5.9}{\cos 27°} = \frac{5.9}{0.8910} = 6.6$$

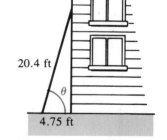

Figure 3.2

Example 2. A 20.4-ft long ladder is placed against a building so that its lower end is 4.75 feet from the base of the building. What angle does the ladder make with the ground?

Solution. The desired angle is designated θ in Figure 3.3. From the figure, we see that

$$\cos \theta = \frac{\text{adjacent side}}{\text{hypotenuse}} = \frac{4.75}{20.4} = 0.233$$

Using a calculator or Table A, we find that $\theta = 76°30′$ to the nearest 10′.

Figure 3.3

Example 3. In Figure 3.4 a radar station tracking a missile determines the angle of elevation to be 20.7° and the line-of-sight distance (called the **slant range**) to be 38.2 km. Determine the altitude and horizontal range of the missile.

Solution. The altitude is

$$h = 38.2 \sin 20.7° = 38.2(0.3535) = 13.5 \text{ km}$$

and the horizontal range is

$$r = 38.2 \cos 20.7° = 38.2(0.9354) = 35.7 \text{ km}$$

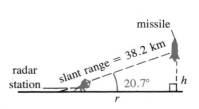

Figure 3.4

Example 4. An airplane flying at a speed of 185 ft/sec starts to descend to the runway on a straight-line glide path that is 7.5° below the horizontal. If the plane is at a 2980-ft altitude at the start of the glide path, how long will it take for the plane to touch down?

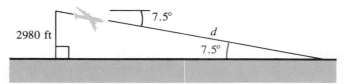

Figure 3.5

Solution. Figure 3.5 shows that the length of the glide path is the hypotenuse of a right triangle. Therefore, we can write

$$\sin 7.5° = \frac{2980}{d}$$

$$d = \frac{2980}{\sin 7.5°} = \frac{2980}{0.1305} = 22{,}800 \text{ ft}$$

Recall that

$$\text{velocity} = \frac{\text{distance}}{\text{time}}$$

Thus, solving for time, we have

$$\text{time} = \frac{\text{distance}}{\text{velocity}} = \frac{22{,}800 \text{ ft}}{185 \text{ ft/sec}} = 123 \text{ sec}$$

The descent takes 123 sec = 2.05 min.

The next example shows how right-triangle trigonometry can be used to find vertical dimensions of objects that appear in reconnaissance and satellite photographs. The analyst needs only the angle of elevation of the sun and the scale of the photograph.

Example 5. A representation of an aerial photograph of a building complex is shown in Figure 3.6(a). If the sun was at an angle of 26.5° when the photograph was taken, how high is the rectangular-shaped building?

Solution. The length of the shadow measures 0.48 cm. To get the real length of the shadow, multiply 0.48 cm by the scale factor given in the photograph. Thus,

$$\text{shadow length} = 0.48(250) = 120 \text{ m}$$

From the model in Figure 3.6(b), we see that $\tan 26.5° = h/120$. Solving for h, we find the height of the building:

$$h = 120 \tan 26.5° = 59.8 \text{ m}$$

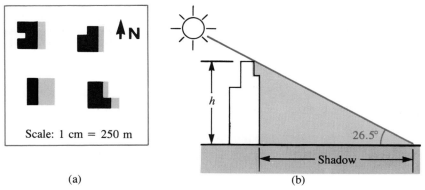

Scale: 1 cm = 250 m

(a)   (b)

Figure 3.6

A parallel of latitude on the Earth's surface is a circle around the Earth in a plane parallel to the equatorial circle. The latitude angle is the angle made by two radii of the Earth, one from the midpoint of the Earth to the equator and one from the midpoint of the Earth to the parallel of latitude. (See Figure 3.7.)

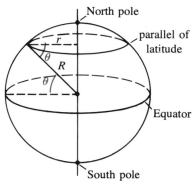

Figure 3.7

Example 6. Show that the length of any parallel of latitude around Earth is equal to the equatorial distance around Earth times the cosine of the latitude angle.

Solution. By the definition of the cosine function,

$$\cos \theta = \frac{r}{R} \quad \text{or} \quad r = R \cos \theta$$

Let the length of the parallel of latitude be C_p. If C_e denotes the average circumference of Earth, then

$$C_p = 2\pi r = 2\pi R \cos \theta = C_e \cos \theta$$

Example 7 shows how right-triangle trigonometry can be used to find the height of an object when a side of the right triangle cannot be obtained.

The procedure is to measure the angle of elevation to the top of the object at two different locations and the distance between the two locations.

Example 7. Two observers who are 4250 ft apart measure the angle of elevation to the top of a mountain to be 18.7° and 25.3°, respectively. (See Figure 3.8.) Find the height of the mountain.

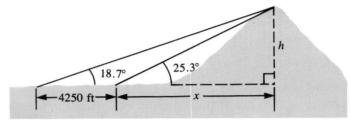

Figure 3.8

Solution. From Figure 3.8 we can write the two equations

$$\tan 18.7° = \frac{h}{x + 4250} \tag{1}$$

$$\tan 25.3° = \frac{h}{x} \tag{2}$$

The only two unknowns in these two equations are x and h. Solving (2) for x, we get

$$x = \frac{h}{\tan 25.3°} = h \cot 25.3°$$

Substituting this expression into Equation (1), we have

$$\tan 18.7° = \frac{h}{h \cot 25.3° + 4250}$$

To solve this equation for h, we proceed as follows.

$$(h \cot 25.3° + 4250) \tan 18.7° = h$$

$$h \cot 25.3° \tan 18.7° + 4250 \tan 18.7° = h$$

$$(1 - \cot 25.3° \tan 18.7°)h = 4250 \tan 18.7°$$

$$h = \frac{4250 \tan 18.7°}{1 - \cot 25.3° \tan 18.7°}$$

$$= \frac{4250(0.3385)}{1 - (2.116)(0.3385)}$$

$$= 5070$$

Thus, the top of the mountain is 5070 ft above the observers.

Exercises for Section 3.1

In Exercises 1–5, solve each right triangle.

1.

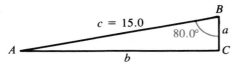

c $a = 4.0$ $b = 7.0$

2.

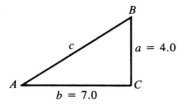

$c = 15.0$ $80.0°$ a b

3.

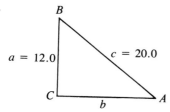

$a = 12.0$ $c = 20.0$ b

4.

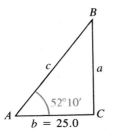

c a $52°10'$ $b = 25.0$

5.

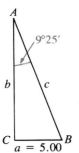

$9°25'$ b c $a = 5.00$

6. One side of a rectangle is half as long as the diagonal. The diagonal is 5.0 m long. How long are the sides of the rectangle? Solve without using the Pythagorean theorem.

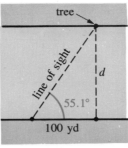

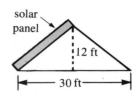

Figure 3.9   Figure 3.10

7. An engineer wishing to know the width of a river walks 100 yd downstream from a point that is directly across from a tree on the opposite bank. If the angle between the river bank and the line of sight to the tree at this second point is 55.1°, what is the width of the river? (See Figure 3.9.)

8. A solar collector is placed on the roof of a house, as shown in Figure 3.10. What angle does the collector make with the vertical?

9. A solar panel (as shown in Figure 3.11) is to be tilted so that angle $\phi = 100°$ when the elevation angle of the sun is 27°. Find h, if the length of the panel is 6.4 m.

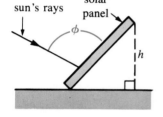

Figure 3.11

10. A television antenna stands on top of a house that is 20 ft tall. The angle subtended by the antenna from a point 30 ft from the base of the building is 15°. Find the height of the antenna.

11. Civil engineers designing a steel truss for a bridge, shown in Figure 3.12, want BC to be 10.0 m and AC to be 7.00 m. What angle will AB make with AC? with BC?

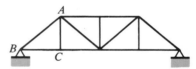

Figure 3.12

12. At noon in the tropics, when the sun is directly overhead, a fisherman holds his 5.0-m pole inclined 30° to the horizontal. How long is the shadow of the pole? How high is the tip of the pole above the level of the other end?

13. A hexagonal bolt head measures 12 mm across the flats. Find the distance c across the corners. (See Figure 3.13.)

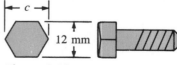

Figure 3.13

14. A 24-in.-wide sheet of aluminum is bent along its centerline to form a V-shaped gutter. Find the angle between the sides of the gutter if it is 7 in. deep.

15. A cylindrical steel bar rests in a V-shaped groove, as shown in Figure 3.14. Find the radius of the bar if $a = 1.5$ in. and $\theta = 45°$.

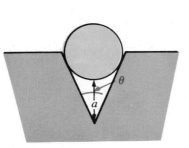

Figure 3.14

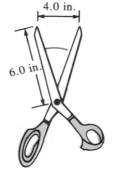

Figure 3.15

16. The length of each blade of a pair of shears from the pivot to the point is 6.0 in. (See Figure 3.15.) When the points of the open shears are 4.0 in. apart, what angle do the blades make with each other?

17. From the top of a building that is 220.0 ft high, an observer looks down on a parking lot. If the lines of sight of the observer to two different cars in the lot are 28°15′ and 36°20′ below the horizontal, respectively, what is the distance between the two cars?

18. Lock-on for a certain automatic landing system, for use when visibility is poor, occurs when the airplane is 4.0 mi (slant range) from the runway and at an altitude of 3800 ft. If the glide path is a straight line to the runway, what angle does it make with the horizontal?

19. An airplane takes off with an airspeed of 265 ft/sec and climbs at an angle of 8.7° with the horizontal until it reaches an altitude of 5800 ft. (See Figure 3.16.) How long does it take the plane to reach this altitude?

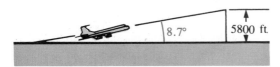

Figure 3.16

20. A civil engineering student is given the sketch of a survey shown in Figure 3.17 and asked to find the distance x. Show how this can be done using right triangles and then compute x.

21. Ten holes are to be drilled in a circular cover plate of a rocket motor. The holes are equally spaced on a circle of radius 12.9 cm, as shown in Figure 3.18. What is the straight-line center-to-center distance between the holes?

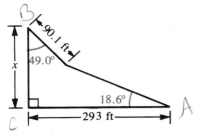

Figure 3.17

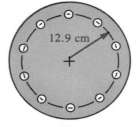

Figure 3.18

22. A 78.0-ft rocket with its base on the ground is elevated at an angle of 69°40'. What is the height of the nose of the rocket above the ground?

23. The triangular wing of a delta-wing airplane is swept back at an angle of 51.5° to the centerline of the fuselage. If the leading edge of the wing is 28.3 ft long and the fuselage is 4.20 ft wide, what is the wingspan of the airplane? (See Figure 3.19.)

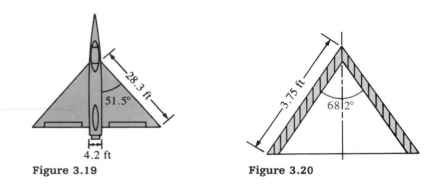

Figure 3.19   **Figure 3.20**

24. The shroud of a nose cone is shown in cross-section in Figure 3.20. What is the diameter of the rocket using this nose cone?

25. An astronomer measures the shadow of a crater on the moon in a photograph and finds its length to be 0.32 cm. If the sun was at an angle of 49.1° to the horizontal when the photograph was taken, how deep is the crater? (See Figure 3.21.) Assume the map scale is 1 cm = 2500 m.

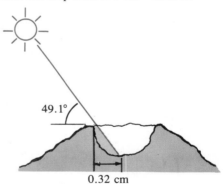

Figure 3.21

26. Figure 3.22 represents an aerial photograph of a cliff in a remote region of Antarctica. Compute the height of the cliff if the elevation angle of the sun was 19.0° when the photograph was taken.

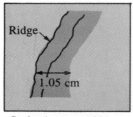
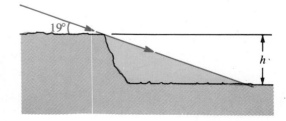

Scale: 1 cm = 1200 m

Figure 3.22

27. A space capsule orbits the moon at an altitude of 100 mi. As shown in Figure 3.23, a sighting from the capsule to the moon's horizon shows an angle of depression of 22.6°. Find the radius of the moon.

28. Determine the length of the Arctic Circle (66°33′N). Assume the circumference of the earth is 25,000 mi.

29. The 40° parallel of latitude passes through the United States. If a citizen of the United States were to travel due east along the 40° parallel, how far would he or she travel before returning home?

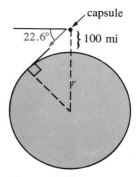

Figure 3.23

An interesting method of measuring the height of clouds is shown in Figure 3.24. A sweeping light beam is placed at point *A*, and a light source detector is placed at point *B*. The axis of the detector is kept vertical, and the light beam is made to sweep from the horizontal ($\alpha = 0°$) to the vertical ($\alpha = 90°$). When the beam illuminates the base of the clouds directly above the detector, as shown in the figure, the detector is activated, and the angle α is read. Since *d* is known, the height *h* can be computed. Use the cloud altitude detector in Exercises 30–32.

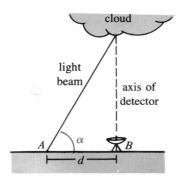

Figure 3.24

30. Compute the height of the cloud if the light source is 100 ft from the detector and the angle is 82°.

31. Find the angle α when the clouds are 1000 ft high and the light source is located 100 ft from the detector.

32. Suppose that a detector that is 1 km from the light beam detects a cloud layer when $\alpha = 60°$ and another layer when $\alpha = 75°$. What is the vertical separation of the cloud layers?

33. An observer at the base of a hill knows that the television antenna on top of the hill is 550 ft high. If the angle of elevation from the observer to the base of the antenna is 16.4° and to the top of the antenna is 29.1°, how high is the hill?

34. As shown in Figure 3.25, a fishing boat sailing due north sites a lighthouse as 16.2° east of north. If 12.7 miles later the lighthouse is sited as 43.7° east of north, how close will the boat come to the lighthouse?

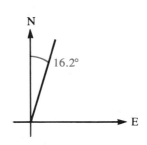

Figure 3.25

35. A woman walking along the prairie stops to measure the angle of elevation to a high mountain. It measures 30°. The woman then walks a kilometer toward the mountain and measures again. This time the angle of elevation is 45°. (See Figure 3.26.) How high is the mountain?

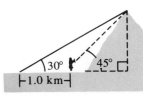

Figure 3.26

Physical quantities that can be described by a single number (such as temperature and volume) are called **scalar** quantities. Other quantities that can only be described by both a magnitude and a direction (such as velocity and force) are called **vector** quantities. For example, velocity is a vector quantity because an object moving 70 miles per hour to the north is quite different from one moving 70 miles per hour to the east.

Vectors are set in boldface type to distinguish them from scalars. The magnitude of a vector **F** is denoted by |**F**| and is always a positive number. The direction of a vector is given in a variety of ways, depending on the application. A vector is represented graphically by an arrow, with the tip of the arrowhead at the **terminal point** of the vector. The length of the arrow corresponds to the magnitude of the vector, and the direction of the arrowhead gives its direction. Thus, in Figure 3.27, one arrow represents a velocity of 100 mph due north and the other a force of 200 lb acting at 45° above the horizontal.

For mathematical purposes two vectors are considered equal if they have the same direction and length, regardless of the location of the initial point of the vector. Thus, in Figure 3.28 all the vectors are mathematically equivalent.

Note that this type of vector equality may not always be the kind you need. For example, if **F** in Figure 3.29 is a 10-lb force pointing down, it does make a difference in which of the three places it is applied. Vectors for which you may ignore the actual point of application are said to be *free*. Physical situations such as those shown in Figure 3.29 cannot be described using free vectors and hence will not be discussed here.

If we first impose a coordinate system on the plane, we can move all the vectors in this plane so that their initial points are at the origin. (See Figure 3.30.) Such vectors are said to be in **standard position**. In effect, a vector at the origin represents all other vectors with the same direction and length.

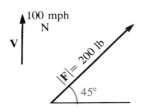

Figure 3.27

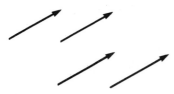

Figure 3.28

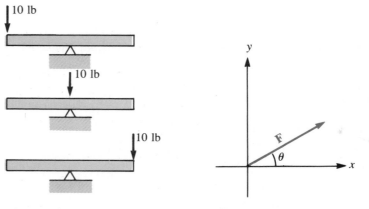

Figure 3.29   **Figure 3.30**

If a vector is placed in standard position, its length can be found from the Pythagorean theorem. The direction of the vector is the angle it makes with the positive x-axis.

Example 1. Find the magnitude and direction of the vector **A** in standard position whose terminal point is at (2, 1).

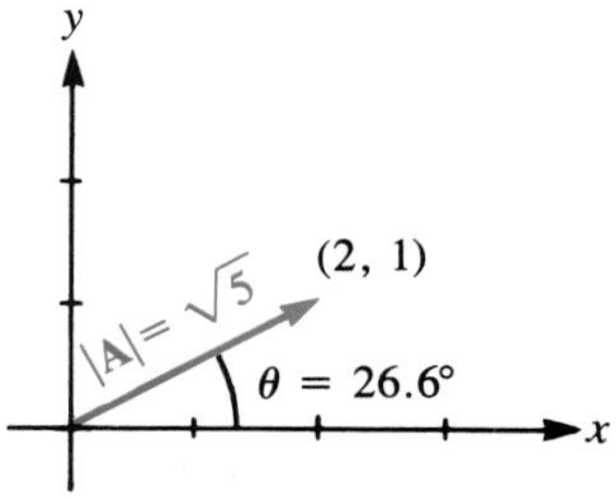

Solution. Figure 3.31 shows that the length of the vector is

$$|A| = \sqrt{2^2 + 1^2} = \sqrt{5}$$

and the angle θ that **A** makes with the positive x-axis is given by $\tan \theta = \frac{1}{2}$. Thus,

$$\theta = 26.6°$$

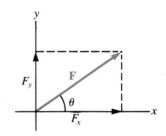

Figure 3.31

The perpendicular projections of the vector onto the x- and y-axes are called the **components** of the vector. We say that a vector **F** is resolved into its x and y components, called the **horizontal** and **vertical components of F**, respectively. Resolving a vector into its x and y components is a simple problem of trigonometry. From Figure 3.32, we get

$$F_x = |F| \cos \theta \quad \text{Horizontal component of F}$$
$$F_y = |F| \sin \theta \quad \text{Vertical component of F}$$
$$F_x^2 + F_y^2 = |F|^2$$

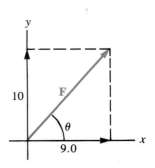

Figure 3.32

Example 2. Find the horizontal and vertical components of a force vector of magnitude 15 lb acting at an angle of 30° to the horizontal.

Solution

$$F_x = |F| \cos \theta = (15)(0.866) = 13$$
$$F_x = |F| \sin \theta = (15)(0.5) = 7.5$$

Example 3. Find the magnitude and direction of the vector whose components are shown in Figure 3.33.

Solution. The magnitude is

$$|F| = \sqrt{10^2 + 9^2} = \sqrt{181} \approx 13.5.$$

The angle that the vector makes with the horizontal is determined as follows:

$$\tan \theta = \frac{10}{9.0} = 1.111$$

$$\theta = 48° \quad \textit{To the nearest degree}$$

Figure 3.33

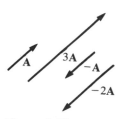

Figure 3.34

Given any vector **A**, we can obtain other vectors in the same direction as **A** (or the direction opposite to that of **A**) by multiplying **A** by a real number c. The resulting vector, denoted by $c\mathbf{A}$, is a vector that points in the same direction as **A** if $c > 0$ or the direction opposite that of **A** if $c < 0$. The magnitude of $c\mathbf{A}$ is $|c||\mathbf{A}|$; that is, it is larger than $|\mathbf{A}|$ if $|c| > 1$ and smaller than $|\mathbf{A}|$ if $|c| < 1$. The vector $c\mathbf{A}$ is called a **scalar** multiple of the vector **A**. Figure 3.34 shows some scalar multiples of a given vector **A**.

The particular scalar multiple of a vector obtained by multiplying a vector **A** by the reciprocal of its magnitude is a vector in the direction of **A** whose length is 1. This is the **unit vector** in the direction of **A**, denoted by $\mathbf{u}_A$. Thus,

$$\mathbf{u_A} = \frac{\mathbf{A}}{|\mathbf{A}|} \tag{3.3}$$

A rule called the **parallelogram rule** gives the procedure for adding two vectors.

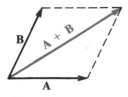

Figure 3.35

> **The Parallelogram Rule:** To add two vectors **A** and **B**, place both with their initial points together. Then form a parallelogram with these vectors as sides. The vector that starts from the initial point of **A** and **B** and forms the diagonal of the parallelogram is called the **sum**, or **resultant**, of **A** and **B**. Figure 3.35 depicts the resultant of two vectors.

Addition of vectors is frequently accomplished by using the component method of representation for each vector in the sum. For instance, the vector sum $\mathbf{A} + \mathbf{B}$ is shown in Figure 3.36, along with the components A_x, A_y, B_x, and B_y. From the figure, we see that the horizontal component of $\mathbf{A} + \mathbf{B}$ is $A_x + B_x$ and the vertical component is $A_y + B_y$. In words, the horizontal component of the sum of two vectors is the sum of the individual horizontal components, and the vertical component is the sum of the individual vertical components. Thus, finding the sum of two vectors can be reduced to finding a vector from its x and y components.

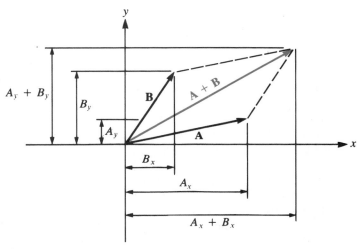

Figure 3.36

Example 4. Find the sum of the two vectors given in Figure 3.37, **A** of magnitude 100 and direction 63.0° and **B** of magnitude 40 and direction 325.0°.

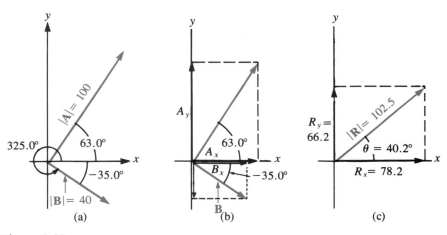

Figure 3.37

Solution. First, we find A_x, A_y, B_x, and B_y.

$$A_x = 100 \cos 63° = 100(0.4540) = 45.4$$
$$A_y = 100 \sin 63° = 100(0.8910) = 89.1$$
$$B_x = 40 \cos (-35°) = 40(0.8192) = 32.8$$
$$B_y = 40 \sin (-35°) = 40(-0.5736) = -22.9$$

Let $\mathbf{A} + \mathbf{B} = \mathbf{R}$. Then,

$$R_x = A_x + B_x = 45.4 + 32.8 = 78.2$$
$$R_y = A_y + B_y = 89.1 + (-22.9) = 66.2$$

Finally, the magnitude of $\mathbf{A} + \mathbf{B}$ is

$$|\mathbf{R}| = \sqrt{R_x^2 + R_y^2} = \sqrt{78.2^2 + 66.2^2} = 102.5$$

and the direction is given by

$$\tan \theta = \frac{R_y}{R_x} = \frac{66.2}{78.2} = 0.8465$$

$$\theta = 40.2°$$

An Alternative Form of the Parallelogram Rule

Vectors $\mathbf{A}$ and $\mathbf{B}$ can also be added by placing the initial point of $\mathbf{B}$ at the terminal point of $\mathbf{A}$. Then, $\mathbf{A} + \mathbf{B}$ is the arrow whose initial point is at the initial point of $\mathbf{A}$ and whose terminal point is at the terminal point of $\mathbf{B}$. [See Figure 3.38(a).] Figure 3.38(b) is included to show that this method is just another way of applying the parallelogram rule.

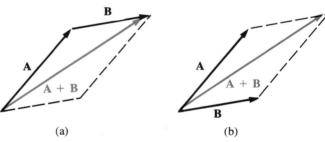

(a)   (b)

Figure 3.38

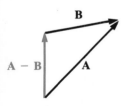

Figure 3.39

For vector subtraction, we reverse the procedure for addition by placing the terminal point of $\mathbf{B}$ at the terminal point of $\mathbf{A}$. Then, $\mathbf{A} - \mathbf{B}$ is the arrow drawn from the initial point of $\mathbf{A}$ to the initial point of $\mathbf{B}$. (See Figure 3.39.)

One advantage of using this rule for vector addition is that we can add several vectors sequentially without computing intermediate resultants. For instance, to add the three vectors shown in Figure 3.40, place the initial point of $\mathbf{B}$ at the terminal point of $\mathbf{A}$ and then place the initial point of $\mathbf{C}$ at the terminal point of $\mathbf{B}$. The resultant vector $\mathbf{A} + \mathbf{B} + \mathbf{C}$ is the vector drawn from the initial point of $\mathbf{A}$ to the terminal point of $\mathbf{C}$. This method is particularly useful when vector addition is being done graphically. Notice that since addition is a commutative operation for vectors ($\mathbf{A} + \mathbf{B} = \mathbf{B} + \mathbf{A}$), the order of the addition is unimportant. In Figure 3.40(b) and (c) the order of addition is different but the resultant vector is the same.

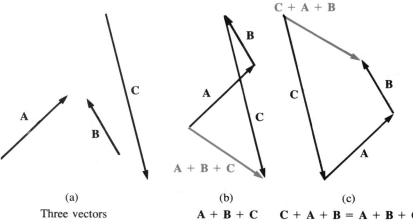

(a)

Three vectors

(b)

A + B + C

(c)

C + A + B = A + B + C

Figure 3.40

In Exercises 1–10, draw the vector whose initial point is at the origin and whose terminal point is at the indicated point. Calculate the magnitude and direction of each vector.

1. $(1, 2)$

2. $(-3, 2)$

3. $(\sqrt{2}, \sqrt{7})$

4. $(1, -1)$

5. $(-4, -4)$

6. $(5, 8)$

7. $(4, -3)$

8. $(-3, -7)$

9. $(-\sqrt{3}, 6)$

10. $(\sqrt{5}, 3)$

In Exercises 11–20, find the horizontal and vertical components of each vector.

11. $|\mathbf{F}| = 10$, $\theta = 50°$

12. $|\mathbf{F}| = 25$, $\theta = 75°$

13. $|\mathbf{F}| = 13.7$, $\theta = 34°10'$

14. $|\mathbf{F}| = 0.751$, $\theta = 56°30'$

15. $|\mathbf{F}| = 158$, $\theta = 125°$

16. $|\mathbf{F}| = 875$, $\theta = 145°$

17. $|\mathbf{F}| = 43.5$, $\theta = 220°$

18. $|\mathbf{F}| = 9.41$, $\theta = 195°$

19. $|\mathbf{F}| = 10.4$, $\theta = 335°$

20. $|\mathbf{F}| = 0.05$, $\theta = 280°$

In Exercises 21–29, find the magnitude and direction of the vector whose components are given.

21. $F_x = 20.0$, $F_y = 15.0$

22. $F_x = 56.0$, $F_y = 13.0$

23. $F_x = 17.5$, $F_y = 69.3$

24. $F_x = 0.012$, $F_y = 0.200$

25. $F_x = 0.130$, $F_y = 0.080$

26. $F_x = 1930$, $F_y = 565$

27. $F_x = 8.0$, $F_y = -7.0$ **28.** $F_x = -3.0$, $F_y = 5.0$

29. $F_x = -2.0$, $F_y = -7.0$

In Exercises 30–37, the vectors are defined in terms of a magnitude and a direction. Find the sum of **A** and **B**.

30. $|\mathbf{A}| = 20$, $\theta_A = 15°$
 $|\mathbf{B}| = 25$, $\theta_B = 50°$

31. $|\mathbf{A}| = 16$, $\theta_A = 25°$
 $|\mathbf{B}| = 22$, $\theta_B = 70°$

32. $|\mathbf{A}| = 15$, $\theta_A = 0°$
 $|\mathbf{B}| = 26$, $\theta_B = 60°$

33. $|\mathbf{A}| = 9.5$, $\theta_A = 90°$
 $|\mathbf{B}| = 5.1$, $\theta_B = 40°$

34. $|\mathbf{A}| = 2.5$, $\theta_A = 35°$
 $|\mathbf{B}| = 3.0$, $\theta_B = 120°$

35. $|\mathbf{A}| = 29.2$, $\theta_A = 15.6°$
 $|\mathbf{B}| = 82.6$, $\theta_B = 150°$

36. $|\mathbf{A}| = 125$, $\theta_A = 145°$
 $|\mathbf{B}| = 92$, $\theta_B = 215°$

37. $|\mathbf{A}| = 550$, $\theta_A = 140°$
 $|\mathbf{B}| = 925$, $\theta_B = 310°$

In Exercises 38–45, perform each indicated operation graphically, using the vectors in Figure 3.41.

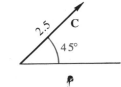

Figure 3.41

38. $\mathbf{A} + \mathbf{B}$ **39.** $3\mathbf{B} - \mathbf{A}$

40. $4\mathbf{A} + \mathbf{B} + 5\mathbf{C}$ **41.** $\mathbf{A} - 2\mathbf{B}$

42. $\mathbf{C} - \mathbf{B} + 2\mathbf{A}$ **43.** $2\mathbf{C} - \mathbf{B} - \mathbf{A}$

44. $\mathbf{A} + 3\mathbf{C} - \mathbf{B}$ **45.** $\mathbf{B} - 2\mathbf{A} + \mathbf{C}$

3.3 Applications of Vectors

Vectors are important tools for solving a wide range of physical problems. We have already mentioned that velocity and force are vector quantities. The examples and exercises in this section illustrate how vectors are used to solve problems involving these quantities.

   Consider a boat moving across a river, as shown in Figure 3.42. If the pilot heads for a point directly on the other shoreline the boat will end up downstream from that point, because as the boat moves across the river it is being carried downstream by the current. The true motion of the boat is given by the vector addition of the velocity vector of the boat and that of the current.

Example 1. A boat that can travel at a rate of 3.5 km/hr in still water is pointed directly across a river having a current of 4.8 km/hr. Calculate the actual velocity of the boat relative to the shoreline.

Solution. Referring to the diagram in Figure 3.42, we see that the velocity vector is the vector sum of the velocity of the boat and that of the current. Since these vectors are at right angles, we use the Pythagorean theorem to get

$$|\mathbf{v}| = \sqrt{3.5^2 + 4.8^2} = 5.9 \text{ km/hr}$$

The angle θ that the velocity vector makes with the shoreline is then given by

$$\tan \theta = \frac{3.5}{4.8} = 0.7292$$

$$\theta = 36°$$

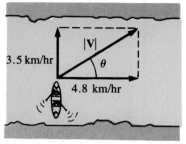

Figure 3.42

Navigation

In navigation it is customary to measure the direction in which a ship or airplane is moving by an angle called the **course** of the vehicle. Specifically, the course of a ship or airplane is the angle measured clockwise from north to the direction in which the ship or plane is moving. (See Figure 3.43.)

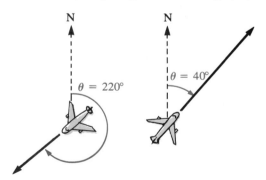

Figure 3.43 Course Angle

To maintain a certain course in a crosswind, the pilot must point the plane slightly into the wind. The wind has the same effect on the plane as the current of the river had on the boat in Example 1. The direction in which a craft is pointed in order to maintain a certain course is called its **heading**. The relationship between heading and course is shown in Figure 3.44. The triangle formed by the three vectors shows that the course vector is the vector sum of the heading vector and the wind vector.

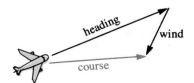

Figure 3.44 Heading vector + wind vector = course vector

Example 2. The pilot of an airplane with a cruising speed of 220 mph wishes to fly a course of 270° in a wind of 15 mph from the south. Calculate the heading the pilot should set for this course.

Solution. Since the wind is from the south and the desired course is due west, the vector sum is the right triangle shown in Figure 3.45. The angle ϕ between the course vector and the heading vector is given by

$$\sin \phi = \frac{15}{220} = 0.0682$$

$$\phi = 3.9°$$

Finally, from the figure we see that the desired heading is

$$\theta = 270° - \phi = 270° - 3.9° = 266.1°$$

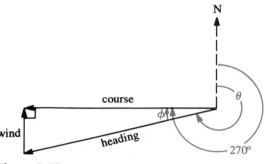

Figure 3.45

An airplane's ground speed depends on the wind field in which the airplane is flying. For instance, if a plane with a cruising speed of 200 mph is flying into a 10-mph head wind, its ground speed is 190 mph; if it is flying with a 10-mph tail wind, its ground speed is 210 mph. In general, the **ground speed** is the vector sum of the airspeed and the wind speed, as shown in Figure 3.46.

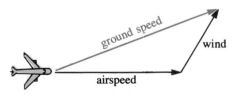

Figure 3.46

Example 3. Consider a flight from Chicago to Boston to be along a west to east direction, with an airline distance of 870 miles. A light plane having an airspeed of 180 mph is to make the trip. How many flying hours will it take for the trip if there is a constant southerly wind* of 23.0 mph? What is the heading for the trip?

---

* Wind direction is conventionally specified by the direction from which the wind is blowing. Thus, a "southerly wind" is a wind blowing from the south.

Solution. Let θ be the angle necessary to compensate for the wind. Then, from Figure 3.47, we have

$$\sin \theta = \frac{23.0}{180} = 0.128$$

$$\theta = 7°20'$$

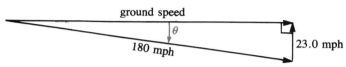

ground speed

θ

180 mph

23.0 mph

Figure 3.47

Hence the ground speed of the plane is

$$(\cos 7°20')(180 \text{ mph}) = 179 \text{ mph}$$

The trip will take

$$\frac{870 \text{ mi}}{179 \text{ mph}} = 4.86 \text{ hr, or about 4 hr 52 min}$$

The heading for the trip is $90° + 7°20'$, or $97°20'$. (See Figure 3.48.)

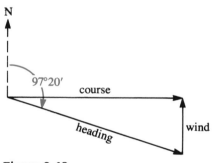

N

97°20'

course

heading

wind

Figure 3.48

If the vector quantities to be added are not at right angles, the sum may be obtained by the method of addition of components discussed in Section 3.2.

Example 4. A ship with a heading of 72° and a speed of 12 knots moves through an ocean current of 2 knots from the north. Calculate the course of the ship.

Solution. Figure 3.49 shows the conditions.

Since the heading vector and the current vector are not at right angles, we use the addition of components method to get the course vector. The north and east components of the ship's speed are

$$v_n = 12 \cos 72° = 3.7 \text{ knots}$$

$$v_e = 12 \sin 72° = 11.4 \text{ knots}$$

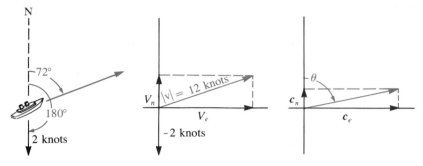

Figure 3.49

The north and east components of the course vector are then

$$c_n = -2 + 3.7 = 1.7 \text{ knots}$$
$$c_e = v_e = 11.4 \text{ knots}$$

The course angle is given by

$$\tan \theta = \frac{c_e}{c_n} = \frac{11.4}{1.7} = 6.706$$
$$\theta = 81.5°$$

Force Analysis

Perhaps the most common use of vectors is in representing forces. Considerable time is spent in fields such as physics and mechanical engineering analyzing the effects of forces. An important step in this process is representing force vectors in component form.

Example 5. A physics lab experiment on components of forces is shown schematically in Figure 3.50(a). Calculate the force parallel to the plane.

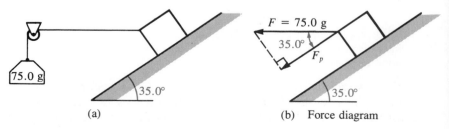

Figure 3.50

Solution. The vector representation of the 75.0-g force is shown in Figure 3.50(b). Since we want the component of force parallel to the plane, we must use

$$\cos 35.0° = \frac{F_p}{75.0}$$

Solving for F_p, we get

$$F_p = 75.0 \cos 35.0° = 75.0(0.8192) = 61.4 \text{ g}$$

as the parallel component of the applied force.

The weight of an astronaut on the moon is one-sixth his or her weight on Earth. This fact has a marked effect on such simple acts as walking, running, jumping, and the like. To study these effects and to train astronauts to work under lunar gravity conditions, scientists at NASA Langley Research Center designed an inclined plane apparatus to simulate reduced gravity.

The apparatus consists of an inclined plane and a sling that holds the astronaut in a position perpendicular to the inclined plane, as shown in Figure 3.51. The sling is attached to one end of a long cable that runs parallel to the inclined plane. The other end of the cable is attached to a trolley that runs along a track high overhead. This device allows the astronaut to move freely in a plane perpendicular to the inclined plane.

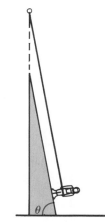

Figure 3.51

Example 6. Make a vector diagram to show the components of the astronaut's weight that are parallel to the plane and perpendicular to the plane. The perpendicular component represents the force exerted by the feet of the astronaut against the plane.

Solution. In order to draw a correct vector diagram it is necessary to know that weight is a force that always acts vertically downward—that is, toward the center of the earth. Figure 3.52 shows the components, where W is the weight of the astronaut. The parallel component is $W \sin \theta$, and the perpendicular component is $W \cos \theta$.

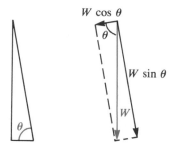

Figure 3.52

Example 7. From the point of view of the astronaut in the sling, the inclined plane is the ground, and the downward force against the inclined plane is $W \cos \theta$. What is the value of θ required to simulate lunar gravity?

Solution. To simulate lunar gravity, we must have $W \cos \theta = W/6$. Thus,

$$\cos \theta = \frac{1}{6} = 0.1667$$

$$\theta = 80°24' \qquad \text{To the nearest minute}$$

The next example deals with the design of a lunar lander. Surprisingly, it is an application of trigonometry that gives the design requirements for construction of the legs of the lander. Since three points determine a plane, lunar landers are designed with three legs.

Figure 3.53

Before discussing the lunar lander problem, we note that an object is said to be in **equilibrium** if the forces acting on the object have a vector sum equal to 0. Thus, to be in equilibrium, a force in one direction must be balanced by an equal force in the opposite direction. For instance, in Example 6 the component of weight parallel to the plane ($W \sin \theta$) must be balanced by the tension in the cable, and the perpendicular component ($W \cos \theta$) must be balanced by the force exerted by the plane.

Example 8. A spacecraft designed to soft-land on the moon has three feet that form an equilateral triangle on level ground. Each of the three legs makes an angle of 37.0° with the vertical. If the impact force of 15,000 lb is evenly distributed, find the axial force on each leg.

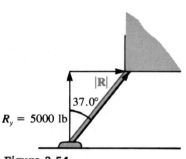

$R_y = 5000$ lb

Figure 3.54

Solution. We assume that each of the three legs shares the impact force equally. Thus, each leg must withstand a vertical force of 5000 lb. Since the legs are in equilibrium, each leg must have an internal force whose vertical component balances the impact force. From the force diagram in Figure 3.54, we have

$$\cos 37.0° = \frac{5000}{|\mathbf{R}|}$$

$$|\mathbf{R}| = \frac{5000}{\cos 37.0°} = 6260 \text{ lb}$$

Exercises for Section 3.3

1. What are the horizontal and vertical components of the velocity of a ball thrown 100 ft/sec at an angle of 40° with the horizontal?

2. A boat that travels at the rate of 5 mph in still water is pointed directly across a stream having a current of 3 mph. What will be the actual speed of the boat, and in which direction will the boat go?

3. In which direction must the boat in Exercise 3 be pointed for the boat to go straight across the stream?

4. An airplane having a speed of 190 knots starts to climb at an angle of 13° above the horizontal. How fast is the airplane rising vertically?

5. A balloon rising at the rate of 20 ft/sec is being carried horizontally by a wind that has a velocity of 25 mph. (See Figure 3.55.) Find the actual velocity of the balloon and the angle that its path makes with the vertical (60 mph = 88 fps).

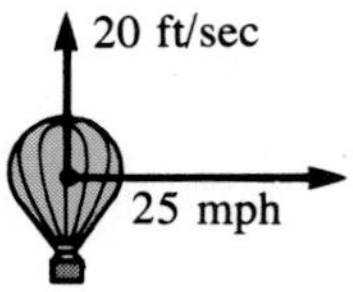

6. An object is thrown vertically downward with a speed of 50 ft/sec from a plane moving horizontally with a speed of 250 ft/sec. What is the velocity of the object as it leaves the plane?

7. A bullet is fired from a plane at an angle of 20° below the horizontal in the direction in which the plane is moving. If the bullet leaves the muzzle of the

20 ft/sec

25 mph

Figure 3.55

gun with a speed of 1200 ft/sec and the plane is flying 500 ft/sec, what is the resultant velocity of the bullet?

8. A rocket is launched from a ship at an elevation angle of 42° in the direction in which the ship is moving. The speed of the rocket is 540 ft/sec, and that of the boat is 45 ft/sec. Find the resultant velocity of the rocket.

9. A plane is headed due north at 300 mph. If the wind is from the east at 50 mph, what is the ground speed of the plane? What is its course?

10. A plane is headed due west with an airspeed of 150 mph. If the wind is from the southwest at 35 mph, what is the ground speed of the plane? What is the course of the plane?

11. An airplane with a heading of 140° and a cruising speed of 185 mph is flying in an 18-mph wind field from due south. Calculate the ground speed of the airplane.

12. Calculate the ground speed of the airplane in Exercise 11 if the wind field is from due west.

13. A pilot is preparing a flight plan from the Juliette airport to Denver, which is 280 miles due north. If the airplane cruises at 160 mph and there is a constant 18.0-mph wind from the west, what heading should the pilot fly? How long will it take to make the trip?

14. If the airplane in Exercise 13 makes the return trip in a west wind of 32.0 mph, determine the heading the pilot should fly.

15. Find the horizontal and vertical components of the force in Figure 3.56.

16. What are the horizontal and vertical components in Figure 3.56 if $|\mathbf{F}| = 250$ lb and $\theta = 21.7°$?

17. The weight of an object is represented by a vector acting vertically downward. (See Figure 3.57.) If a 15.3-lb block rests on a plane inclined at 28.4° above the horizontal, what is the component of force acting perpendicular to the plane?

18. In Exercise 17, find the component of force acting parallel to the plane.

19. A 75.0-lb block rests on an inclined plane. If the block exerts a force of 67.0 lb perpendicular to the plane, what is the angle of inclination of the plane?

20. A horizontal force of 750 lb is applied to a block resting on a plane inclined at 17.9° above the horizontal. Find the component of the force parallel to the plane.

21. What is the angle of the inclined plane used to simulate gravity for a 200-lb astronaut walking on an asteroid whose gravity is $\frac{1}{8}$ that of Earth?

22. If the inclined plane used to simulate gravity were inclined at 60° (see Figure 3.58), what percentage of the astronaut's weight would bear against the plane?

23. Suppose that each of the three legs of a lunar lander will withstand an axial load of 1000 lb. What is the maximum angle that the legs can make with the vertical and still support a total impact force on the lander of 2500 lb?

24. Suppose that a lunar lander is to be designed for a total impact force of 3500 lb and that each of the three legs makes an angle of 28° with the vertical. Find the design impact force in each leg.

25. In Figure 3.59, an object weighing 120 lb hangs at the end of a rope. The object

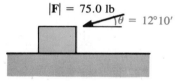

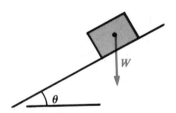

Figure 3.56

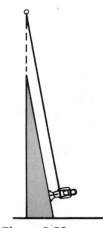

Figure 3.57

Figure 3.58

is pulled sideways by a horizontal force of 30 lb. What angle does the rope make with the vertical? (*Hint*: Weight is a vector that is always considered to be acting vertically downward, and the system is in equilibrium.)

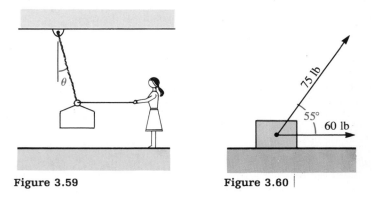

Figure 3.59   Figure 3.60

26. A force of 60 lb acts horizontally on an object. Another force of 75 lb acts on the object at an angle of 55° with the horizontal. What is the resultant of these forces? (See Figure 3.60.)

27. Find the resultant force in Figure 3.60, if the 75-lb force is replaced by a 115-lb force.

Figure 3.61(a) shows a schematic diagram of an electric circuit containing an alternating current generator, a resistance R, and an inductive reactance X_L, connected in series. A quantity Z, called the impedance of the circuit and used in a.c. circuit theory, is related to the resistance and the inductive reactance by the right-triangle relationship shown in Figure 3.61(b). The angle θ is called the phase angle. The following exercises are illustrative of the use of the impedance triangle in a.c. circuit theory. In Exercises 28–33, solve for the missing components in the impedance triangle.

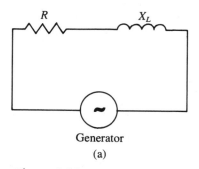

Generator
(a)

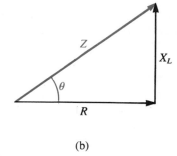

(b)

Figure 3.61

28. $R = 30.0$, $\theta = 60.0°$

29. $R = 30.0$, $Z = 60.0$

30. $Z = 100$, $\theta = 20.0°$

31. $R = 200$, $X = 100$

32. $Z = 1000$, $X = 800$

33. $X = 25.0$, $\theta = 30.0°$

3.4 Oblique Triangles: The Law of Cosines

Any triangle that is not a right triangle is called **oblique**. Hence, in an oblique triangle, no angle is equal to 90°. In the rest of this chapter we will consider conditions under which oblique triangles can be solved. As noted in the discussion of right triangles, knowledge of at least three of the six parts is necessary to solve a triangle. The three parts needed are not arbitrary. For example, three angles will not define a unique solution, because many similar triangles can be constructed with these angles but different side lengths. There are four different combinations of side lengths and angles that will describe a triangle.

Case 1: Two sides and the included angle are given.

Case 2: Three sides are given.

Case 3: Two angles and one side are given.

Of course, as was noted earlier, the given information must be such that a triangle can be formed. For instance, no triangle can be formed from a given side and two angles of 88° and 95°, because the sum of the angles of a triangle cannot exceed 180°. It is important that you be aware of inconsistencies of this type when solving triangles. A fourth case that arises in solving oblique triangles is important, even though the information given may yield two different triangles.

Case 4: Two sides and an angle opposite one of the sides are given.

This case is sometimes referred to as the **ambiguous** case since two triangles, one triangle, or no triangles may result from data given in this form. For instance, Figure 3.62 shows two triangles that can be obtained from the information $a = 15$, $b = 20$, and $A = 20°$. In Section 3.6, we will discuss this case in detail. For the present, you should remember that it is possible for two noncongruent triangles to have the same Case 4 data.

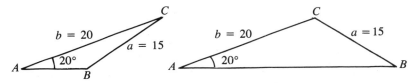

Figure 3.62

The **Law of Cosines** is a formula that enables us to solve an oblique triangle when two sides and the included angle are given, as in Case 1, or when three sides are given, as in Case 2. To derive the law of cosines, we subdivide a general oblique triangle into two right triangles.

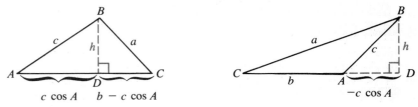

Figure 3.63   Law of Cosines

Consider any oblique triangle ABC; either of the triangles shown in Figure 3.63 will do. Drop a perpendicular from the vertex B to side AC or its extension. Call the length of this perpendicular h. In either case, we obtain $h = c \sin A$, and hence

$$h^2 = c^2 \sin^2 A$$

Using

$$\sin^2 A + \cos^2 A = 1$$

we have

$$h^2 = c^2(1 - \cos^2 A)$$

Referring to Figure 3.63, we also know, from right triangle BCD, that

$$h^2 = a^2 - (b - c \cos A)^2$$

Equating the right-hand sides of the two preceding equations, we get

$$c^2(1 - \cos^2 A) = a^2 - (b - c \cos A)^2$$
$$c^2 - c^2 \cos^2 A = a^2 - b^2 + 2bc \cos A - c^2 \cos^2 A$$

Simplifying this expression, we arrive at

$$a^2 = b^2 + c^2 - 2bc \cos A \qquad (3.4)$$

In a similar manner, it can be shown that

$$b^2 = a^2 + c^2 - 2ac \cos B$$

and

$$c^2 = a^2 + b^2 - 2ab \cos C$$

Each of these formulas is a statement of the Law of Cosines. This law says that the square of any side of a triangle is equal to the sum of the squares of the other two sides minus twice the product of these sides times the cosine of the angle between them. If the angle is 90°, the Law of Cosines reduces to the Pythagorean theorem, so it is quite properly considered an extension of that famous theorem.

A calculator can be used to make the arithmetic computations that arise when the Law of Cosines is applied to a specific triangle. For example, if b, c, and A are given parts of a triangle, the following keystrokes can be used to calculate the length of side a. Referring to Equation (3.4), we have

Some calculators require a grouping symbol before and after the expression 2bc cos A. In this case the keystroke sequence following the minus sign is

Check to see how your calculator works with Equation (3.4).

The Law of Cosines gives the relationship between three sides and one of the angles of any triangle. Thus, if any three of these parts are given, you can compute the remaining parts or show that such a triangle cannot exist. For example, if three sides of a triangle are given, any angle may be found from Equation (3.4) by solving for $\cos A$, $\cos B$, or $\cos C$. Alternative forms of the Law of Cosines are given in Equation (3.5):

$$\cos A = \frac{b^2 + c^2 - a^2}{2bc}$$

$$\cos B = \frac{a^2 + c^2 - b^2}{2ac} \qquad (3.5)$$

$$\cos C = \frac{a^2 + b^2 - c^2}{2ab}$$

Note that in this form the law says that the cosine of an angle may be found by computing a fraction whose numerator is the sum of squares of the adjacent sides minus the square of the opposite side and whose denominator is twice the product of the adjacent sides.

To solve for angle A when sides a, b, and c are known, use Equation (3.5) and the following keystrokes on your calculator.

Example 1. Solve the triangle with side $a = 5.18$, side $b = 6.00$, and angle $C = 60.0°$, as shown in Figure 3.64.

Solution

$$c^2 = a^2 + b^2 - 2ab \cos C$$
$$= 5.18^2 + 6.00^2 - 2(5.18)(6.00) \cos 60.0°$$
$$= 31.7524$$

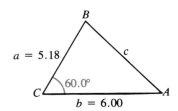

Figure 3.64

Therefore,

$$c = \sqrt{31.7524} = 5.63$$

To solve for angle A, we use Equation (3.5) of the Law of Cosines:

$$\cos A = \frac{b^2 + c^2 - a^2}{2bc}$$

$$= \frac{6.00^2 + 5.63^2 - 5.18^2}{2(6.00)(5.63)}$$

$$= 0.6049$$

Therefore,

$$A = 52.8°$$

The remaining angle could also be found from the Law of Cosines, but since the sum of the angles is 180°,

$$B = 180° - 60° - 52.8°$$

$$= 67.2°$$

Example 2. Two airplanes leave an airport at the same time, one going northeast at 400 mph and the other going directly west at 300 mph. How far apart are they 2 hr after leaving?

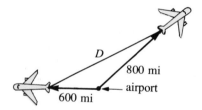

Figure 3.65

Solution. From Figure 3.65 and the Law of Cosines,

$$D = \sqrt{600^2 + 800^2 - 2(600)(800) \cos 135.0°}$$

Notice that $\cos 135.0° = -\cos 45°$, so

$$D = \sqrt{360,000 + 640,000 - (960,000)(-\cos 45°)}$$

$$= \sqrt{1,678,822}$$

$$= 1300 \text{ mi}$$

Example 3. In a steel bridge, one part of a truss is in the form of an isosceles triangle, as shown in Figure 3.66. At what angles do the sides of the truss meet?

Solution

$$\cos A = \frac{20^2 + 30^2 - 20^2}{(2)(20)(30)} = \frac{900}{1200} = 0.75$$

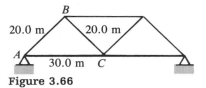

Figure 3.66

Hence,

$$A = 41.4°$$

This is also the value of angle C, since the triangle is isosceles. Then,

$$B = 180° - 2(41.4°) = 97.2°$$

Suppose we had decided to use the Law of Cosines to find angle B. Then,

$$\cos B = \frac{20^2 + 20^2 - 30^2}{(2)(20)(20)} = \frac{800 - 900}{800} = -\frac{1}{8} = -0.125$$

The fact that $\cos B$ is negative tells us that angle B is greater than 90° and less than 180°. The angle whose cosine is 0.125 is 82.8°, and therefore

$$B = 180° - 82.8° = 97.2°$$

This agrees with the previous result. Notice how the determination of angle B was affected by the fact that $\cos B$ was negative. Of course, if you use a calculator to evaluate $\cos B = -0.125$, the display shows 97.2° directly.

Example 4. A satellite traveling in a circular orbit 1000 mi above Earth passes directly over a tracking station at noon. Assume that the satellite takes 2.0 hr to make an orbit and that the radius of Earth is 4000 mi. Find the distance between the satellite and tracking station at 12:03 P.M.

Solution. From Figure 3.67, we see that the angle β must be computed. Since the satellite takes 2 hr (or 120 min) for an orbit of 360°, it moves $360°/120 = 3°$ during each minute. Thus, the satellite travels a total of 9.0° in 3 minutes. Hence, $\beta = 9.0°$. By the Law of Cosines,

$$x = \sqrt{(4000)^2 + (5000)^2 - 2(4000)(5000) \cos 9.0°}$$

$$= \sqrt{1,492,466}$$

$$= 1220$$

The distance between the satellite and the tracking station is about 1220 mi.

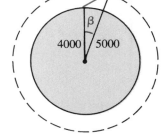

Figure 3.67

Comment: The form of the Law of Cosines in Equation (3.5) confirms that the sides of triangles cannot have just any lengths. For example, for $a = 1$, $b = 3$, and $c = 1$, Equation (3.5) gives

$$\cos A = \frac{(3)^2 + (1)^2 - (1)^2}{2(3)(1)} = \frac{9}{6} = \frac{3}{2}$$

which is impossible since $|\cos A| \leq 1$. Hence, no such triangle exists. The same conclusion can be reached by noting that $a + c < b$.

Exercises for Section 3.4

In Exercises 1–6, use the Law of Cosines to find the unknown side.

1. $a = 45.0, b = 67.0, C = 35°$
2. $a = 20.0, b = 40.0, C = 28°$
3. $a = 10.5, b = 40.8, C = 120°$
4. $b = 12.9, c = 15.3, A = 104.2°$
5. $b = 38.0, c = 42.0, A = 135.3°$
6. $a = 3.49, b = 3.54, C = 5°24'$

In Exercises 7–12, use the Law of Cosines to find the largest angle.

7. $a = 7.23, b = 6.00, c = 8.61$
8. $a = 16.0, b = 17.0, c = 18.0$
9. $a = 18.0, b = 14.0, c = 10.0$
10. $a = 300, b = 500, c = 600$
11. $a = 170, b = 250, c = 120$
12. $a = 56.0, b = 67.0, c = 82.0$

In Exercises 13–18, use the Law of Cosines to solve the triangles.

13. $a = 4.21, b = 1.84, C = 30.7°$
14. $a = 5.92, b = 7.11, C = 60.6°$
15. $a = 120, b = 145, C = 94°25'$
16. $a = 900, b = 700, c = 500$
17. $a = 2.00, b = 3.00, c = 4.00$
18. $a = 5.01, c = 5.88, B = 28°40'$

19. A boy starting from point A walks 2.50 km due west to point B. He then takes a path that is 25.4° south of west and walks 1.40 km to point C. What is the distance between A and C?

20. If the sides of the triangular sections of a geodesic dome, as shown in Figure 3.68, are 1.65, 1.65, and 1.92 m, what are the interior angles of the triangular sections?

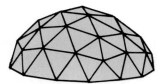

Geodesic Dome
Figure 3.68

Historical note: The geodesic dome was popularized in the 1950s by architect R. Buckminster Fuller.

21. The stake that marked corner C of a triangular lot ABC has been lost. Consulting her deed to the property, the owner finds that $AB = 80.0$ ft, $BC = 50.0$ ft, and $CA = 40.0$ ft. At what angle with AB should she run a line so that by laying off 40.0 ft along this line she can locate corner C?

22. A solar collector is placed on a roof that makes an angle of 24.0° with the horizontal. If the upper end of the collector is supported as shown in Figure 3.69, how long is the collector?

23. A reflector used in a solar furnace is composed of triangular sections having side lengths of 5.50, 5.50, and 1.30 ft. Find the interior angle between the sides of equal length.

24. An airplane flying directly north toward a city C alters its course toward the northeast at a point 100 km from C and heads for city B, approximately 50.0 km away. If B and C are 60.0 km apart, what course should the airplane fly to get to B?

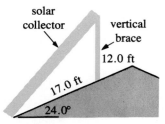

Figure 3.69

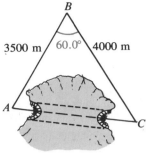

Figure 3.70

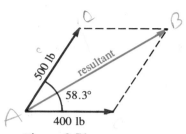

Figure 3.71

25. In planning a tunnel under a hill, an engineer lays out the triangle ABC, shown in Figure 3.70, in order to determine the course of the tunnel. If $AB = 3500$ m, $BC = 4000$ m, and angle $B = 60.0°$, determine the sizes of the angles A and C and the length of AC.

26. If two forces, one of 500 lb and the other of 400 lb, act from a point at an angle of 58.3° with each other, what is the size of the resultant? (See Figure 3.71.)

27. In order to measure the distance between two points A and B on opposite sides of a building, a third point C is chosen from which the following measurements are made: $CA = 200$ m, $CB = 400$ m, and angle $ACB = 60.0°$. What is the distance between A and B?

28. Show that for any triangle ABC,

$$\frac{a^2 + b^2 + c^2}{2abc} = \frac{\cos A}{a} + \frac{\cos B}{b} + \frac{\cos C}{c}$$

29. Two hikers leave camp at noon. One walks due west at 3.2 mph, and the other walks a line 75° north of east at 2.4 mph. Approximately how far apart will the two hikers be at 1 : 30 P.M.? Assume they walk in straight lines and at constant rates.

3.5 The Law of Sines

To solve triangles for which two angles and one side are given, we use a formula known as the **Law of Sines**. This formula, in conjunction with the Law of Cosines, enables us to solve any triangle for which we are given three parts, or at least to declare that no solution is possible.

Consider either triangle shown in Figure 3.72. We see from Figure 3.72(a) that if we draw a perpendicular h from the vertex B to side b or its extension,

$$h = c \sin A \quad \text{and} \quad h = a \sin C$$

Equating these two expressions, we get

$$c \sin A = a \sin C$$

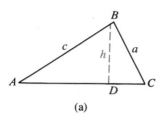

(a)

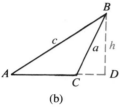

(b)

Figure 3.72   Law of Sines

or, rearranging factors,

$$\frac{a}{\sin A} = \frac{c}{\sin C}$$

In a similar manner, we can show that

$$\frac{a}{\sin A} = \frac{b}{\sin B}$$

Hence,

$$\frac{a}{\sin A} = \frac{b}{\sin B} = \frac{c}{\sin C} \qquad (3.6)$$

Equation (3.6) is called the **Law of Sines**. The Law of Sines states that, in any triangle, the ratios formed by dividing each side by the sine of the angle opposite it are equal.

Combinations of any two of the three ratios given in Equation (3.6) will yield an equation with four parts. Obviously, if we know three of these parts, we can find the fourth.

If the Law of Sines is written

$$\frac{a}{\sin A} = \frac{b}{\sin B}$$

then side a is given by

$$a = \frac{b \sin A}{\sin B}$$

The keystroke sequence for side a is then

Similarly, if we solve the Law of Sines for sin A, we get

$$\sin A = \frac{a \sin B}{b}$$

Angle A can then be found by the following keystrokes:

Example 1.   Given that $c = 10$, $A = 40.0°$, and $B = 60.0°$, find a, b, and C. (See Figure 3.73.)

Solution.   We begin by observing that

$$C = 180° - (A + B) = 180° - (40.0° + 60.0°) = 80.0°$$

Using the Law of Sines, we have

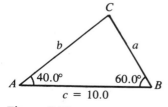

Figure 3.73

$$\frac{a}{\sin 40°} = \frac{10.0}{\sin 80°} \quad \text{or} \quad a = \frac{10.0(0.6428)}{0.9848} = 6.53$$

and

$$\frac{b}{\sin 60°} = \frac{10.0}{\sin 80°} \quad \text{or} \quad b = \frac{10.0(0.8660)}{0.9848} = 8.79$$

Warning: You should be aware of the fact that since $\sin A = \sin (180° - A)$ the formula

$$\sin A = \frac{a \sin B}{b}$$

gives two possible values for angle A. Therefore, you must be careful when using this form of the Law of Sines. The next example illustrates the problem.

Example 2. Given $A = 33.7°$, $b = 2.17$, and $c = 1.09$, find angle B. (See Figure 3.74.)

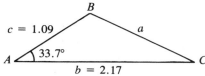

Figure 3.74

Solution. Before we can solve for angle B, we must find side a. By the Law of Cosines, we have

$$a = \sqrt{2.17^2 + 1.09^2 - 2(2.17)(1.09) \cos 33.7°} = 1.40$$

Now, if we use the Law of Sines to find angle B, we get

$$\sin B = \frac{2.17 \sin 33.7°}{1.40} = 0.860$$

$$B = 59.3° \quad \text{or} \quad B = 120.7°$$

However, if we use the Law of Cosines to find B, we get

$$\cos B = \frac{1.40^2 + 1.09^2 - 2.17^2}{2(1.40)(1.09)} = -0.5114$$

$$B = 120.7°$$

We note that the correct value $120.7°$ is the supplement of $59.3°$. However, the Law of Sines alone gives us no clue that the supplement of $59.3°$ is the required answer. A rough drawing of the triangle will usually be sufficient to tell you which of the two angles to use.

Example 3. A surveyor desires to run a straight line from A in the direction AB, as shown in Figure 3.75, but finds that an obstruction interferes with the line of sight. Therefore, the crew lays off the line segment $\overline{BX}$ for a distance of 150.0 m in such a way that angle $ABX = 135.0°$ and runs XY at an angle of $75.0°$ with $\overline{BX}$. At

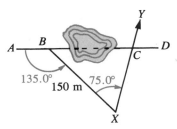

Figure 3.75

what distance from X on this line should a stake be placed so that A, B, and C are on the same straight line?

Solution. Since angle $CBX = 45.0°$, we have, from the Law of Sines,

$$\frac{\overline{CX}}{\sin 45.0°} = \frac{150.0}{\sin 60.0°}$$

Hence,

$$\overline{CX} = \frac{\sin 45.0°}{\sin 60.0°}\,150 = \frac{\sqrt{2}/2}{\sqrt{3}/2}\,150 = \frac{\sqrt{2}}{\sqrt{3}}\,150 \approx \frac{1.414}{1.732}\,150 = 122\text{ m}$$

The next two examples involve cases in which two sides of a triangle and the angle opposite one of them are given (Case 4). You will recall that such data may define one, two, or no triangles. For simplicity, we restrict the following discussion to examples in which a unique triangle is defined by the given data. The discussion of the ambiguous case is presented in the next section.

Example 4. A satellite traveling in a circular orbit 1000 mi above Earth is due to pass directly over a tracking station at noon. Assume that the satellite takes 2 hr to make an orbit and that the radius of Earth is 4000 mi. If the tracking antenna is aimed $30.0°$ above the horizon, at what time will the satellite pass through the beam of the antenna? (See Figure 3.76.) Assume that the beam is directed to intercept the satellite before it is overhead.

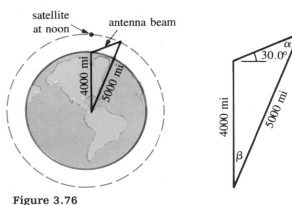

Figure 3.76

Solution. From the Law of Sines,

$$\frac{\sin \alpha}{4000} = \frac{\sin 120°}{5000}$$

$$\sin \alpha = \frac{4000 \sin 120°}{5000} = 0.6928$$

Hence,

$$\alpha = 43.85° \quad \text{and} \quad \alpha' = 180° - 43.85° = 136.15°$$

Since $120° + α' > 180°$, there is only one solution. Thus,

$$β = 180° - (120° + 43.85°) = 16.15°$$

The time involved in the change from $β = 16.15°$ to $β = 0.0°$ is given by $(16.15°/360°)\ (120\ \text{min}) = 5.38\ \text{min} = 5\ \text{min}\ 23\ \text{sec}$. Thus, the satellite will pass through the beam of the antenna at $12:00 - 5.38\ \text{min}$, or $11:54:37$ A.M.

Example 5. Consider a round trip flight from Chicago to Boston, with a one-way airline distance of 870 mi. A light plane having an airspeed of 180 mph is to make the round trip. How many flying hours will it take for the round trip if there is a constant southwest wind of 23.0 mph? What headings will the pilot use for the two parts of the trip? (Heading and ground speed are defined in Section 3.3.)

Solution. For the eastbound trip, shown in Figure 3.77, the Law of Sines is applied to determine $θ$:

$$\sin θ = \frac{23.0}{180} \sin 45.0° = 0.0904$$

$$θ = 5.2°$$

Applying the Law of Sines again, we determine the ground speed represented by $\overline{AB}$:

$$\frac{\overline{AB}}{\sin C} = \frac{180}{\sin B} \quad \text{or} \quad \overline{AB} = \frac{\sin 129.8°}{\sin 45°} \cdot 180 = 196\ \text{mph}$$

Thus, the time required for the eastbound trip is

$$\text{time} = \frac{\text{distance}}{\text{velocity}} = \frac{870\ \text{mi}}{196\ \text{mph}} = 4.44\ \text{hr}$$

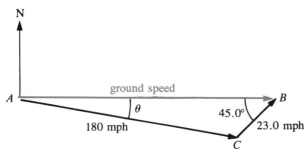

Figure 3.77 Eastbound Trip

For the westbound trip, shown in Figure 3.78, $θ$ is again $5.2°$, and the ground speed is found by use of the Law of Sines.

$$\overline{AB} = \frac{\sin 39.8°}{\sin 135°} \cdot 180 = 163\ \text{mph}$$

The time required is

$$\text{time} = \frac{870\ \text{mi}}{163\ \text{mph}} = 5.34\ \text{hr}$$

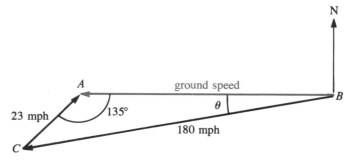

Figure 3.78 Westbound Trip

Thus, the total time for the round trip is 9.78 hr, or 9 hr 47 min. The heading for the eastbound trip is 90° + 5.2°, or 95.2°, and the heading for the westbound trip is 270° − 5.2°, or 264.8°.

Exercises for Section 3.5

In Exercises 1–15, solve the given oblique triangles.

1. $A = 32.0°$, $B = 48.0°$, $a = 10.0$
2. $A = 60.0°$, $B = 45.0°$, $b = 3.00$

3. $A = 45.2°$, $a = 8.82$, $b = 5.15$
4. $A = 75.0°$, $a = 27.7$, $b = 11.8$

5. $A = 35.6°$, $b = 12.2$, $a = 17.5$
6. $A = 82.1°$, $b = 7.21$, $a = 29.0$

7. $A = 120°50'$, $a = 6.61$, $b = 5.09$
8. $A = 51°10'$, $a = 59.2$, $b = 53.5$

9. $C = 53.0°$, $b = 18.3$, $c = 30.2$
10. $C = 58.0°$, $c = 83.0$, $b = 51.0$

11. $B = 122.0°$, $b = 30.0$, $a = 25.0$
12. $B = 63.0°$, $b = 5.00$, $c = 4.00$

13. $C = 110.0°$, $B = 50.0°$, $b = 40.0$
14. $C = 73.2°$, $A = 13.7°$, $c = 20.5$

15. $B = 48.0°$, $A = 43.4°$, $c = 61.3$

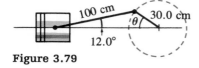

Figure 3.79

16. The crank and connecting rod of an engine, illustrated in Figure 3.79, are 30.0 cm and 100 cm long, respectively. What angle does the crank make with the horizontal when the angle made by the connecting rod is 12.0°?

17. One end of a 15.5-ft plank is placed on the ground at a point 10.8 ft from the start of a 42.7° incline, and the other end is allowed to rest on the incline. How far up the incline does the plank extend?

18. A 300-ft broadcast antenna stands at the top of a hill whose sides are inclined at 18.6° to the horizontal. How far down the hill will a 250-ft support cable extend if it is attached halfway up the antenna?

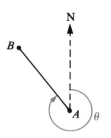

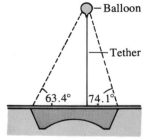

Figure 3.80 Bearing of *B* from *A*   **Figure 3.81**

19. Coast guard station Bravo is located 230 mi due north of an automated search and rescue station. Station Bravo receives a distress message from an oil tanker at a bearing of 124.6°, and the automated station receives the same message at a bearing of 52.1°. How long will it take a helicopter from Bravo to reach the ship if the helicopter can fly at 125 mph? The **bearing** of *B* from *A* is the angle measured clockwise from north at *A* to the line segment from *A* to *B*. (See Figure 3.80.)

20. A balloon is tethered above a bridge by a cord. To find the height of the balloon above the surface of the bridge, a girl measures the length of the bridge and the angles of elevation to the balloon at each end of the bridge. If she finds the length of the bridge to be 263 ft and the angles of elevation to be 64.3° and 74.1°, what is the height of the balloon? (See Figure 3.81.)

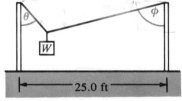

Figure 3.82

21. A weight is attached to two vertical poles, as shown in Figure 3.82. How far is the weight from the left post if $\theta = 41.0°$ and $\phi = 75.0°$?

22. How far will the weight in Exercise 21 be from the left post if $\theta = 38.7°$ and $\phi = 69.1°$?

23. From a position at the base of a hill, an observer notes that the angle of elevation to the top of an antenna is 43.5°. (See Figure 3.83.) After walking 1500 ft toward the base of the antenna up a slope of 30.0° the observer finds the angle of elevation to be 75.4°. Find the height of the antenna and the height of the hill.

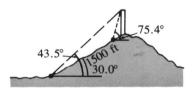

Figure 3.83

24. A satellite traveling in a circular orbit 1500 mi above Earth is due to pass directly over a tracking station at noon. Assume that the satellite takes 90 min to make an orbit and that the radius of the Earth is 4000 mi. If the tracking antenna is aimed 20.0° above the horizon, at what time will the satellite pass through the beam of the antenna?

25. Determine at what angle above the horizon the antenna in Exercise 24 must be pointed in order for its beam to intercept the satellite at 12:05 P.M.

26. Consider a flight from Miami to New York to be along a north-south direction, with an airline distance of approximately 1000 mi. A jetliner having an airspeed of 500 mph makes the round trip. If there is a constant northwest wind of 100 mph, how long will it take for the round trip? What headings will the pilot use for the two parts of the trip?

3.6 The Ambiguous Case

We will now analyze how to solve those triangles for which the measures of two sides and the angle opposite one of them is given. For ease of discussion, suppose that two sides a and b and an angle A are given. As you will see, there may be one, two, or no triangles with these measurements.

$A < 90°$: Perhaps the best way to clarify the situation in which angle A is acute is to draw a figure. To locate the vertex C, let us construct a line segment having a length of b units along one side of angle A. Then it is obvious from Figure 3.84 that the length of side a will determine whether there are two, one, or no triangles. In Figure 3.84(a), only one triangle is possible since $a \geq b$. In Figure 3.84(b), two triangles can be made by swinging an arc of length a from C so that it intersects side c at two points. In Figure 3.84(c), the length of side a is such that it intersects side c at one point to form a single right triangle. Finally, in Figure 3.84(d), side a is too short to intersect side c, and therefore no triangle can be formed.

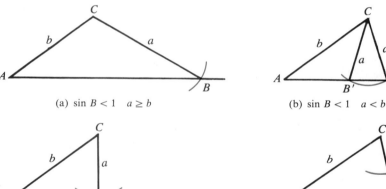

(a) $\sin B < 1$ $a \geq b$ (b) $\sin B < 1$ $a < b$

(c) $\sin B = 1$ (d) $\sin B > 1$ (Impossible)

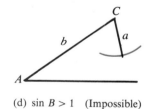

Figure 3.84   Ambiguous Case, $A < 90°$

The situation into which a particular set of data falls can be determined analytically by solving

$$\frac{a}{\sin A} = \frac{b}{\sin B}$$

for $\sin B$.

(1) If $a < b$ and $\sin B > 1$, then B cannot exist, so there is no triangle.

(2) If $a < b$ and $\sin B = 1$, then $B = 90°$, so there is one right triangle.

(3) If $a < b$ and $\sin B < 1$, then two angles at B are possible; the acute angle B and the obtuse angle $B' = 180° - B$.

(4) If $a \geq b$, then there is only one triangle.

$A \geq 90°$: The case in which angle A is greater than or equal to $90°$ is much easier to analyze than the case in which $A < 90°$. As you can see in Figure 3.85(a), if $a \leq b$, there is no triangle. If $a > b$, there is one triangle, as shown in Figure 3.85(b).

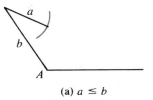

(a) $a \leq b$ (b) $a > b$

Figure 3.85   Ambiguous Case, $A > 90°$

Example 1.   How many triangles can be formed if $a = 4.0$, $b = 10$, and $A = 30°$?

Solution.   Using the Law of Sines, we have

$$\frac{\sin B}{10} = \frac{\sin 30°}{4.0} \quad \text{or} \quad \sin B = \frac{10(0.5)}{4.0} = 1.25$$

Since $\sin B > 1$, there is no triangle corresponding to the given information. (See Figure 3.86.)

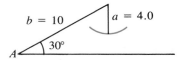

Figure 3.86

Example 2.   Solve the triangle with $B = 134.7°$, $b = 526$, and $c = 481$. (See Figure 3.87.)

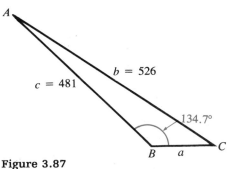

Figure 3.87

Solution.   In this triangle, the given angle is obtuse and the side opposite the given angle is greater than the side adjacent to it. This means that there is only one triangle that can be formed with the given information. Using the Law of Sines, we have

$$\frac{\sin C}{481} = \frac{\sin 134.7°}{526} \quad \text{or} \quad \sin C = \frac{481(0.7108)}{526} = 0.6500$$

Since B is obtuse, both A and C are acute. Consequently, angle $C = 40.5°$, and $C' = 180° - 40.5° = 139.5°$. Since $B' + C' > 180°$, we conclude that there is only one solution. Angle A is then given by

$$A = 180° - (134.7° + 40.5°) = 4.8°$$

Finally, side a is given by

$$\frac{a}{\sin 4.8°} = \frac{526}{\sin 134.7°} \quad \text{or} \quad a = \frac{526(0.0837)}{0.7108} = 61.9$$

Example 3. Verify that two triangles can be drawn for $a = 9.00$, $b = 10.0$, and $A = 60.0°$, and then solve each triangle.

Solution. Substituting the given values in the Law of Sines, we get

$$\frac{\sin B}{10.0} = \frac{\sin 60°}{9.00} \quad \text{or} \quad \sin B = \frac{10.0(0.8660)}{9.00} = 0.9622$$

Consequently, $B = 74.2°$ and $B' = 180° - 74.2° = 105.8°$. Since $A + B' < 180°$, there are two possible solutions. These are shown in Figure 3.88. One triangle is ABC, and the other is $AB'C'$.

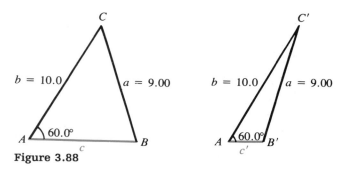

Figure 3.88

To solve triangle ABC, we note that angle C is given by

$$C = 180° - (A + B) = 180° - (60° + 74.2°) = 45.8°$$

Hence,

$$\frac{c}{\sin 45.8°} = \frac{9.00}{\sin 60°} \quad \text{or} \quad c = \frac{9.00(0.7169)}{0.8660} = 7.45$$

Similarly, in triangle $AB'C'$, we have

$$C' = 180° - (A + B') = 180° - (60° + 105.8°) = 14.2°$$

So

$$\frac{c'}{\sin 14.2°} = \frac{9.00}{\sin 60°} \quad \text{or} \quad c' = \frac{9.00(0.2453)}{0.8660} = 2.55$$

Example 4. Verify that only one triangle can be drawn for the case in which $b = 10.0$, $A = 60.0°$, and $a = 11.0$.

Solution. Figure 3.89 depicts the information given. Applying the Law of Sines, we get

$$\frac{11.0}{0.866} = \frac{10.0}{\sin B}$$

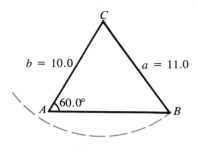

Figure 3.89

from which sin $B = 0.7873$. Since the side opposite the given angle is greater than the side adjacent to it, there is only one solution. Notice that had we missed the fact that $a > b$, we would have gotten

$$B = 51.9° \quad \text{and} \quad B' = 128.1°$$

However, the "solution" $B' = 128.1°$ is impossible because a triangle cannot have angles of $60.0°$ and $128.1°$. Thus, we reach the same conclusion as before.

Comment: This last example shows how important it is to continually check that both given and calculated data are consistent and do not produce impossible situations.

The various subcases within the ambiguous case need not be memorized, since, for any specific problem, the ambiguity or insolubility will become evident when the solution is attempted. For example, if in the course of solving a triangle you find that sin $B > 1$, this means that no triangle corresponding to the given data exists. If sin $B < 1$, there are two angles that may be solutions if they do not cause the sum of the angles in the triangle to exceed $180°$.

Exercises for Section 3.6

In Exercises 1–5, state whether the triangle has one solution, two solutions, or no solutions, given that $A = 30°$ and $b = 4$.

1. $a = 1$ 2. $a = 2$ 3. $a = 3$ 4. $a = 4$ 5. $a = 5$

For the triangles in Exercises 6–19, find all the unknown measurements or show that no such triangle exists.

6. $a = 20.0$, $b = 10.0$, $A = 35°40'$ 7. $a = 2.00$, $b = 6.00$, $A = 26°20'$

8. $a = 4.00$, $b = 8.00$, $A = 30.0°$ 9. $a = 15.0$, $c = 8.00$, $A = 150.0°$

10. $a = 50.0$, $b = 19.0$, $B = 22°30'$ 11. $b = 60.0$, $c = 74.0$, $B = 140.0°$

12. $a = 50.0$, $c = 10.0$, $A = 48.0°$ 13. $C = 28.0°$, $a = 20.0$, $c = 15.0$

14. $B = 40.0°$, $a = 12.0$, $b = 10.0$ 15. $A = 30.0°$, $b = 400$, $a = 300$

16. $B = 100.0°$, $a = 10.0$, $b = 12.0$ 17. $C = 70.0°$, $b = 100$, $c = 100$

18. $B = 41.2°$, $a = 4.20$, $b = 3.20$ 19. $a = 0.900$, $b = 0.700$, $A = 72°15'$

20. If $b = 12$ and $A = 30.0°$, for what values of a will there be two triangles?

21. If $c = 15$ and $B = 25.0°$, for what values of b will there be two triangles?

3.7 Analysis of the General Triangle

In Section 3.4 we mentioned that most of the time you could expect three parts of a triangle to be sufficient to determine it uniquely. With the aid of the two fundamental laws derived in Sections 3.4 and 3.5, we are now in a position to summarize the various approaches to solving a triangle.

Case 1:
Two Sides and an Included Angle Are Given: Use the Law of Cosines to obtain the third side. A second angle may be obtained using either the Law of Cosines or the Law of Sines. The third angle is computed by subtracting the sum of the other two from 180°.

Case 2:
Three Sides Are Given: Use the Law of Cosines to obtain one of the angles, preferably the largest one. A second angle may be obtained using either the Law of Cosines or the Law of Sines, and the third one may be obtained from the fact that the sum of the angles must be 180°. If the sum of any two sides exceeds the length of the third side, no triangle can be formed.

Case 3:
Two Angles and A Side Are Given: The two given angles must have a sum of less than 180°; otherwise no triangle is possible. Use the Law of Sines to determine the two unknown sides.

Case 4:
Two Sides and a Nonincluded Angle Are Given: If two sides a and b and an angle A are given, there may be two, one, or no triangles with these measurements. A carefully drawn figure will usually make the situation clear. Analytically, we have the following rules:

(1) If A is acute, there are no, one, or two solutions, depending on whether $a < b \sin A$, $a = b \sin A$, or $a > b \sin A$, unless $a \geq b$, in which case there is only one solution.

(2) If A is obtuse, there is either no solution or one solution corresponding to $a \leq b$ or $a > b$.

Exercises for Section 3.7

In Exercises 1–17, solve each triangle or show that no such triangle exists.

1. $A = 60.0°$, $B = 75.0°$, $a = 600$ 2. $A = 75.0°$, $a = 120$, $b = 75.0$

3. $B = 15.0°$, $C = 105.0°$, $a = 4.00$ 4. $A = 30.0°$, $b = 60.0$, $c = 50.0$

5. $a = 8$, $b = 2$, $c = 6$ 6. $C = 15.0°$, $b = 15.0$, $c = 10.0$

7. $C = 30.0°$, $a = 300$, $b = 500$ 8. $a = 2000$, $b = 1000$, $c = 2500$

9. $B = 120.0°$, $a = 60.0$, $b = 25.0$ 10. $B = 30.0°$, $a = 500$, $b = 400$

11. $B = 70.0°$, $a = 12.0$, $b = 6.00$ 12. $A = 20.0°$, $a = 2.00$, $b = 3.00$

13. $A = 60.0°$, $B = 100°$, $b = 2.00$ 14. $B = 125.2°$, $a = 2.20$, $b = 1.30$

15. $A = 100.0°$, $a = 2.00$, $b = 1.00$ 16. $A = 42.3°$, $a = 20.0$, $c = 30.0$

17. $a = 2.00$, $b = 4.00$, $c = 8.00$

18. From a window 35.0 ft above the street, the angle of depression to the curb on the other side of the street is 15.0°, and to the curb on the near side is 45.0°. How wide is the street?

19. At successive milestones on a straight road leading to a mountain, readings of 30.0° and 45.0°, respectively, are made for the angle of elevation to the top of the mountain. What is the line-of-sight distance to the top of the mountain from the milestone nearest to it?

20. Two forces, one of 75.0 lb and the other of 100 lb, act at a point. If the angle between the forces is 60.0°, find the magnitude and direction of the resultant force. Give the direction as the angle between the resultant and the 100-lb force.

21. The airspeed of a plane is 400 mph, and there is a 75.0-mph wind from the northeast at a time when the heading of the plane is due east. Find the ground speed and the direction of the path of the plane.

3.8 Area Formulas

Recall that the area of a triangle is equal to one-half the product of any base with the corresponding altitude. With the use of the Law of Cosines and the Law of Sines, we can derive some equivalent expressions for area for which the height need not be specifically computed. In this section we will examine three such formulas.

Consider the triangle in Figure 3.90. The altitude from the vertex B to side b is given by $h = c \sin A$. Hence, the area S is given by

$$S = \tfrac{1}{2} bc \sin A \qquad (3.7)$$

This procedure could be repeated for each vertex to derive equivalent

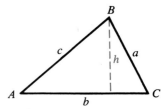

Figure 3.90

formulas. In general, *the area of a triangle is equal to one-half the product of the lengths of any two sides and the sine of the included angle.*

Example 1. Find the area of a triangle with $a = 3.8$ in., $b = 5.1$ in., and $C = 48°$.

Solution. Since C is the angle between sides a and b, the area is given by

$$S = \tfrac{1}{2}(3.8)(5.1) \sin 48° = 7.2 \text{ in.}^2$$

The expression for area may be altered further by using the Law of Sines. Since $c = b \sin C / \sin B$, we have from Equation (3.7) that

$$S = \frac{b^2}{2} \frac{\sin A \sin C}{\sin B} \tag{3.8}$$

Thus, the area may be calculated if one of the sides and two of the angles are given (since the third angle may then be found easily).

Example 2. Find the area of a triangle with two angles and an included side given by 30°, 45°, and 2.7 cm, respectively.

Solution. The remaining angle is 105°. Hence,

$$S = \frac{(2.7)^2}{2} \frac{\sin 30° \sin 45°}{\sin 105°}$$

$$= \frac{7.29(0.5)(0.7071)}{2(0.9659)}$$

$$= 1.3 \text{ cm}^2$$

A formula for the area may be derived from Equation (3.7) in order to obtain an expression in terms of the length of the sides.

$$S^2 = \tfrac{1}{4}b^2c^2 \sin^2 A$$

$$= \tfrac{1}{4}b^2c^2(1 - \cos^2 A)$$

$$= (\tfrac{1}{2}bc)(\tfrac{1}{2}bc)(1 - \cos A)(1 + \cos A)$$

Now, from the Law of Cosines, we have that

$$\tfrac{1}{2}bc(1 + \cos A) = \tfrac{1}{2}bc\left(1 + \frac{b^2 + c^2 - a^2}{2bc}\right)$$

$$= \frac{2bc + b^2 + c^2 - a^2}{4}$$

$$= \frac{(b + c)^2 - a^2}{4}$$

$$= \frac{(b + c - a)(b + c + a)}{4}$$

Similarly, we could obtain

$$\tfrac{1}{2}bc(1 - \cos A) = \frac{(a - b + c)(a + b - c)}{4}$$

Therefore, the expression S^2 becomes

$$S^2 = \frac{(b + c - a)(a + b + c)(a - b + c)(a + b - c)}{16}$$

$$S = \tfrac{1}{4}\sqrt{(b + c - a)(a + b + c)(a - b + c)(a + b - c)}$$

Sometimes we express this formula in terms of the perimeter, $P = a + b + c$:

$$S = \tfrac{1}{4}\sqrt{P(P - 2a)(P - 2b)(P - 2c)} \qquad (3.9)$$

(handwritten annotations: $S = \frac{a+b+c}{2}$, area $= \sqrt{(s-a)(s-b)(s-c)}$)

Example 3. Find the area of the triangle whose sides are 2, 2, and 3.

Solution. Since the perimeter is 7, we have

$$S = \tfrac{1}{4}\sqrt{7(7 - 4)(7 - 4)(7 - 6)} = \tfrac{1}{4}\sqrt{63} = \tfrac{3}{4}\sqrt{7} \approx 1.98 \text{ square units}$$

Note that Equation (3.9) defines the inherent limitations on the lengths of the sides of a triangle. For example, no triangle exists whose sides are 1, 2, and 3.

Exercises for Section 3.8

In Exercises 1–10, find the areas of the triangles with the given measurements.

1. $a = 15.0$, $b = 5.00$, $C = 30.0°$
2. $a = 12.0$, $b = 10.0$, $c = 5.00$
3. $a = 10.0$, $A = 60.0°$, $B = 45.0°$
4. $a = 10.0$, $A = 120°$, $B = 30.0°$
5. $a = 3.0$, $b = 4.0$, $c = 5.0$
6. $a = 1.0$, $b = 4.0$, $c = 5.0$
7. $a = 1.22$, $b = 1.39$, $c = 2.51$
8. $b = 4.30$, $B = 37.2°$, $C = 68.3°$
9. $c = 2.42$, $A = 108.3°$, $B = 31.4°$
10. $b = 25.6$, $A = 100.3°$, $B = 30.6°$

11. Why is it impossible to express the area of a triangle only in terms of its angles?

12. How are the areas of similar triangles related?

13. Workers need a triangular steel plate that is 12.0, 16.0, and 24.0 in., respectively, along the sides. What is the area of such a plate?

14. If the plate mentioned in Exercise 13 is made of sheet steel weighing 0.90 oz/sq in. of surface, how much does the plate weigh?

15. A farmer has a field shaped as shown in Figure 3.91. Find the area of the field.

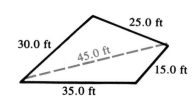

Figure 3.91

16. A home is built on a lot in the shape of a quadrilateral, with measurements 35, 100, 28, and 83 m. If the diagonal measures 110 m, what is the area of the lot in hectares? (1 hectare = 10,000 m².)

17. Find the perimeter of a triangle if two of the sides are 100 m and 150 m and the area is 600 m².

18. Find the area of a quadrilateral whose sides are successively 3, 5, 6, and 4 m if the angle between the sides of length 3 and 5 is 100°.

19. Find all parts of a triangle whose area is 25.0 m² and two of whose sides are 15.3 and 11.3 m.

20. Find all possible triangles with an area of 96.5 m² and two angles 68.2° and 58.3°.

21. A solar reflector is composed of 36 triangular sections having side lengths of 6.2, 6.2, and 1.1 m. Find the total area of the reflector.

22. A triangular-shaped mirror used in a solar furnace has side lengths of 25, 25, and 3.5 cm. What is the area of the mirror?

Key Topics for Chapter 3

Define and/or discuss each of the following.

Right Triangles
Vectors
Components of a Vector
Resultant
Course
Bearing

Heading
Law of Cosines
Law of Sines
Ambiguous and Impossible Cases
Solution to the General Triangle
Area of a Triangle

Review Exercises for Chapter 3

1. Solve the right triangle ABC if $A = 32°$ and $a = 3.0$.

2. Solve the right triangle ABC if $A = 17°23'$ and $b = 5.8$.

3. Solve the right triangle ABC if $a = 29$ and $c = 41$.

4. Solve the right triangle ABC if $b = 7.5$ and $c = 11.3$.

5. Find the horizontal and vertical components of $\mathbf{F}$ if $|\mathbf{F}| = 50$ lb and $\theta = 36.7°$.

6. Find the horizontal and vertical components of $\mathbf{F}$ if $|\mathbf{F}| = 8$ lb and $\theta = 17.9°$.

7. Find the resultant force for components $F_x = 43$ lb and $F_y = 72$ lb.

8. Find the resultant force for components $F_x = 9.4$ lb and $F_y = 3.7$ lb.

9. An engineer's drawing of a component of a trimming die is shown in Figure 3.92. Find the measure of angle θ.

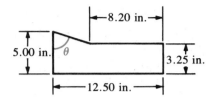

Figure 3.92

10. An airplane heading due east at 350 mph experiences a 40-mph crosswind out of the north. In what direction is the plane moving relative to an observer on the ground?

11. A tunnel through a mountain ascends at an angle of 5°25' with the horizontal. If the tunnel is 5000 ft long, what is the vertical rise of the tunnel?

12. Eight holes are equally spaced on the circumference of a circle. If the center-to-center distance between the holes is 3.8 cm, what is the radius of the circle?

13. An ant starts at the origin of the coordinate plane and travels on a straight line making an angle of 23.0° with the positive x-axis. What are the coordinates of the ant after it has traveled 15 in.?

14. The angle of elevation to a weather balloon from two tracking stations is 30° and 45°. If the balloon is at an altitude of 50,000 ft, how far apart are the two tracking stations?

In Exercises 15–25, solve each triangle or show that no triangle exists.

Law of Cosines:

15. $A = 29.0°$, $b = 17.0$, $c = 28.0$ 16. $a = 7.80$, $c = 9.10$, $B = 38°18'$

17. $a = 11.2$, $b = 7.90$, $c = 15.4$ 18. $a = 210$, $b = 175$, $c = 78.0$

19. $a = 23.0$, $b = 5.88$, $c = 17.8$

Law of Sines:

20. $A = 39°12'$, $B = 17°42'$, $c = 20.8$ 21. $C = 27.6°$, $A = 112.2°$, $a = 3120$

22. $b = 75.0$, $B = 11.5°$, $C = 40°$ 23. $a = 15.0$, $b = 12.0$, $A = 25.0°$

24. $A = 42.0°$, $c = 25.0$, $a = 17.0$ 25. $B = 63°54'$, $a = 23.0$, $b = 12.0$

In Exercises 26–28, find the area of the indicated triangle.

26. $a = 15.6$, $b = 19.2$, $c = 27.8$ 27. $b = 7.5$, $c = 3.9$, $A = 42.5°$

28. $A = 72°$, $B = 37°$, $b = 29$

29. Solve the triangle $a = 9.06$, $c = 6.68$, $B = 138.0°$ and find its area.

30. Two light planes leave Kennedy airport at the same time, one flying a course of 265.0° at 175 mph and the other flying a course of 300.0° at 190 mph. How far apart will the two planes be at the end of $2\frac{1}{2}$ hr?

31. At a certain point, the angle of elevation to the top of a tower that stands on level ground is 30.0°. At a point 100 m nearer the tower, the angle of elevation is 58.0°. How high is the tower?

32. From a helicopter, the angles of depression to two successive milestones on a level road below are 15.0° and 30.0°, respectively. Find the height of the helicopter.

33. In measuring the height of a bell tower with a transit set 5.00 ft above the ground, a student finds that from a point A the angle of elevation to the top of the tower is 45.0°. After moving the transit 50.0 ft in a straight line toward the tower, the student finds the angle to be 60.0°. Find the height of the bell tower.

34. A plane flies due east out of Atlanta at 250 mph for 1 hr and then turns and flies 35.0° north of east at 300 mph for 1.5 hr. How far is the plane from Atlanta at the end of 2.5 hr?

35. An airplane heading due east at 350 mph experiences a 40-mph crosswind out of the north. What direction is the plane moving relative to an observer on the ground?

36. How much would it cost to lay a 4-in. thick slab of concrete in the shape of a triangle whose sides are 15, 20, and 22 ft if the cost of the concrete is $30/yd^3?

37. A 25-ft ladder leans against a building built on a slope. If the foot of the ladder is 11 ft from the base of the building and the angle between the side of the building and the ground is 128°, how high up the side of the building does the ladder reach? (See Figure 3.93.)

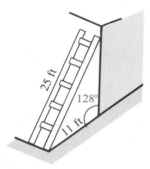

Figure 3.93

4

Radian Measure

4.1 The Radian

In Chapter 1 we defined the degree measure for comparing angles of different sizes. In this section we will define another unit of angular measure called the **radian**. Although less familiar than the degree, the radian is used extensively in advanced mathematics and is the standard unit of angular measurement in the International System.

> **Definition 4.1:** One **radian** is the measure of an angle whose vertex is at the center of a circle and whose sides subtend an arc on the circle equal in length to the radius of the circle. (See Figure 4.1.)

The definition of the radian is independent of the size of the circle. The radian measure of a central angle is found by determining how many times the length of the radius is contained in the subtended arc length.

To establish the relationship between the degree measure and the radian measure, we observe that the circumference, C, of a circle of radius r is given by $C = 2\pi r$. The number of radians in one circumference is thus $C/r = 2\pi$. Since there are $360°$ in one circumference, it follows that there are 2π radians in $360°$, or, in equation form,

$$2\pi \text{ radians} = 360°$$

from which we get the following important conversion formulas.

$$1 \text{ degree} = \frac{\pi}{180} \text{ radians} \approx 0.0175 \text{ radians}$$

$$1 \text{ radian} = \frac{180}{\pi} \text{ degrees} \approx 57.3 \text{ degrees}$$

$$(4.1)$$

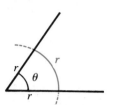

Figure 4.1   One Radian

Sometimes Formulas (4.1) are combined as

$$\frac{\text{degree measure of an angle}}{180°} = \frac{\text{radian measure}}{\pi}$$

From Formulas (4.1) we obtain the following two rules.

> **Rule 1:**   To convert from radian measure to degree measure, multiply the radian measure by $180/\pi$.
>
> **Rule 2:**   To convert from degree measure to radian measure, multiply the degree measure by $\pi/180$.

The next two examples show how to use these rules to convert from degrees to radians and vice versa.

Example 1.   Express (a) 60°, (b) 225°, and (c) 24.8° in radian measure.

Solution

(a) $60° = 60\left(\dfrac{\pi}{180}\right)$ radians $= \dfrac{\pi}{3}$ radians

(b) $225° = 225\left(\dfrac{\pi}{180}\right)$ radians $= \dfrac{5\pi}{4}$ radians

(c) $24.8° = 24.8(0.0175)$ radians $= 0.434$ radian

When the radian measure is a convenient multiple of π, you will usually find it beneficial *not* to convert it to a decimal fraction. When such a conversion is necessary, 3.1416 is often used as a decimal approximation to π.

Example 2.   Express (a) $\dfrac{\pi}{6}$ radians, (b) $\dfrac{3\pi}{4}$ radians, and (c) 1.13 radians in degrees.

Solution

(a) $\dfrac{\pi}{6}$ radians $= \left(\dfrac{\pi}{6}\right)\left(\dfrac{180}{\pi}\right)$ degrees $= 30°$

(b) $\dfrac{3\pi}{4}$ radians $= \left(\dfrac{3\pi}{4}\right)\left(\dfrac{180}{\pi}\right)$ degrees $= 135°$

(c) 1.13 radians $= 1.13(57.3) = 64.7°$

Some calculators have a **d↔r** *key that allows for direct conversion from degrees to radians and from radians to degrees. Thus, the entry 180 followed by* **d↔r** *will yield 3.1416 radians. If you do not have a* **d↔r** *key, you will have to multiply by $\pi/180$ to convert from degrees to radians and by $180/\pi$ to convert from radians to degrees.*

Most calculators have a special button for the number π. If yours does not have a **π** *button, you can use 3.1416 as an approximation, but your answers may be slightly different from those given in the answer section.*

Table 4.1 is a conversion table showing frequently occurring angles with both their degree measure and their radian measure. *You should memorize the entries in this table so that you don't have to make the conversion calculation.*

Table 4.1 Table of Degree and Radian Measures for Commonly Occurring Angles

Angle in Degrees	Angle in Radians
0	0
30	$\pi/6$
45	$\pi/4$
60	$\pi/3$
90	$\pi/2$
120	$2\pi/3$
135	$3\pi/4$
150	$5\pi/6$
180	π
270	$3\pi/2$
360	2π

The word "radian" is understood; it need not be written. This is not the case with degree measurement; its units must always be included. The next example emphasizes the difference between these two angular measures.

Example 3. Compare the angle of 60° with that of 60 radians. (See Figure 4.2.)

Solution. The angle of 60 radians is obtained by 9 repeated revolutions of the terminal side (each revolution being 2π radians) plus an additional 3.45 radians. To obtain the value 3.45, divide 60 by 2π; the remainder is 0.549, which when multiplied by 2π is 3.45. Thus the angle of 60 radians is coterminal with an angle of 3.45 radians.

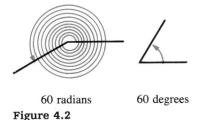

60 radians   60 degrees

Figure 4.2

Trigonometric Functions for Angles in Radians

The six trigonometric functions may be found for angles measured in radians by using either a calculator or Table C in the appendix. When using a calculator, remember to use the radian mode. The functional values in the following example were evaluated using a calculator.

Example 4. Evaluate (a) sin 1.38, (b) sec 3, and (c) tan $\frac{1}{2}$.

Solution
(a) sin 1.38 = 0.98185
(b) sec 3 = −1.0101 (Note: sec 3 = 1/cos 3.)
(c) tan $\frac{1}{2}$ = 0.54630

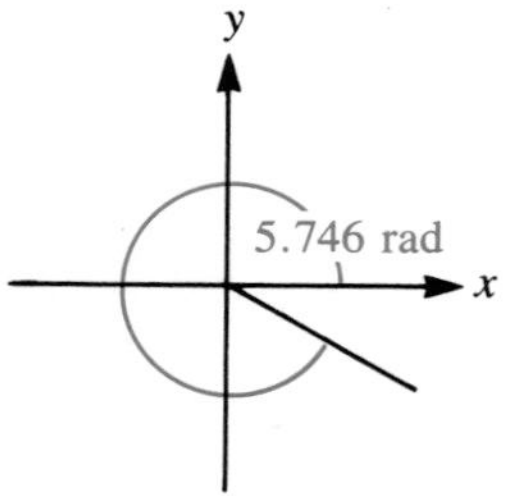

Example 5. Find angle θ, where $0 \le \theta < 2\pi$ radians, if (a) sin θ = −0.5113 and cos θ > 0 and (b) sec θ = −2.2332 and tan θ > 0.

Solution
(a) We note that θ is a fourth-quadrant angle, since sin θ < 0 and cos θ > 0. The reference angle for θ is the acute angle α whose sine is 0.5113. Using a calculator, we get

$$\alpha = 0.5113 \quad \boxed{\text{inv}} \quad \boxed{\text{sin}} \quad = 0.537 \text{ radian}$$

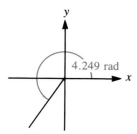

Figure 4.3

The desired fourth-quadrant angle is

$$\theta_4 = 2\pi - 0.537 = 5.746 \text{ radians}$$

(b) In this case, angle θ is a third-quadrant angle, since sec θ < 0 and tan θ > 0. The reference angle for θ is the acute angle α whose secant is 2.2332. Using a calculator, we get

$$\alpha = 2.2332 \quad \boxed{\text{1/x}} \quad \boxed{\text{inv}} \quad \boxed{\text{cos}} \quad = 1.107 \text{ radians}$$

The value of θ is then

$$\theta_3 = \pi + 1.107 = 4.249 \text{ radians}$$

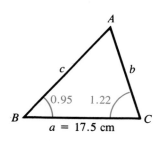

Figure 4.4

Example 6. The interior angles of a triangle can be stated in radians as well as in degrees. In Figure 4.5, $a = 17.5$ cm, $B = 0.95$ radians, and $C = 1.22$ radians. Find angle A and sides b and c.

Solution. The sum of the interior angles of a triangle is 180°, or π radians. Using 3.14 to approximate π, we have

$$A = 3.14 - (0.95 + 1.22) = 0.97 \text{ radian}$$

Now, using the Law of Sines and a calculator (or Table C in the appendix), we have

$$\frac{b}{\sin 0.95} = \frac{17.5}{\sin 0.97} \quad \text{or} \quad b = \frac{17.5 \sin 0.95}{\sin 0.97} = \frac{17.5(0.8134)}{0.8249} = 17.3 \text{ cm}$$

$$\frac{c}{\sin 1.22} = \frac{17.5}{\sin 0.97} \quad \text{or} \quad c = \frac{17.5 \sin 1.22}{\sin 0.97} = \frac{17.5(0.9391)}{0.8249} = 19.9 \text{ cm}$$

Figure 4.5

Interpolation

Sometimes interpolation must be used when the tabulated values in Table C do not correspond exactly to the desired radian measure.

Example 7. Use Table C to find $\tan \sqrt{2}$.

Solution. Table C lists angles in radians to two decimal places. We can therefore use interpolation to find trigonometric functions of angles to three decimal places. Using $\sqrt{2} = 1.414$, we seek $\tan 1.414$.

Angle		Tangent	
$0.010 \begin{bmatrix} 0.004 \begin{bmatrix} 1.410 \\ 1.414 \\ 1.420 \end{bmatrix} \end{bmatrix}$		$\begin{bmatrix} 6.165 \\ \ldots\ldots \\ 6.581 \end{bmatrix} c \end{bmatrix} 0.416$	

$$\frac{c}{0.416} = \frac{0.004}{0.010}$$

$$c = 0.166$$

Therefore, $\tan \sqrt{2} = 6.165 + 0.166 = 6.331$.

Example 8. Use Table C to estimate the value of the acute angle θ if $\cos \theta = 0.8145$.

Solution. From the table, we have $\cos 0.61 = 0.8196$ and $\cos 0.62 = 0.8139$.

Angle		Cosine	
$0.010 \begin{bmatrix} c \begin{bmatrix} 0.610 \\ \ldots\ldots \\ 0.620 \end{bmatrix} \end{bmatrix}$		$\begin{bmatrix} 0.8196 \\ 0.8145 \\ 0.8139 \end{bmatrix} 0.0051 \end{bmatrix} 0.0057$	

$$\frac{c}{0.010} = \frac{0.0051}{0.0057}$$

$$c = 0.009$$

The correction c is rounded off to the nearest thousandth for consistency with the accuracy of the table. Therefore, $\theta = 0.619$ radian.

Exercises for Section 4.1

1. What is the difference in radian measure of two coterminal angles?

2. What is the radian measure of a right angle? of a straight angle?

In Exercises 3–20, express each angle in radian measure. Express the measure in multiples of π when convenient.

3. $75°$	4. $-30°$	5. $480°$	6. $210°$
7. $-240°$	8. $42°$	9. $95°$	10. $1485°$
11. $750°$	12. $-300°$	13. $92.1°$	14. $105.7°$
15. $0.092°$	16. $34°39'$	17. $253°36'$	18. $311°48'$
19. $0°27'$	20. $400°40'$		

In Exercises 21–28, give the radian measure of an angle between $-\pi$ and π whose terminal side passes through the given point.

21. (2, 2)   **22.** (0, 3)   **23.** $(-\pi, 0)$ **24.** (5, -5)

25. $(-1, -1)$ **26.** $(-25, 0)$ **27.** (0, -1)   **28.** $(-3, -3)$

In Exercises 29–38, draw the angle and name the initial and terminal sides. If the given angle is greater than 2π, indicate an angle between 0 and 2π coterminal with the one given. Express each of the given angles in degrees.

29. 4   **30.** 2π **31.** π **32.** $-\dfrac{7\pi}{6}$

33. -3π **34.** -100 **35.** 100   **36.** 30

37. 100π **38.** -100π

In Exercises 39–46, use a calculator or Table C to find the value of each trigonometric function.

39. tan 1.23   **40.** sin 0.54   **41.** cot(-3)   **42.** sec 0.02

43. cot 3.78   **44.** csc 0.5   **45.** cos 2   **46.** tan π

In Exercises 47–54, use a calculator or Table C to find θ where $0 \le \theta < 2\pi$.

47. $\sin \theta = 0.4331$ and $\cos \theta < 0$ **48.** $\sin \theta = -0.4253$ and $\tan \theta < 0$

49. $\cos \theta = -0.8675$ and $\cot \theta > 0$ **50.** $\cos \theta = 0.3326$ and $\csc \theta < 0$

51. $\cos \theta = 0.9012$ and $\tan \theta > 0$ **52.** $\tan \theta = -6.8269$ and $\sin \theta > 0$

53. $\cot \theta = -1.0502$ and $\sin \theta < 0$ **54.** $\sec \theta = 1.113$ and $\cot \theta < 0$

In Exercises 55–66, find the value of each trigonometric function without using a calculator or table.

55. $\sin \dfrac{\pi}{4}$ **56.** $\cos \dfrac{\pi}{3}$ **57.** $\tan \dfrac{5\pi}{3}$

58. $\cos 7\pi$ **59.** $\tan \dfrac{7\pi}{4}$ **60.** $\sin \dfrac{-5\pi}{6}$

61. $\cos \dfrac{13\pi}{6}$ **62.** $\sin \dfrac{7\pi}{3}$ **63.** $\csc \dfrac{3\pi}{4}$

64. $\cot \dfrac{5\pi}{3}$ **65.** $\sec 5\pi$ **66.** $\cot \dfrac{19\pi}{6}$

In Exercises 67–70, use the Law of Sines to solve each triangle.

67. $A = 0.88$ radian, $B = 1.29$ radians, $c = 20.7$

68. $A = 1.31$ radians, $C = 1.00$ radian, $b = 0.652$

69. $B = 0.59$ radian, $C = 1.27$ radians, $a = 274$

70. $B = 0.72$ radian, $C = 0.93$ radian, $a = 13,500$

71. Construct a table showing the exact values of the trigonometric functions for angles with radian measures of 0, π/6, π/4, and π/3 and angles around the unit circle having these angles as reference angles.

72. Through how many radians does a person standing on the equator rotate in 8 hr?

73. Determine an angle with a measure between 0 and 2π that is coterminal with an angle whose measure is 80 radians.

74. A roulette wheel consists of 38 equal compartments. Those numbered from 1 to 36 are alternately black and red. There are also two green pockets numbered 0 and 00 on opposite sides of the wheel. (See Figure 4.6.)
    (a) How many radians are represented by red or black?
    (b) How many radians are represented by green?
    (c) How many radians are represented by any one number?

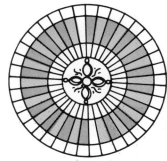

Figure 4.6

4.2 Arc Length and Area of a Sector

The decision as to whether to measure an angle in degrees or radians is sometimes a matter of personal preference, but more often than not the unit is dictated by the circumstances of the problem. In this section, we present several problems that are best solved using radian measure.

Arc Length

Figure 4.7 shows a central angle subtending an arc of length s. The radian measure of angle θ is found by determining how many times the radius of the circle is contained in the length of the arc—that is, $\theta = s/r$. If this formula is solved for s, we get a formula for the length of an arc.

> **Arc Length Formula:** The length of an arc, s, subtended by a central angle θ of a circle of radius r is given by
>
> $$s = r\theta \qquad (4.2)$$
>
> where θ is measured in radians.

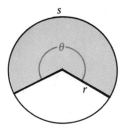

Figure 4.7

Example 1. Find the length of an arc on a circle of radius 5.0 cm that subtends a central angle of 38°.

Solution. To use Formula (4.2) you must first convert the degree measure to radian measure. Thus

$$38° \left(\frac{\pi}{180} \text{ radians} \right) = \frac{19\pi}{90} \text{ radians}$$

Therefore, $s = 5.0 \times 19\pi/90$ cm $= 3.3$ cm.

The formula $s = r\theta$ is used in physics to relate linear displacement, s, to angular displacement, θ. The next example illustrates this application of the arc length formula.

Example 2. As the drum in Figure 4.8 rotates counterclockwise, the cord is wound around the drum. How far will the weight on the end of the cord be moved when the drum is rotated through an angle of 53.8°, if $r = 4.5$ ft?

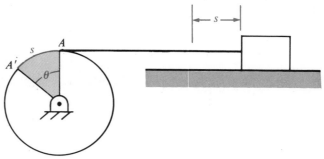

Figure 4.8

Solution. As the drum rotates, the cord will be wound around the drum, so the distance that the weight will move is equal to the arc length along the edge of the drum formed by a rotation of 53.8°. To use Formula (4.2), we must specify the angle of rotation in radians. Thus,

$$\theta = 53.8 \left(\frac{\pi}{180} \right) = 0.939$$

The distance s is then given by

$$s = r\theta = 4.5(0.939) = 4.2 \text{ ft}$$

Area of a Sector of a Circle

Another application of radian measure is in finding areas of sectors of circles (see Figure 4.9). To obtain a formula for computing the area of a circular sector with central angle θ, we note that the area of the entire circle is given by $A = \pi r^2$. This formula can be written as $A = \frac{1}{2}(2\pi)r^2$, where 2π is the central angle of the entire circle. Therefore, the area of a circle can be expressed in terms of its central angle and the square of its radius. The area of a sector of a circle is the same proportion of the total area of the circle as its central angle is of 2π. For instance, a circular sector with a central angle of $\frac{1}{4}\pi$ has an area that is $\frac{1}{4}\pi/2\pi = \frac{1}{8}$ of the total area. This relationship is the basis for the formula for the area of a circular sector.

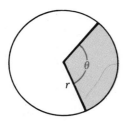

Figure 4.9

> **Area of a Circular Sector:** The area of a circular sector with a central angle θ and radius r is given by
>
> $$A = \tfrac{1}{2}\theta r^2 \qquad (4.3)$$
>
> where θ is measured in radians.

Example 3. Determine the area of the sector of a circle with a radius of 15.0 ft if the central angle of the sector is 52.7°.

Solution. To use Formula (4.3), we must convert 52.7° to radians. Thus,

$$\theta = 52.7°\left(\frac{\pi}{180}\right) = 0.920 \text{ radian}$$

The area of the sector is then

$$A = \tfrac{1}{2}(0.920)(15.0)^2 = 103 \text{ ft}^2$$

Exercises for Section 4.2

1. A circle has a radius of 25.3 cm. Calculate the length of the arc on the circumference subtended by a central angle of 78.6°.

2. Calculate the length of an arc on the circumference of a circle of radius 9.75 in., if the arc is subtended by a central angle of 187°.

3. Calculate the central angle of a circle with a 4.5-ft radius if the angle subtends an arc of 2.0 ft.

4. Calculate the central angle of a circle with a 21.4-cm radius if the angle subtends an arc of 1.9 cm.

5. A 92°15′ central angle of a circle subtends an arc of length 33 ft on the circumference of the circle. Find the radius of the circle.

6. Find the radius of a circle if a 310°30′ central angle subtends an arc of 0.720 m on the circumference.

7. A 10.0-ft-long pendulum swings through an arc of 30°. How long is the arc described by its midpoint?

8. A racing car travels a circular course about the judges' stand. If the angle subtended by the line of sight while the car travels 1.0 mi is 120°, how large is the entire track?

9. How high will the weight in Figure 4.10 be lifted if the drum is rotated through an angle of 81.5°?

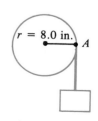

$r = 8.0$ in.

A

Figure 4.10

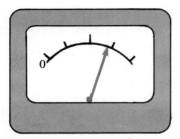

Figure 4.11

10. Through what angle, in degrees, must the drum in Figure 4.10 be rotated to raise the weight 10 in.?

11. Compute the radius of the drum in Figure 4.10 if the weight is raised 5.2 in. by a rotation of 70.7°.

12. The scale on an ammeter is 8.0 cm long. If the scale is an arc of a circle having a radius of 3.2 cm, what angle in degrees will the needle make between the zero reading and a full scale reading? (See Figure 4.11.)

13. A voltage of 6.2 V causes the needle on a voltmeter to deflect through an angle of 48°. If the needle is 1.75 in. long, how far does the tip of the needle move in indicating the applied voltage?

14. The diameter of the earth is approximately 8000 mi. Find the distance between two points on the equator whose longitude differs by 3°.

15. The rotation of either of the wheels shown in Figure 4.12 causes the other wheel to rotate also. How much rotation occurs in the larger wheel when the smaller one rotates through an angle of 100°?

16. How much rotation must occur in the smaller wheel in Figure 4.12 to cause the larger one to rotate a quarter of a revolution?

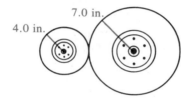

Figure 4.12

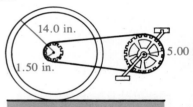

Figure 4.13

17. A schematic drawing of a typical bicycle drive chain is shown in Figure 4.13. How far will the bicycle move forward for each complete revolution of the drive sprocket?

18. How much rotation of the drive sprocket must occur to move the bicycle in Figure 4.13 a total of 5.00 ft forward?

19. A cable is looped over a pulley, as shown in Figure 4.14. Given that the radius of the pulley is 7.0 in. and angle θ is 78.6°, determine how much of the cable is in contact with the pulley.

20. Referring to Figure 4.14, find angle θ if 5.8 in. of cable contacts the pulley.

21. Find the area of a sector of a circle, given that the central angle is 85.9° and the diameter is 24.0 ft.

22. Find the area of a sector of a circle, given that the central angle is 46° and the radius is 30.25 m.

23. The area of a circular sector is 4.05 in.², and its central angle is 15.2°. Find the radius of the circle.

24. The area of a circular sector is 44.0 cm², and its radius is 12.0 cm. Find the central angle of the sector. Express the answer to the nearest tenth of a degree.

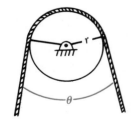

Figure 4.14

25. The area of a sector of a circle is 100 ft^2, and its bounding arc has a length of 5.75 ft. Find the central angle of the sector in degrees and the radius of the circle in feet.

26. The area of a circular sector is 75.4 mi^2, and its bounding arc has a length of 1.25 mi. Find the radius of the circle.

27. The area of a sector of a circular flower bed is found to be about 125 ft^2. If the radius of the field is 84 in., what is the central angle of the sector in degrees? (See Figure 4.15.)

Figure 4.15

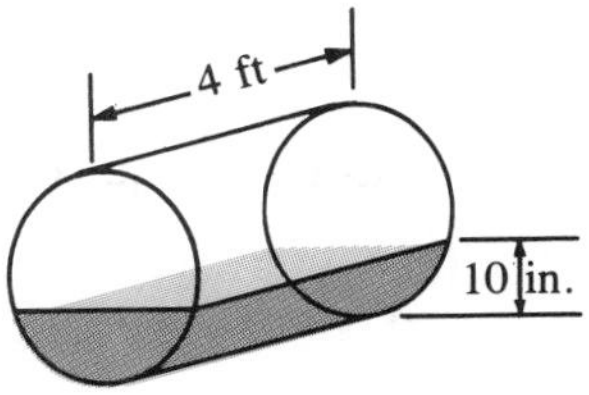

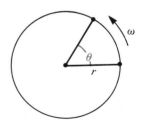

Figure 4.16

28. A circle of radius 3 ft is divided unevenly into red and blue sectors. What is the total central angle for the blue sectors if the area is 100 in.^{2}?

29. A cylindrical tank with a horizontal axis is filled with oil to a depth of 10 in. Given that the tank is 4.0 ft long and has a diameter of 3.0 ft, find the volume of oil in the tank. (See Figure 4.16.)

30. Derive a formula for the radius, r, of a circle that has an arc of s units bounding a sector of area A.

4.3 Angular Velocity

Consider a point on the circumference of a circle of radius r, as shown in Figure 4.17. If the circle rotates at a constant rate about its center, the point is said to have an angular velocity. Angular velocity, which is a measure of the rate of change of angular displacement, is denoted by ω and defined as follows.

Figure 4.17

Definition 4.1: The **angular velocity** of a rotating object is the change in its angular displacement divided by the time required for the change to occur. Symbolically,

$$\omega = \frac{\theta}{t} \qquad (4.4)$$

Notice that angular velocity is independent of the radius of the circle.

Two common units of angular velocity are revolutions/minute (rpm) and rad/sec, but any ratio of angular displacement units to time units can be used. Conversion from one unit of angular velocity to another is sometimes required. The next example shows how this is done.

Example 1
(a) Change 80 rpm to rad/sec.

(b) Change 100 rad/sec to rpm.

Solution

(a) $80 \dfrac{\text{rev}}{\text{min}} \times 2\pi \dfrac{\text{rad}}{\text{rev}} \times \dfrac{1}{60} \dfrac{\text{min}}{\text{sec}} = \dfrac{8\pi}{3} \text{ rad/sec} = 8.4 \text{ rad/sec}$

(b) $100 \dfrac{\text{rad}}{\text{sec}} \times \dfrac{1}{2\pi} \dfrac{\text{rev}}{\text{rad}} \times 60 \dfrac{\text{sec}}{\text{min}} = \dfrac{3000}{\pi} \text{ rpm} = 955 \text{ rpm}$

The angular velocity of a particle on the circumference of a rotating circle is not the same as its linear velocity. Linear velocity is denoted by v and defined as follows.

Definition 4.2: The **linear velocity** of an object is the change in displacement divided by the time required for the change to occur. Symbolically,

$$v = \frac{s}{t} \qquad (4.5)$$

where s is the distance traveled by the particle along the circumference of the circle and t is the elapsed time.

Some typical units for v are ft/sec, m/sec, and mi/hr.

If s is the distance measured along the circumference of a circle of radius r, we can relate ω to v as follows. From the previous section, we know that $s = r\theta$, if θ is measured in radians. Thus, we can substitute $r\theta$ for s in Formula (4.5) to obtain

$$v = \frac{s}{t} = \frac{r\theta}{t}$$

Since $\omega = \theta/t$, the relation between linear velocity and angular velocity is

$$v = r\omega \qquad (4.6)$$

Caution: Formula (4.6) is only valid if ω is measured in radians/unit time. If ω is given in other units, such as rpm, the units must be converted to rad/unit time before Formula (4.6) is used.

Comment: Formula (4.6) shows the dependency of linear velocity on the radius of the circle of rotation. For a given angular velocity, the larger the radius the larger the linear velocity of a point on the circle. Thus, if P_1 and P_2 are points on a rotating wheel, as shown in Figure 4.18, P_2 will travel faster in the linear sense than P_1, even though both have the same angular velocity.

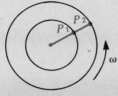

Figure 4.18

Example 2.

Consider a point P on the circumference of a wheel whose radius is 18 cm. Suppose the wheel rotates at $\pi/9$ rad/sec. Determine (a) the linear velocity of point P and (b) the linear distance traveled by P in 10 sec.

Solution

(a) Substituting $\omega = \pi/9$ rad/sec and $r = 18$ cm in Formula (4.6), we get

$$v = r\omega = 18\left(\frac{\pi}{9}\right) = 2\pi \text{ cm/sec}$$

(b) The distance traveled by the point in 10 sec is found by using the formula $s = vt$:

$$s = 2\pi(10) = 20\pi \text{ cm}$$

Example 3.

A belt runs a pulley of radius 5.0 cm at 100 rpm. Find the angular velocity of the pulley in rad/sec and the linear velocity of a point on the belt. (See Figure 4.19.)

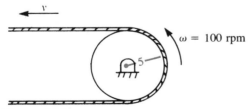

Figure 4.19

Solution.

To write the angular velocity of the pulley in rad/sec, change revolutions to radians and minutes to seconds. Thus,

$$\omega = 100 \, \frac{\text{rev}}{\text{min}} \times 2\pi \, \frac{\text{rad}}{\text{rev}} \times \frac{1 \text{ min}}{60 \text{ sec}} = \frac{10\pi}{3} \text{ rad/sec}$$

The linear velocity of a point on the belt will be the same as that of a point on the circumference of the pulley, which is given by

$$v = r\omega = 5\left(\frac{10\pi}{3}\right) = \frac{50\pi}{3} \text{ cm/sec} = 52.4 \text{ cm/sec}$$

Example 4. A bicycle travels 15 mi at a constant speed for 1 hr. What is the angular velocity of a point on the wheel in revolutions per minute and in radians per second? Assume the bicycle has 26-in. diameter wheels.

Solution. In 1 hr a point on the tip of a wheel will travel 15 mi. Thus,

$$s = 15 \text{ mi} \times 5280 \frac{\text{ft}}{\text{mi}} \times 12 \frac{\text{in}}{\text{ft}}$$

$$= 950,400 \text{ in.}$$

(Note that we had to change either the 15 mi to inches or the 26-in. diameter to miles.)

The circumference of the wheel is $\pi d = 26\pi$ in. Thus, the number of revolutions made by the wheel of the bicycle in 1 hr is

$$\frac{950,400}{26\pi} = 11,635 \text{ rev/hr}$$

To find ω in rpm, we have

$$\omega = 11,635 \frac{\text{rev}}{\text{hr}} \times \frac{1 \text{ hour}}{60 \text{ min}}$$

$$= 194 \text{ rpm}$$

Finally, to find ω in rad/sec, we have

$$\omega = 194 \frac{\text{rev}}{\text{min}} \times 2\pi \frac{\text{rad}}{\text{rev}} \times \frac{1 \text{ min}}{60 \text{ sec}}$$

$$= 20.3 \text{ rad/sec}$$

Exercises for Section 4.3

In Exercises 1–5, find the linear velocity in ft/sec of a point on the rim of a wheel of the given radius with the given angular velocity.

1. $r = 1$ ft, $\omega = 12.5$ rad/sec

2. $r = 3$ ft, $\omega = 25$ rpm

3. $r = 5$ cm, $\omega = 100$ rpm

4. $r = 4.3$ m, $\omega = 5$ rpm

5. $r = 7$ in., $\omega = 250$ rpm

In Exercises 6–10, find the angular velocity in rad/sec that corresponds to the given linear velocity.

6. $r = 10$ in., $v = 150$ in./sec

7. $r = 3$ ft, $v = 2$ ft/min

8. $r = 10$ ft, $v = 30$ mph

9. $r = 1.3$ m, $v = 140$ km/hr

10. $r = 10$ cm, $v = 100$ km/hr

In Exercises 11 and 12, find the distance traveled by a point on the rim of a wheel in the given period of time.

11. $r = 10$ cm, $\omega = \pi$ rad/sec, $t = 20$ sec

12. $r = 2$ ft, $\omega = 5$ rpm, $t = 3$ min

13. A 10-in.-diameter pulley rotates with angular velocity of 200 rpm. What is the linear velocity in ft/sec of a belt attached to the pulley? Approximately how many miles does a point on the belt travel in 8 hr?

14. (a) What is the angular velocity in rad/sec of the hour hand of a clock?
    (b) How far does the tip of the hour hand move in 1 hr if it is 8 in. long?

15. How far does the tip of a 4-in. minute hand move in 90 sec?

16. What is the linear velocity in ft/sec of the tip of a 6-in. second hand?

17. A bicycle is ridden at a constant speed of 12 mph for 2 hr. What is the angular velocity in rad/sec of a 26-in.-diameter wheel?

18. A bicycle travels 10 mi in 1 hr. Find the angular velocity of the drive sprocket in rpm if the drive sprocket has a 5-in. radius and is connected to a 14-in.-radius rear wheel with a sprocket of 1.50 in. (See Figure 4.20.)

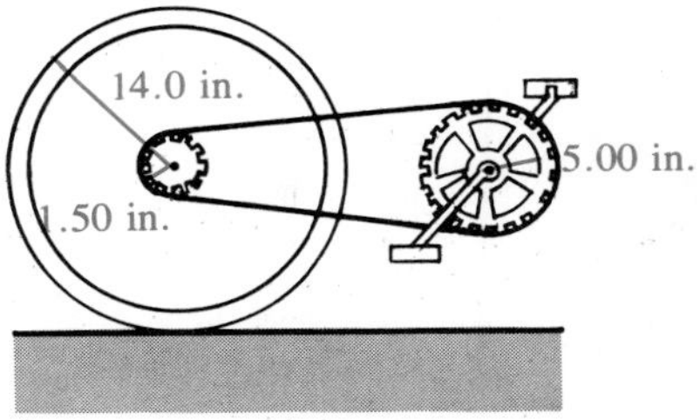

19. Two pulleys, one with a radius of 1 ft and the other with a radius of 2 ft, are connected by a belt. If the larger pulley has an angular velocity of 100 rpm, find the speed of the belt in ft/sec and the angular velocity of the other pulley. (See Figure 4.21.)

For Exercises 20 and 21, refer to Figure 4.22.

20. Find the linear velocity in mph of a person standing on the equator.

21. Find the linear velocity in mph of a person standing at a latitude of 35°.

22. A satellite circles the Earth at a height of 100 mi. If the satellite circles the earth every 90 min, what is the linear velocity in mph of the satellite? (See Figure 4.23.)

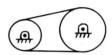

14.0 in.

5.00 in.

1.50 in.

Figure 4.20

Figure 4.21

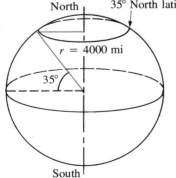

North

35° North latitude

$r = 4000$ mi

35°

South

Figure 4.22

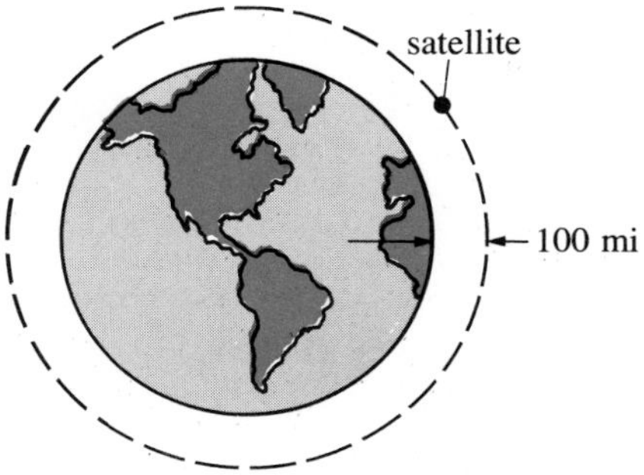

satellite

100 mi

Figure 4.23

Key Topics for Chapter 4

Define and/or discuss each of the following.

Radian Measure
Trigonometric Functions Using Radians
Arc Length

Area of a Sector
Angular Velocity
Linear Velocity

Review Exercises for Chapter 4

In Exercises 1–5, express each angle in radian measure. Use multiples of π.

1. $36°$ 2. $25°$ 3. $110°$ 4. $3000°$ 5. $-85°$

In Exercises 6–10, express each angle in degrees.

6. $\dfrac{5\pi}{12}$ 7. $\dfrac{\pi}{18}$ 8. $\dfrac{13\pi}{6}$ 9. 435 radians

10. 180π

In Exercises 11–20, evaluate the given function. If the radian measure is given as a multiple or fractional part of π, give the exact value. Otherwise use a calculator or Table C and interpolation when necessary.

11. $\sin \dfrac{7\pi}{6}$ 12. $\cos \dfrac{16\pi}{3}$ 13. $\tan \dfrac{25\pi}{6}$ 14. $\cot \dfrac{-5\pi}{4}$

15. $\sec \pi$ 16. $\sin 4.16$ 17. $\cos 2.73$ 18. $\tan 5.27$

19. $\sin -3.07$ 20. $\sec 0.3$

In Exercises 21–30, find the value of x for $0 \le x < 2\pi$. When appropriate, give the exact value.

21. $\sin x = -\dfrac{\sqrt{3}}{2}$, $\cos x > 0$ 22. $\cos x = \dfrac{1}{2}$, $\sin x > 0$

23. $\tan x = 3$, $\cos x > 0$ 24. $\cos x = \dfrac{\sqrt{2}}{2}$, $\tan x < 0$

25. $\sec x = -2$, $\sin x > 0$ 26. $\sin x = -0.8102$, $\cos x < 0$

27. $\tan x = 1.202$, $\sin x < 0$ 28. $\sin x = 0.6032$, $\tan x > 0$

29. $\tan x = 2$, $\cos x < 0$ 30. $\csc x = -1.118$, $\tan x > 0$

31. Determine θ if $\tan \theta = 2$, $2\pi < \theta < 4\pi$.

32. Determine θ if $\sin \theta = 0.6773$, $-\pi < \theta < \pi$.

33. A central angle of 47° subtends an arc of 5.5 cm on the circumference of a circle. Determine the radius of the circle.

34. A central angle of a circle subtends an arc on its circumference of 15 in. If the radius of the circle is 26 in., what is the measure of the central angle?

35. The drum shown in Figure 4.24 rotates clockwise through an angle of 200°. Determine how far the block is moved if the radius of the drum is 0.75 ft.

36. Compute the radius of the drum in Figure 4.24 if the weight is moved 3.5 m by a rotation of 35.2°.

37. Determine the area of the sector of a circle if the angle of the circular sector is 40° and the diameter of the circle is 28 cm.

38. The area of a circular sector is 100 ft² and the subtended arc is 9.5 ft. Find the radius of the circle.

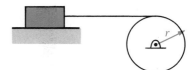

Figure 4.24

39. A bicycle wheel moves along the pavement at 14 mph. Determine the angular velocity of the wheel in rad/sec and in rpm if it has a diameter of 27 in.

40. Compute the velocity with which the weight in Figure 4.24 will move if the drum is rotating with an angular velocity of 2.5 rpm and the radius of the drum is 8.0 in. Express the velocity in ft/sec.

41. A conveyor belt rolls on cylindrical bearings that are 2.5 in. in diameter. If the belt has a velocity of 4.6 ft/sec, what is the angular velocity of the roller bearings?

42. The two wheels shown in Figure 4.25 are connected by a belt. Determine the angular velocity of the smaller wheel if the angular velocity of the larger one is 25 rev/sec. What is the velocity of the belt?

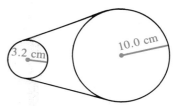

Figure 4.25

43. Assume that the radius of the larger wheel in Figure 4.25 is R and the radius of the smaller one is r. If the angular velocity of the larger wheel is ω, show that the angular velocity of the smaller one is ωR/r.

5

Analytic Trigonometry

5.1 Trigonometric Functions of Real Numbers

Modern trigonometry consists of two more or less distinct branches. The study of the six ratios and their applications to problems involving triangles is one branch, called **triangle trigonometry**. The other branch, which is concerned with the general functional behavior of the six ratios, is called **analytic trigonometry**.

In analytic trigonometry we consider the six trigonometric functions to be functions of real numbers in addition to being functions of angles. The discussion in this section shows that the reinterpretation of the domain of the trigonometric functions to include real numbers is a matter of matching real numbers with the radian measures of angles.

We begin by locating the unit circle in the rectangular plane so that its center is at the origin and it passes through (1, 0), as shown in Figure 5.1(a). The circumference of the unit circle is 2π, or approximately 6.28.

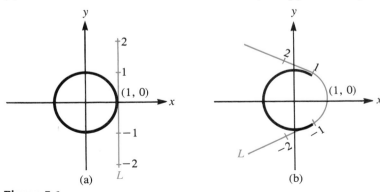

Figure 5.1

Let L denote a real number line that is parallel to the y-axis with its origin at $(1, 0)$. Now if L is wrapped around the unit circle, each point on the line is mapped onto a point on the circle, as in Figure 5.1(b). The positive half-line is wound in a counterclockwise direction, whereas the negative half-line is wound in a clockwise direction.

Each real number u mapped onto the unit circle determines both a point P in the plane and an angle α in standard position, as shown in Figure 5.2. The point P and the angle α are said to be *associated* with the real number u.

A relationship can be established between the number u and the angle α based on the fact that the radian measure of an angle is the ratio of the length of the arc of a circle subtended by the angle to the radius of the circle; that is,

$$\alpha \text{ (radians)} = \frac{u \text{ (arc length)}}{r \text{ (radius)}}$$

Since $r = 1$ for the unit circle, the measure of angle α is numerically equal to the arc length u. Symbolically,

$$\alpha \text{ (radians)} = u$$

Example 1 illustrates this fact.

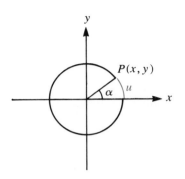

Figure 5.2

Example 1. Sketch the points and angles associated with the real numbers 2, 10, -3.6, and 6.

Solution. See Figure 5.3.

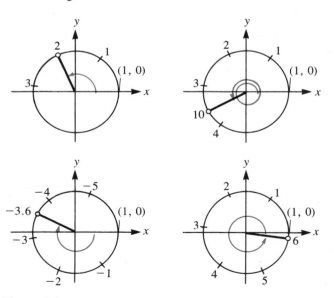

Figure 5.3

Using the natural association of real numbers with angles in standard position, we can define the trigonometric functions for any real number u. If α, in radians, is the angle associated with the real number u, then

$$\sin u = \sin \alpha \qquad \cos u = \cos \alpha \qquad \tan u = \tan \alpha$$

$$\sec u = \sec \alpha \qquad \csc u = \csc \alpha \qquad \cot u = \cot \alpha$$

To clearly understand what has been accomplished by these definitions, you must keep in mind that u is a real number and not an angle. These definitions permit discussion of trigonometric functions without reference to the concept of angle.

The trigonometric functions were defined in Chapter 2 as ratios of numerical values. The extension to real numbers shows how the same functions are related to a circle. Because of their relationship to a circle, the trigonometric functions are often called the *circular functions*.

Warning: Since it is customary to delete the dimension of radians, it will be impossible to distinguish between an argument that is a real number and an argument that is an angle in radians. Take care to indicate the correct units of measurements if the argument is an angle measured in degrees, minutes, and seconds. Otherwise, by convention, the argument is an angle measured in radians or is a real number. Thus, sin 30° means the sine of an angle of 30°, but sin 30 means either of the following, both of which have the same numerical value:

(1) the sine of an angle of 30 radians.

(2) the sine of the real number 30.

When the unit circle is located in the rectangular plane, the point P associated with the number u has coordinates (x, y). This association is shown in Figure 5.2. An interesting and important relationship exists between the coordinates of the point associated with u and $\cos u$ and $\sin u$. Since we are using the unit circle, we have

$$\cos u = \frac{x}{r} = \frac{x}{1} = x \qquad \sin u = \frac{y}{r} = \frac{y}{1} = y$$

That is, **the x- and y-coordinates of the point associated with the real number u are equal to $\cos u$ and $\sin u$, respectively.**

Figure 5.4 shows a unit circle with real numbers from 0 to 2π marked off on the circumference. Using this circle, we can approximate values of $\cos u$ and $\sin u$ for any real number u in the interval $(0, 2\pi)$. As u increases, the values of $\sin u$ and $\cos u$ repeat themselves every circumference, or 2π units. Thus,

$$\cos u = \cos(u + 2n\pi) \qquad \sin u = \sin(u + 2n\pi)$$

where n is an integer.

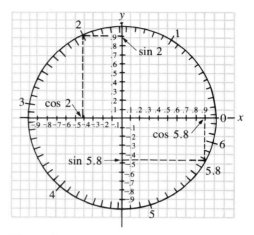

Figure 5.4

Example 2. Using Figure 5.4, find the cosine and sine of (a) 2, (b) 5.8, (c) 8.28, and (d) −4.28.

Solution. Using Figure 5.4, we estimate as follows.
(a) cos 2 = −0.42, sin 2 = 0.91.
(b) cos 5.8 = 0.89, sin 5.8 = −0.46.
(c) To find cos 8.28 and sin 8.28, note that 8.28 = 6.28 + 2. Thus, cos 8.28 = cos 2 = −0.42, sin 8.28 = sin 2 = 0.91.
(d) To find cos (−4.28) and sin (−4.28), note that −4.28 = 2 − 6.28. Thus, cos (−4.28) = cos 2 = −0.42, sin (−4.28) = sin 2 = 0.91.

The values of the trigonometric functions for real numbers may be found from Table C in the Appendix or from a calculator. Table C may be interpreted as a table of trigonometric functions of either real numbers or angles measured in radians. Similarly, when numbers are entered into a calculator in radian mode, the numbers may be considered either real numbers or radian measures of angles.

Example 3. Use a calculator or Table C to find cos 0.5, sin 1.4, and tan 0.714 to four decimal places.

Solution. From Table C or a calculator,

$$\cos 0.5 = 0.8776$$

$$\sin 1.4 = 0.9854$$

If tan 0.714 must be approximated by interpolation, we have, from Table C,

$$\tan 0.71 = 0.8595$$

$$\tan 0.72 = 0.8771$$

and thus

$$\tan 0.714 \approx (0.4)(.0176) + 0.8595 \approx 0.8665$$

Example 4. Find the values of x for which $\tan x = 1$.

Solution. Some of the numbers in the solution set are $-7\pi/4$, $-3\pi/4$, $\pi/4$, and $5\pi/4$. Notice that each of these numbers is an integral multiple of $\pi/4$. Using this fact, we can write the desired solution set in the form $(4n + 1)(\pi/4)$, where n is an integer.

The names of the trigonometric functions are the same whether they are used in the sense of ratios or in a functional sense. However, by writing $y = \sin x$ or $f(x) = \sin x$, we emphasize the functional concept of the sine function. Do not be misled into believing that only one letter may be used for the argument of the trigonometric functions. Any convenient letter or symbol suffices. Thus, $\sin x$, $\sin u$, $\sin \theta$, and $\sin y$ all mean the same thing. Only the application reveals whether the argument is to be interpreted as an angle or as a real number.

The following list illustrates some practical applications of trigonometric functions in which the argument has nothing to do with angles:

■ A weight hanging on a certain vibrating spring has a velocity described by $\sin 3t$. In this case, the argument is $3t$, where t is the time in seconds.

■ The instantaneous voltage for certain electrical systems is given by $156 \sin 377t$, where t is the value of the time in seconds.

■ The equation of motion of a shaft with flexible bearings is given by

$$x = x_0 \sin \frac{\pi}{2L} x$$

where x and L are given in centimeters.

Exercises for Section 5.1

1. Sketch the point on the unit circle associated with each real number. Find the cosine and sine of each number using Figure 5.4, and check the answer using Table C or a calculator.

(a) 1 (b) -2 (c) 3 (d) 10
(e) 3π (f) -4 (g) -4π (h) $\frac{1}{3}\pi$
(i) $\frac{1}{3}$ (j) $\frac{1}{2}$ (k) $\sqrt{7}$ (l) 5.15

2. In calculus the ratio $\dfrac{\sin x}{x}$ is important. Use Table C or a calculator to find the value of the ratio for each of the following values of x.
   (a) 0.3   (b) 0.2   (c) 0.1   (d) 0.05   (e) 0.01

   What value do you think $\dfrac{\sin x}{x}$ approaches as x approaches 0? What is $\dfrac{\sin x}{x}$ when $x = 0$?

3. If $f(x) = \sin x$ and $g(x) = \cos x$, show that $[f(x)]^2 + [g(x)]^2 = 1$.

Given that $f(x) = \sin x$, find the values in Exercises 4–10. (Use Table C or a calculator if necessary.)

4. $f(\tfrac{1}{2}\pi)$ 5. $f(\pi)$ 6. $f(50)$ 7. $f(-10)$

8. $f(3\pi)$ 9. $f(\tfrac{1}{6}\pi)$ 10. $f(2\pi)$

Given that $g(x) = \cos x$, find the values in Exercises 11–17. (Use Table C or a calculator if necessary.)

11. $g(0)$ 12. $g(\tfrac{1}{3}\pi)$ 13. $g(\pi)$ 14. $g(25)$

15. $g(-10)$ 16. $g(5\pi)$ 17. $g(5)$

18. If $f(x) = \sin x$, what is the solution to the equation $f(x) = 0$?

19. If $g(x) = \cos x$, what is the solution to the equation $g(x) = 0$?

20. For which values of x is $\sin x = 1$?

21. For which values of x is $\cos x = 1$?

22. For which values of x is $\sec x = \tfrac{1}{2}$?

23. Solve the inequality $\cos x \le \sec x$.

24. Solve the inequality $\sin x > \csc x$.

25. Solve the inequality $\sin x \ge 1$.

In Exercises 26–28, recall that a zero of a function $f(x)$ is a value of $x = x_0$ such that $f(x_0) = 0$.

26. What are the zeros of $\sin x$?

27. What are the zeros of $\cos x$?

28. Which pairs of trigonometric functions have the same zeros?

29. The velocity of a certain weight attached to a spring is given by $v = \sin 2t$, where v is in fps when t is measured in seconds. Find the velocity for $t = 0.7$ sec.

30. The instantaneous voltage applied to a circuit is given by $V = 220 \sin 377t$, where t is measured in seconds. Find the voltage being applied when $t = 0.01$ sec.

31. In calculus we show that if an object has motion described by $y = \tan t$, its velocity is given by $v = \sec^2 t$. Find the time between 0 and 1 sec for which the velocity is 2 m/sec.

We begin this section with a discussion of periodicity and boundedness. Each property is defined and discussed in a general setting and is then related to the trigonometric functions.

Periodicity

Something is *periodic* when it repeats itself at regular intervals. Examples of periodic phenomena occur throughout the physical sciences—the motion of a pendulum, alternating current, the vibration of a tuning fork. The trigonometric functions play an important role in describing periodic phenomena.

Before discussing the periodic nature of the trigonometric functions, we state the following precise but general definition of a periodic function.

Definition 5.1: A function f is said to be *periodic* if there exists a number p, where $p > 0$, such that

$$f(x) = f(x + p)$$

for all x in the domain of the function. If there is a smallest number p for which this expression is true, then p is called the **period** (or the *fundamental period*) of the function.

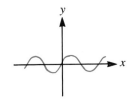

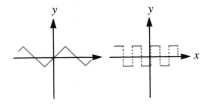

Figure 5.5 Periodic functions

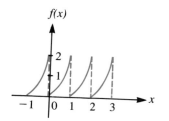

Figure 5.6

Figure 5.5 contains graphs of several periodic functions. A periodic function is completely described by its period and the functional values over one period interval, as the next example illustrates.

Example 1. Sketch the graph of the function that is defined over $0 \leq x < 1$ by $f(x) = 2x^2$ and is periodic with period one.

Solution. We sketch the graph of the function on $0 \leq x < 1$, and then repeat this pattern for all other intervals. (See Figure 5.6.)

If a function is periodic with period p, then it is also periodic with periods $2p, 3p, \ldots, np, \ldots$; that is,

$$f(x) = f(x + p) = f(x + 2p) = f(x + 3p) = \cdots = f(x + np) = \cdots$$

However, only the value p is considered to be *the* period of the function.

Of particular interest to us is the periodic nature of the trigonometric functions. Recall from Section 5.1 that for any real number u, the values of $\cos u$ and $\sin u$ are equal to the x- and y-coordinates, respectively, of the point on a unit circle corresponding to the number u. (See Example 2, Section 5.1.) Further, the values of $\sin u$ and $\cos u$ repeat themselves every 2π units. Thus, $\sin u$ and $\cos u$ are periodic functions with period 2π. Since $\csc u = 1/\sin u$ and $\sec u = 1/\cos u$, it follows that $\csc u$ and $\sec u$ are also periodic with period 2π. We indicate this periodicity by writing

$$\sin u = \sin (u + 2\pi)$$
$$\cos u = \cos (u + 2\pi)$$
$$\sec u = \sec (u + 2\pi)$$
$$\csc u = \csc (u + 2\pi)$$

The values of $\tan u$ in the first quadrant are repeated in the third quadrant. Similarly, the values of $\tan u$ in the second quadrant are repeated in the fourth quadrant. Thus, $\tan u$ is a periodic function with period π. The same is true for $\cot u$, and thus we write

$$\tan u = \tan (u + \pi)$$
$$\cot u = \cot (u + \pi)$$

Boundedness

Many realistic physical phenomena tend to be constrained within certain limits or bounds. Such phenomena are described mathematically by functions called bounded functions. A function is **bounded** if there is a number M for which $|f(x)| \leq M$ for all x in the domain of the function.

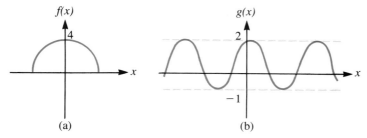

Figure 5.7   Bounded Functions

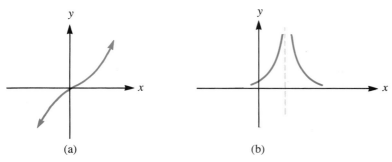

Figure 5.8   Unbounded Functions

For example, the functions graphed in Figure 5.7 are bounded, whereas those in Figure 5.8 are unbounded.

If a function is bounded, it will have many bounds. But in describing the property of boundedness, we usually give only the lower and upper extremes of the functional values. For example, function f in Figure 5.7(a) varies between 0 and 4 so we write $0 \leq f(x) \leq 4$, and the boundedness of g in Figure 5.7(b) is described by $-1 \leq g(x) \leq 2$.

The graph of a function that is bounded between c and d is constrained between the lines $y = c$ and $y = d$, whereas the graph of an unbounded function is unlimited in its vertical extent.

Since the sine and cosine functions may be conceived of as coordinates of points on a unit circle, and since these coordinates are constrained to lie between -1 and 1, the sine and cosine functions are bounded by -1 and 1. Thus, the graphs of the sine and cosine lie between the lines $y = 1$ and $y = -1$.

Even and Odd Functions

Some functions have the property that their functional values do not change when the sign of the independent variable is changed. For example, if $f(x) = x^2$, then $f(-x) = (-x)^2 = x^2$. In general, we denote this by

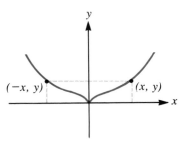

Figure 5.9 Symmetry with Respect to y-axis

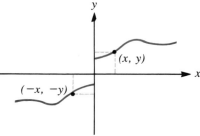

Figure 5.10 Symmetry with Respect to Origin

$$f(x) = f(-x)$$

Functions that have this property are called *even* functions.

Other functions are such that when the sign of the independent variable changes, so does the sign of the dependent variable. For example, if $f(x) = x^3$, then $f(-x) = (-x)^3 = -x^3$. This is written in general as

$$f(-x) = -f(x)$$

Functions that obey this rule are called *odd* functions.

The graphs of even and odd functions have an interesting kind of geometric symmetry.

(1) For an even function $y = f(x)$, if the point (x, y) is on the graph of the function, so is $(-x, y)$. Graphs with this property are said to be *symmetric with respect to the y-axis*. (See Figure 5.9.)

(2) For an odd function $y = f(x)$, if the point with coordinates (x, y) is on the graph of the function, so is the point $(-x, -y)$. Graphs with this property are said to be *symmetric with respect to the origin*. (See Figure 5.10.)

Usually, of course, a function is neither even nor odd. However, it can be proved that every function may be expressed as the sum of an even and an odd function.

By referring to Figure 5.4, Section 5.1, we can demonstrate that $\sin x$ is an odd function; that is,

$$\sin(-x) = -\sin x$$

Similarly, we can show that $\cos x$ is an even function; that is,

$$\cos(-x) = \cos x$$

Thus, the graphs of these two functions have the symmetry properties of odd and even functions, respectively.

Variation of the Trigonometric Functions

As an angle increases from 0 to 2π, its terminal side rotates in a counterclockwise sense. During this interval all values of the trigonometric functions are included. Table 5.1 shows the variation in the values of $\sin u$ and $\cos u$ during one revolution of the terminal side. Because of the property of periodicity, the pattern in Table 5.1 will be repeated for $2\pi \leq u \leq 4\pi$, for $4\pi \leq u \leq 6\pi$, etc.

Table 5.1 Variation of sin u and cos u

When x increases from:	sin u	cos u
0 to $\frac{1}{2}\pi$	increases from 0 to 1	decreases from 1 to 0
$\frac{1}{2}\pi$ to π	decreases from 1 to 0	decreases from 0 to -1
π to $\frac{3}{2}\pi$	decreases from 0 to -1	increases from -1 to 0
$\frac{3}{2}\pi$ to 2π	increases from -1 to 0	increases from 0 to 1

Graphs of the Sine and Cosine Functions

The graph of any function $y = f(x)$ is the set of points in the plane with the coordinates $(x, f(x))$. In general, such a graph is obtained by tabulating values and plotting the corresponding points. But, in particular cases, the task can be significantly shortened by using some general properties of the given function.

We first examine the graphs of the sine and cosine functions considered as functions of *real numbers*. The procedures for graphing the sine and the cosine are almost identical.

We first summarize some of the analytical properties previously discussed.

(1) Both sin x and cos x are *bounded*, above by 1 and below by -1. Thus, the graph of each of the functions lies between the lines $y = 1$ and $y = -1$.

(2) Both sin x and cos x are *periodic* with period 2π. Thus only one interval of length 2π need be considered when the two functions under consideration are being graphed. Outside this interval, the graph repeats itself.

(3) The sine function is odd; that is, $\sin x = -\sin(-x)$. Thus the graph of sin x is symmetric about the origin.
The cosine function is even; that is, $\cos x = \cos(-x)$. Thus, its graph is symmetric about the y-axis.

(4) sin $x = 0$ for $x = 0, \pm\pi, \pm 2\pi, \pm 3\pi$, etc.
cos $x = 0$ for $x = \pm\frac{1}{2}\pi, \pm\frac{3}{2}\pi, \pm\frac{5}{2}\pi$, etc.
In each case, these are the intercepts on the x-axis.

(5) The numerical values of the sine and cosine functions for $0 \leq x \leq \frac{1}{2}\pi$ correspond to the values of sin x and cos x in the first quadrant. The other three quadrants yield values that are numerically the same (though there may be a difference in sign).

To obtain the specific graph, we need to calculate a reasonable number of points for $0 < x < \frac{1}{2}\pi$. Then we can draw a smooth curve through these points, after which we can use the general properties to obtain the remainder of the curve. Usually we emphasize certain "special" points corresponding to $x = 0$, $\frac{1}{2}\pi$, π, $\frac{3}{2}\pi$, and 2π.

Table 5.2 Values of sin x and cos x from $x = 0$ to $x = 2\pi$ in increments of $\frac{1}{6}\pi$

x	0	$\frac{1}{6}\pi$	$\frac{1}{3}\pi$	$\frac{1}{2}\pi$	$\frac{2}{3}\pi$	$\frac{5}{6}\pi$	π	$\frac{7}{6}\pi$	$\frac{4}{3}\pi$	$\frac{3}{2}\pi$	$\frac{5}{3}\pi$	$\frac{11}{6}\pi$	2π
sin x	0	0.5	0.87	1	0.87	0.5	0	-0.5	-0.87	-1	-0.87	-0.5	0
cos x	1	0.87	0.5	0	-0.5	-0.87	-1	-0.87	-0.5	0	0.5	0.87	1

Figure 5.11 shows the graph of one period of the sine function. To the left of the graph is a circle of radius 1. The sine function has values numerically equal to the y coordinates of points on this circle. This figure displays the relationship between points on the unit circle and points on the graph of the function.

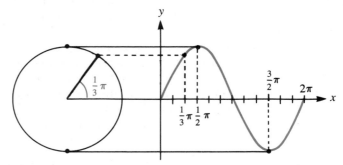

Figure 5.11

Figure 5.12 is a graph of several periods of the function $y = \sin x$, and Figure 5.13 is a graph of several periods of the cosine function. Although in the figures the graphs terminate, in reality they continue indefinitely. A statement of the period is often included with the graph to emphasize this indefinite continuation.

The graph of the sine function is called a sine wave, or a **sinusoid**. The term *sinusoid* may be applied to any curve having the same shape as that of the sine function. For example, the graph of the cosine function is properly

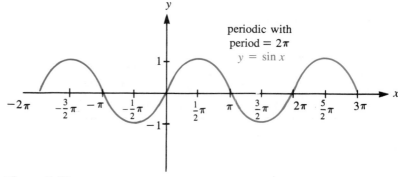

Figure 5.12

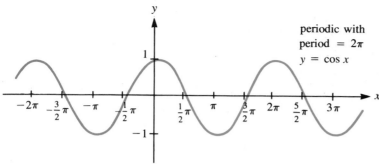

Figure 5.13

called a sinusoid, since it is obtained simply by shifting the graph of the sine function $\frac{1}{2}\pi$ units to the left.

A **cycle** is the shortest segment of the graph that includes one period. The **frequency** of the sinusoid is defined to be the reciprocal of the period. It represents the number of cycles of the function in each unit interval. The graph of any sinusoid should clearly demonstrate the boundedness, the periodicity, and the intercepts.

A sketch of a sinusoid may be obtained by connecting known points on the curve with a smooth line. Although there is no rule stating how many points should be plotted for a given sinusoid, the idea is to choose just enough points to make the shape obvious. In this section we suggest you use 6 to 10 points per period of the graph. Example 2 shows the point-plotting approach to graphing $y = \cos \frac{1}{2}x$, using multiples of $\frac{1}{2}\pi$ for x.

Example 2. Sketch the graph of $y = \cos \frac{1}{2}x$ on the interval $0 \le x \le 4\pi$, using multiples of $\frac{1}{2}\pi$ for x.

Solution. The table shows the various values for this interval. The graph (Figure 5.14) is then obtained by plotting these points. Notice that we have completed one period on the interval $[0, 4\pi]$. We conclude from this that the period of $y = \cos \frac{1}{2}x$ is 4π.

x	0	$\frac{1}{2}\pi$	π	$\frac{3}{2}\pi$	2π	$\frac{5}{2}\pi$	3π	$\frac{7}{2}\pi$	4π
y	1	$\frac{\sqrt{2}}{2}$	0	$\frac{-\sqrt{2}}{2}$	-1	$\frac{-\sqrt{2}}{2}$	0	$\frac{\sqrt{2}}{2}$	1

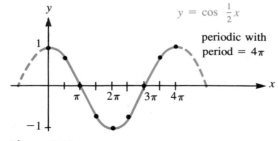

$$y = \cos \frac{1}{2}x$$

periodic with period $= 4\pi$

Figure 5.14

Example 3. Sketch the graph of $y = 3 \sin 2\pi x$ on the interval $0 \le x \le 1$, using multiples of $\frac{1}{6}$ for x.

Solution. We first construct a table of values. The corresponding graph is shown in Figure 5.15. Notice that the period is 1.

x	0	$\frac{1}{6}$	$\frac{1}{3}$	$\frac{1}{2}$	$\frac{2}{3}$	$\frac{5}{6}$	1
y	0	$\frac{3\sqrt{3}}{2}$	$\frac{3\sqrt{3}}{2}$	0	$\frac{-3\sqrt{3}}{2}$	$\frac{-3\sqrt{3}}{2}$	0

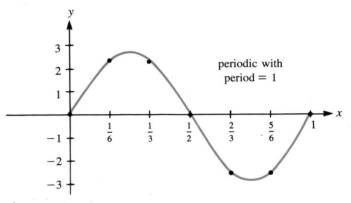

Figure 5.15

Exercises for Section 5.2

1. What are the domain and the range of sin x?

2. What are the domain and the range of cos x?

In Exercises 3–6, indicate the period of each sinusoid whose graph is shown, and whether it is a sine or cosine wave.

3.
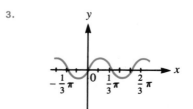

4.
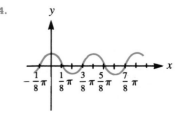

5.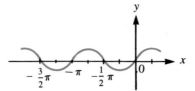

y

$-\pi$ π 3π 5π x

6.

y

$-\frac{3}{2}\pi$ $-\pi$ $-\frac{1}{2}\pi$ 0   x

In Exercises 7–10, determine whether the given function is even or odd.

7. $x \sin x$

8. $x \cos x$

9. $\sin x \cos x$

10. $\sin^2 x$

11. Sketch one period of $y = \sin x$, using multiples of $\frac{1}{4}\pi$ for x.

12. Sketch one period of $y = \cos x$, using multiples of $\frac{1}{4}\pi$ for x.

13. Sketch $y = 2 \sin x$ on the interval $0 \le x \le 2\pi$, using multiples of $\frac{1}{4}\pi$ for x.

14. Sketch $y = 3 \cos x$ on the interval $0 \le x \le 2\pi$, using multiples of $\frac{1}{4}\pi$ for x.

15. Sketch $y = \cos 2x$ on the interval $0 \le x \le \pi$, using multiples of $\frac{1}{8}\pi$. What is its period?

16. Sketch $y = \sin 2x$ on the interval $0 \le x \le \pi$, using multiples of $\frac{1}{8}\pi$. What is its period?

The convention of using multiples of π to plot graphs of the trigonometric functions is convenient but not essential. We can, of course, evaluate $\sin x$ and $\cos x$ for any real number x. Sketch the functions in Exercises 17–22 on the interval $0 \le x \le 7$, using multiples of 1. A calculator will be helpful for these exercises.

17. $y = \cos x$

18. $y = \sin x$

19. $y = 1.5 \sin x$

20. $y = 3 \cos x$

21. $y = \cos \frac{1}{2}x$

22. $y = \sin \frac{1}{2}x$

23. The function $f(x) = |\sin x|$ is called the **full-wave rectified sine wave**. Sketch this function on the interval $0 \le x \le 2\pi$, using multiples of $\frac{1}{4}\pi$. What is its period?

24. The function defined by

$$f(x) = \begin{cases} \sin x, & 0 \le x \le \pi \\ 0, & \pi \le x \le 2\pi \end{cases}$$

and periodic with period 2π is called the **half-wave rectified sine wave**. Sketch this function on the interval $0 \le x \le 4\pi$, using multiples of $\frac{1}{4}\pi$.

25. Sketch $y = |\cos x|$ on the interval $-\frac{1}{2}\pi \le x \le \frac{3}{2}\pi$, using multiples of $\frac{1}{4}\pi$.

26. Make a sketch of $y = x$ and $y = \sin x$ on the same axes and convince yourself that $\sin x < x$ for $x > 0$.

27. The velocity of a particle is given by the equation $v = \cos t$. Sketch the graph of velocity as a function of time. What is the initial velocity? For which times is the velocity equal to zero? How could you describe these points graphically?

28. The vertical component of the motion of a particle that moves in a circular orbit is given by $v = R \sin t$, where R is the radius of the orbit. Sketch the variation in the value of the vertical component for $0 \le t \le \pi$ for a radius of 3 ft.

29. The output of an electrical circuit designed to produce triangular waves is given by

$$E = \begin{cases} \cos t, & 0 \le t \le \frac{1}{2}\pi \\ \sin t, & \frac{1}{2}\pi < t \le \pi \\ -\cos t, & \pi < t \le \frac{3}{2}\pi \\ -\sin t, & \frac{3}{2}\pi < t \le 2\pi \end{cases}$$

Sketch the graph of this output for the given interval.

5.3 Expansion and Contraction of Sine and Cosine Graphs

In Section 5.2 you saw how to graph the sine and cosine functions by point plotting. Graphing modified functions such as $\sin 2x$, $4 \cos x$, and $\cos (x + 3)$ with the point-plotting process can be a tedious process. In this section you will learn how to simplify the procedure for graphing the sine and cosine functions.

We are interested in the following four modifications of the sine and cosine functions:

(1) Multiplication of the function by a constant.
(2) Multiplication of the argument by a constant.
(3) Addition or subtraction of a constant to the argument.
(4) Addition or subtraction of a constant to the functional value.

Multiplication of the Function by a Constant

We have seen that the values of the sine function oscillate between $+1$ and -1. Consider the equation $y = A \sin x$. Since

$$-1 \le \sin x \le 1$$

it follows that

$$-A \le A \sin x \le A$$

The absolute value of A is called the **amplitude** of the sine wave. If $|A|$ is greater than 1, the amplitude of the sine wave is increased; if $|A|$ is less than 1, the amplitude is decreased. Sometimes $|A|$ is called the **maximum**, or

peak, value of the function. Figure 5.16 shows the graph of $y = A \sin x$ for $A = 1, \frac{1}{2},$ and 2.

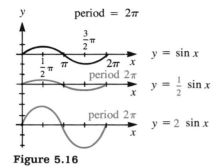

Figure 5.16

Example 1. Sketch the graph of one period of $y = \sqrt{3} \cos x$.

Solution. The function $y = \sqrt{3} \cos x$ has an amplitude of $\sqrt{3}$ and a period of 2π. The graph of $y = \cos x$ is at its maximum at $x = 0$, has a zero at $x = \frac{1}{2}\pi$, is at its minimum at $x = \pi$, has a zero at $x = \frac{3}{2}\pi$, and is at its maximum again at $x = 2\pi$. The graph of the given function is shown in Figure 5.17.

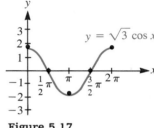

Figure 5.17

Example 2. Sketch one period of $y = -3 \sin x$.

Solution. The amplitude coefficient in this case is negative, which means that the sign of each functional value will be opposite what it is for $y = 3 \sin x$; $y < 0$ when $\sin x > 0$ and $y > 0$ when $\sin x < 0$. Thus the graph of $y = -3 \sin x$ is the reflection in the x-axis of the graph of $y = 3 \sin x$. Both graphs are shown in Figure 5.18 to emphasize the relationship between the two curves.

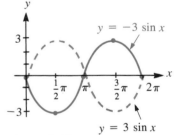

Figure 5.18

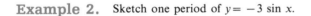

Multiplication of the Argument by a Constant

If we multiply the argument of $\sin x$ by a positive constant B, the function becomes $\sin Bx$. The graph of this function is sinusoidal, but since the argument is Bx one period of $\sin Bx$ is contained in the interval

$$0 \le Bx \le 2\pi$$

or

$$0 \le x \le \frac{2\pi}{B}$$

Therefore, the period of $\sin Bx$ is $2\pi/B$. For example, the period of $\sin 2x$ is π; the period of $\sin \frac{1}{2}x$ is 4π. Graphically, increasing B has the effect of squeezing the sine curve together like an accordion, and decreasing B has the effect of pulling it apart. (See Figure 5.19.)

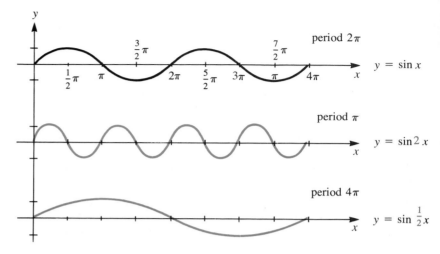

Figure 5.19

Similarly, the period of cos Bx is $2\pi/B$. Thus,

The period of both sin Bx and cos Bx is $\dfrac{2\pi}{B}$, $(B > 0)$.

Example 3

(a) sin $3x$ has a period of $\dfrac{2}{3}\pi$.

(b) cos $\dfrac{1}{4}x$ has a period of $\dfrac{2\pi}{1/4} = 8\pi$.

(c) cos $10x$ has a period of $\dfrac{2\pi}{10} = \dfrac{1}{5}\pi$.

(d) sin $4\pi x$ has a period of $\dfrac{2\pi}{4\pi} = \dfrac{1}{2}$.

Following is a general procedure for sketching sine and cosine functions, given the amplitude and the period.

General Procedure for Sketching Graphs of Sine and Cosine Functions

- For the independent variable, choose units that are equal to $\frac{1}{4}$ of the period of the function.
- Recall that the zeros and the maximum and minimum values occur at multiples of $\frac{1}{4}$ of the period. For example, if the period is 2π, use $\frac{1}{4}(2\pi) = \frac{1}{2}\pi$ as a basic unit on the x-axis; if the period is 5, use $\frac{1}{4}(5)$ or $\frac{5}{4}$.
- The graph of the sine function over one period crosses the x-axis at its initial point, reaches a maximum at $\frac{1}{4}$ period, crosses the x-axis at $\frac{1}{2}$

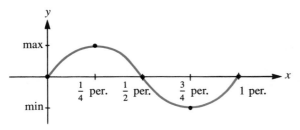

Figure 5.20 Typical Sine Curve

period, reaches a minimum at $\frac{3}{4}$ period, and crosses the x-axis at the end of the period. (See Figure 5.20.)

■ The graph of the cosine function over one period begins at a maximum, crosses the x-axis at $\frac{1}{4}$ period, reaches a minimum at $\frac{1}{2}$ period, crosses the x-axis at $\frac{3}{4}$ period, and reaches a maximum at the end of the period. (See Figure 5.21.)

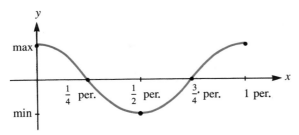

Figure 5.21 Typical Cosine Curve

Example 4. Sketch the graph of one period of $y = 2 \cos 5x$.

Solution. Here the amplitude is 2 and the period is $\frac{2}{5}\pi$. The graph is shown in Figure 5.22. Notice that the units along the x-axis are in multiples of $\frac{1}{4}(\frac{2}{5}\pi) = \frac{1}{10}\pi$.

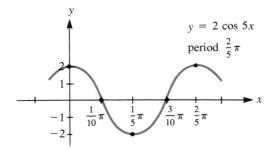

Figure 5.22

Example 5. Sketch one period of the graph of $s = 2.1 \sin 3\pi t$.

Solution. In this case the amplitude is 2.1 and the period is $\frac{2\pi}{3\pi} = \frac{2}{3}$. The graph appears in Figure 5.23.

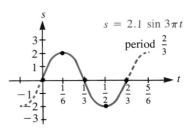

Figure 5.23

Example 6. Write the equation of the sine function whose amplitude is 3 and whose period is 7.

Solution. The general form of the desired sine function is $y = A \sin Bx$. The amplitude 3 yields $A = 3$. The period is given by $\dfrac{2\pi}{B}$, so we must have $\dfrac{2\pi}{B} = 7$, from which $B = \dfrac{2\pi}{7}$. The desired function is

$$y = 3 \sin \tfrac{2}{7}\pi x$$

Exercises for Section 5.3

In Exercises 1–20, sketch the graphs of the given functions. In each case give the amplitude and the period.

1. $y = 3 \sin x$
2. $y = \frac{1}{2} \sin x$
3. $y = 6 \cos x$
4. $y = \frac{1}{3} \sin x$
5. $y = \sin \frac{2}{3}x$
6. $y = 0.3 \sin 3x$
7. $y = \sin \pi x$
8. $y = \cos 0.1x$
9. $s = \frac{1}{2} \cos 2t$
10. $y = 100 \cos 3x$
11. $y = 8.2 \sin 0.4x$
12. $v = 3 \sin \frac{7}{6}t$
13. $y = -\cos \frac{2}{5}x$
14. $y = -5 \cos 7x$
15. $p = \pi \cos 100t$
16. $i = -0.02 \sin \pi t$
17. $y = -12 \sin 0.2x$
18. $P = 10^6 \cos \dfrac{\pi}{1000}x$
19. $v = \dfrac{\cos 1000\pi x}{50}$
20. $y = \dfrac{3 \sin 25\pi t}{200}$

In Exercises 21–28, write the equation of the sine function having the indicated amplitude and period.

21. Amplitude $= \frac{1}{3}$, period $= 12$
22. Amplitude $= \frac{1}{2}$, period $= 15$
23. Amplitude $= 20$, period $= \frac{3}{8}$
24. Amplitude $= \sqrt{5}$, period $= \frac{1}{3}$
25. Amplitude $= 2.4$, period $= \frac{1}{3}\pi$
26. Amplitude $= 0.94$, period $= \frac{1}{6}\pi$
27. Amplitude $= \pi$, period $= 3\pi$
28. Amplitude $= 2/\pi$, period $= 7\pi$

29. The motion of a pendulum can be represented by the equation $x = A \sin Bt$. Write the equation of motion of a pendulum oscillating with an amplitude of 3.2 ft and a period of 2.5 sec. (See Figure 5.24.)

30. The equation for the voltage drop across the terminals of an ordinary electric outlet is given approximately by

$$E = 156 \sin (110 \, \pi t)$$

Sketch the voltage curve for several cycles.

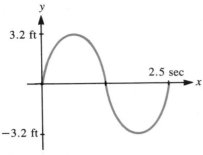

y

3.2 ft

2.5 sec

x

−3.2 ft

Figure 5.24

31. If B is small, the equation $y = \sin Bx$ approximates the shape of ocean waves. Sketch several cycles of an ocean wave described by

$$y = \sin \tfrac{1}{20}\,\pi x$$

5.4 Vertical and Horizontal Translation

A vertical or horizontal relocation of the graph of a function in which the graph's shape is not changed or distorted is called a **translation**. A function is translated vertically if a constant is added to the value of the function, and it is translated horizontally if a constant is added to the argument of the function. Both of these translations, which are important in applied work, are explained here in the context of sine and cosine functions.

Addition of a Constant to the Value of the Function

Consider the graph of $y = D + \sin x$; that is, $y - D = \sin x$. Since D is added to $\sin x$ for each x, the graph of $D + \sin x$ is simply the graph of $\sin x$ displaced D units vertically. The graph is translated up if D is positive and down if D is negative. D is called the **mean**, or **average**, **value** of the function.

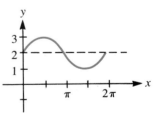

Figure 5.25

Example 1. Sketch the graph of $y = 2 + \sin x$.

Solution. The graph is shown in Figure 5.25. The mean value is 2, the amplitude is 1, and the period is 2π.

Example 2. Sketch the graph of $s = -1 + 3 \cos 2t$.

Solution. The graph is shown in Figure 5.26. The mean value is -1, the amplitude is 3, and the period is π.

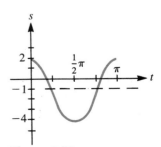

Figure 5.26

Addition of a Constant to the Argument

The addition of a constant to the argument of $\sin x$ is written $\sin (x - C)$. The constant C has the effect of shifting the graph of the sine function to the right or to the left. Notice that $\sin (x - C)$ is zero when $x - C = 0$ (that is, for $x = C$). The value of x for which the argument of the sine function is zero is called the **phase shift**. If C is positive the shift is to the right, and if C is negative the shift is to the left. Figure 5.27 shows three sine waves with phase shifts of 0, $-\frac{1}{4}\pi$, and $\frac{1}{4}\pi$.

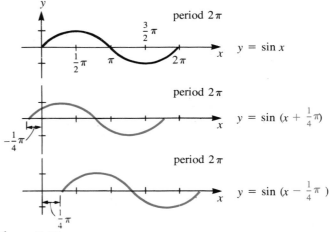

Figure 5.27

In the more general case, the effects of changes in amplitude, period, and phase shift are all combined. The function

$$y = A \sin (Bx - C) = A \sin B\left(x - \frac{C}{B}\right)$$

has an amplitude of A, a period of $2\pi/B$ and a phase shift corresponding to the value of x given by $Bx - C = 0$, that is $x = C/B$. Figure 5.28 shows a graph of the basic sine curve and the graph of $y = 3 \sin (2x - \frac{1}{3}\pi)$.

Example 3. Sketch the graph of $y = 2 \sin (\frac{1}{3}x + \frac{1}{9}\pi)$.

Solution. The amplitude of the graph is 2, the period is $2\pi/(\frac{1}{3}) = 6\pi$, and the phase shift is $-\frac{1}{3}\pi$. (See Figure 5.29.)

In all cases, the distinctive shape of the sine curve remains unaltered. This basic shape is expanded or contracted vertically because of multiplication by the amplitude constant A, expanded or contracted horizontally by the constant B, and shifted to the right or left by the constant C/B.

A similar analysis could be made for the cosine function. We will not

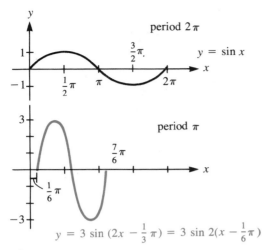

$$y = 3 \sin (2x - \tfrac{1}{3}\pi) = 3 \sin 2(x - \tfrac{1}{6}\pi)$$

Figure 5.28

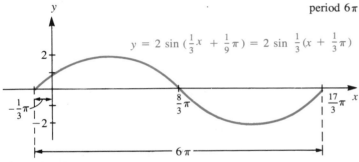

Figure 5.29

discuss in detail the function $y = D + A \cos (Bx - C)$, but the constants A, B, C, and D alter the basic cosine graph in the same manner as they do the sine graph.

Example 4. Sketch the graph of $y = 3 \cos (\tfrac{1}{2}x + \tfrac{1}{4}\pi)$.

Solution. The amplitude is 3, since the basic cosine function is multiplied by 3. The period is $2\pi/(\tfrac{1}{2}) = 4\pi$. The phase shift is found from the equation $\tfrac{1}{2}x + \tfrac{1}{4}\pi = 0$; that is, $x = -\tfrac{1}{2}\pi$. Hence the phase shift is $\tfrac{1}{2}\pi$ units to the left. The graph is shown in Figure 5.30.

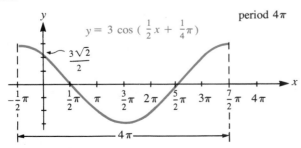

Figure 5.30

If the function is multiplied by a negative constant, you may use one of the relationships in Equation (2.5) of Section 2.3 to put the expression into a more standard form.

Example 5
(a) Sketch the graph of $y = -2 \sin (3x + 1)$.
(b) Express the given function in the form $y = A \sin (Bx + C)$, where A and B are positive constants.

Solution
(a) To graph this function, sketch the function $y = 2 \sin (3x + 1)$ and then reflect this graph in the x-axis. See Figure 5.31.

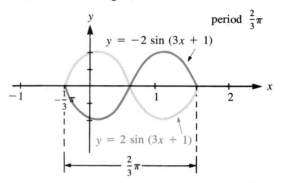

Figure 5.31

(b) Since $\sin x = -\sin (-x)$,
$$y = -2 \sin (3x + 1) = 2 \sin (-3x - 1)$$
Then, since $\sin (x + \pi) = \sin (-x)$,
$$y = 2 \sin [(3x + 1) + \pi] = 2 \sin [3x + (1 + \pi)]$$

Exercises for Section 5.4

In Exercises 21–25, write each expression in the form $A \sin (Bx + C)$, where A and B are positive, and then sketch.

1. $y = 3 + \cos 2x$
2. $s = 4 + \sin 3x$
3. $v = 6 + 8 \sin t$
4. $y = -2 + 3 \cos 6x$
5. $y = -2 + \sin \frac{1}{2}x$
6. $i = 0.2 + 1.3 \cos 0.2t$
7. $M = 3 - 3 \sin 3x$
8. $y = -2 - \sin \pi x$
9. $y = \cos (x + \frac{1}{3}\pi)$
10. $y = 2 \sin \frac{1}{3}x$
11. $y = 2 \cos (\frac{1}{2}x - \frac{1}{2}\pi)$
12. $y = \sin 2(x + \frac{1}{6}\pi)$

13. $y = \cos (2x + \pi)$

14. $y = 3 \cos (3x - \pi)$

15. $y = 4 \sin (\frac{1}{3}x + \frac{1}{3}\pi)$

16. $y = 0.2 \sin (0.25x - \pi)$

17. $y = \cos (\pi x - \frac{1}{4}\pi)$

18. $y = \sqrt{3} \cos (\pi x + \pi)$

19. $y = 4 + \sin (4x - \pi)$

20. $y = 3 - 2 \cos (\frac{1}{2}x + \frac{1}{8}\pi)$

In Exercises 21–25, write each expression in the form $A \sin (Bx + C)$, where A and B are positive, and then sketch.

21. $-\sin (x + 1)$

22. $-\sin (-2x + 3)$

23. $-\sin (2\pi x + \frac{1}{2})$

24. $3 \cos (2\pi x + \pi)$

25. $-\cos (\pi x + 1)$

In Exercises 26–29, write the equation over one period of each sinusoid whose graph is shown.

26.

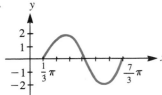

27.

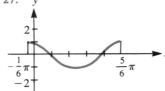

28.

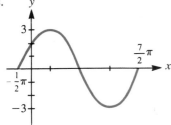

29.

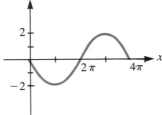

30. It is always possible to express functions of the type $A \sin (Bx + C)$ or $A \cos (Bx + C)$ in the form $A' \sin (B'x + C')$ or $A' \cos (B'x + C')$, where A', B', and C' are positive. Prove this.

31. How are the graphs of $y = \sin (-t)$ and $y = \sin (t)$ related?

32. How are the graphs of $\cos (-t)$ and $\cos (t)$ related?

33. How are the graphs of $\sin (t)$ and $\cos (\frac{1}{2}\pi - t)$ related?

34. A block is attached to a spring as shown in Figure 5.32. The mathematical description of the motion is

$$s = \begin{cases} \frac{7}{8} \cos 2t + \frac{1}{8}, & 0 \le t \le \frac{1}{2}\pi \\ \frac{5}{8} \cos 2t - \frac{1}{8}, & \frac{1}{2}\pi \le t \le \pi \end{cases}$$

The block stops at the position given by $s(\pi)$. Make a sketch of the graph of the motion.

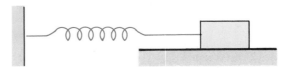

Figure 5.32

35. Repeat Exercise 34 given that the motion is described by

$$s = \begin{cases} \frac{5}{2}\cos 3t + \frac{1}{2}, & 0 \leq t \leq \frac{1}{3}\pi \\ \frac{3}{2}\cos 3t - \frac{1}{2}, & \frac{1}{3}\pi \leq t \leq \frac{2}{3}\pi \\ \frac{1}{2}\cos 3t + \frac{1}{2}, & \frac{2}{3}\pi \leq t \leq \pi \end{cases}$$

The block stops at the position given by $s(\pi)$.

36. The transverse (vertical) displacement of a traveling wave on a stretched string is represented by the equation

$$y = A \cos \frac{2\pi}{\lambda} (x - Vt)$$

where A is the amplitude, λ is the wavelength, x is the distance along the string, V is the velocity of the wave, and t is the time in seconds. Graph this equation on the interval $0 \leq x \leq 4$ in., given that $A = 2$, $\lambda = 2$ in., $V = \frac{1}{2}$ in./sec, and $t = 1$ sec.

37. The output of an alternating current generator is given by

$$I = I_{max} \sin (2\pi ft + \phi)$$

where f is the frequency and ϕ is a constant. Sketch the output current for two periods, given that $f = 60$ Hz, $I_{max} = 157$, and $\phi = \pi/4$.

5.5 Sinusoidal Modeling

In Section 5.1 we established a correspondence between the x- and y-coordinates of a point on a unit circle and the sine and cosine of a real number u mapped onto the unit circle. (See Figure 5.33.) Since $\cos u = x$ and $\sin u = y$, the circular functions are excellent mathematical models of physical problems that involve circular motion. In this section we will describe some typical models.

Imagine an object hanging on a spring as shown in Figure 5.34. If the object is pulled down and released, it will oscillate up and down about its rest point. Assuming there is no frictional force, this oscillatory motion will continue indefinitely. Vibratory motion of this type is called **simple harmonic motion** and can be described mathematically using sine or cosine functions or combinations of the two. In this section we shall restrict ourselves to the use of cosine functions to model harmonic motion.

Simple harmonic motion describes other phenomena besides an oscilla-

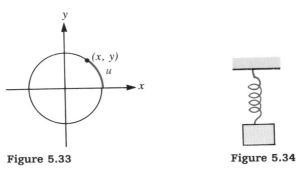

Figure 5.33   Figure 5.34

tion of an object on a spring. These phenomena include the motion of a point on a guitar string that has been plucked and the motion of air brought about by certain sound waves and some radio and television devices.

Simple Harmonic Motion

Consider a point Q moving at a constant angular velocity around the circumference of a circle, as depicted in Figure 5.35. Assume that our first observation is made when Q is at position (x, y). The point P directly below Q is called the projection of Q on the x-axis. As Q revolves, this projection moves back and forth along the x-axis between the extremes of the diameter of the circle. Note that the horizontal motion of P is the same as the vertical motion of the weight hanging on a spring; that is, P is in simple harmonic motion. The method used to describe the motion of P can also be used to describe any other simple harmonic motion.

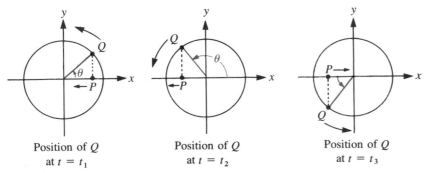

Position of Q Position of Q Position of Q
at $t = t_1$ at $t = t_2$ at $t = t_3$

Figure 5.35

To construct a mathematical model of simple harmonic motion, we proceed as follows. The displacement of P from the origin is labeled $x = \overline{OP}$, the angle made by $\overline{OQ}$ and the positive x-axis is called θ, and the radius is $r = \overline{OQ}$. Then $r = \cos \theta$. More generally, if Q has a constant angular velocity ω, we can write the model as a function of time by letting

$\theta = t - t_0$, where t_0 is some initial time for the motion, sometimes called the phase shift. Then, we have

$$x = r \cos \omega(t - t_0)$$ *Mathematical model of*   (5.1)
simple harmonic motion

In this context, the amplitude of the motion is r, the period is $2\pi/\omega$ time units, and the phase shift is t_0 time units.

Figure 5.36

Example 1.   The motion in the spring-mass system shown in Figure 5.36 is modeled by the equation $y = 4 \cos 0.7t$, where y is the displacement in meters from the rest point and t is the elapsed time. How long does it take for one oscillation of the mass? Where is the mass relative to the rest point when (a) $t = 0.5$ sec, (b) $t = 2.0$ sec, and (c) $t = 3.0$ sec? Notice that when $t = 0$ the object is 4 m above the rest point.

Solution.   The angular velocity is $\omega = 0.7$ rad/sec. Therefore, the time required to complete one oscillation is $2\pi/0.7 \approx 9$ sec. To find the location of the weight at the indicated times, we evaluate $y = 4 \cos 0.7t$.
(a) At $t = 0.5$,

$$y = 4 \cos (0.7)(0.5) \approx 3.8$$

This means that the object is 3.8 m above the rest position.
(b) At $t = 2.0$,

$$y = 4 \cos 0.7(2.0) \approx 0.68$$

The object is now 0.68 m above the rest position.
(c) At $t = 3.0$,

$$y = 4 \cos 0.7(3.0) \approx -2.0$$

The object is approximately 2 m below the rest position at this time. Figure 5.37 shows the position of the object at the various times.

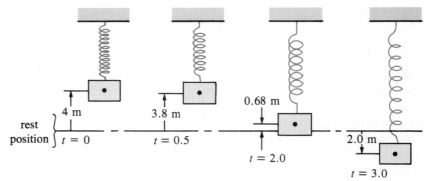

Figure 5.37

Example 2.   The water wheel shown in Figure 5.38 rotates at 3 rpm. Twelve seconds after we make our first observation, point Q is at its greatest height.

(a) Model the distance h of point Q from the surface of the water in terms of the elapsed time.

(b) Determine h when $t = 15$ sec.

Solution

(a) Here,

$$\omega = 3 \, \frac{\text{rev}}{\text{min}} \cdot 2\pi \, \frac{\text{rad}}{\text{rev}} \cdot \frac{1}{60} \, \frac{\text{min}}{\text{sec}} = \frac{1}{10} \, \pi \, \frac{\text{rad}}{\text{sec}}$$

and $r = 7$ ft. The phase shift is $t_0 = 12$ sec, the time required for Q to reach its maximum height. Using Equation (5.1), we find the distance $\overline{OP}$ to be

$$\overline{OP} = 7 \cos \frac{1}{10} \pi(t - 12)$$

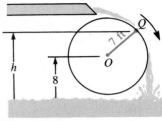

Figure 5.38

Since the center of the wheel is 8 ft above the water level, our model of the distance, h, is

$$h = 8 + \overline{OP}$$
$$= 8 + 7 \cos \tfrac{1}{10}\pi(t - 12) \text{ ft}$$

(b) The value of h for $t = 15$ sec is

$$h = 8 + 7 \cos \tfrac{1}{10}\pi(3) = 12.1$$

Therefore, the point Q is approximately 12 ft above the water level 15 sec after our first observation.

Sinusoids are also used to model physical phenomena other than circular motion. The following examples will give you some idea of the variety of applications involving sinusoids.

The Predator-Prey Problem

Certain ecological systems can be represented by periodic functions. For instance, the sine function is frequently used to describe the predator-prey relationship in a balanced ecological system. (Coyotes are predators; rabbits are prey.) If the number of predators in a region is relatively small, the number of prey will increase. But then, as the prey become more plentiful, the number of predators will increase because food is easy to find. As the number of predators continues to increase, the number of prey will eventually begin to decrease, so food for the predators will become scarce, which will cause the predator population to decrease, which will allow more prey to survive, and so on. This cycle will be repeated over and over again, with the two populations oscillating about their respective mean values.

Example 3. The population of rabbits in a certain region is given by

$$N = 500 + 150 \sin 2t$$

where t is time in years. Discuss the variation in the rabbit population.

Solution. The constant term represents the mean population of rabbits, and the coefficient of the sine function represents the variation in the population. Thus, the mean population is 500, and it varies from a high of 650 to a low of 350. The period of variation is $2\pi/2 = 3.14$ years. The population is shown graphically in Figure 5.39.

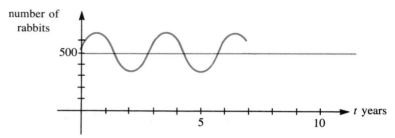

Figure 5.39

Seasonal Temperature Variation

Sine and cosine curves can be used to describe physical conditions represented by meteorological data. Figure 5.40 shows a representation of the daily mean temperatures for Dayton, Ohio from January 1978 to September 1979; the sinusoidal shape is unmistakable. The temperature variation can be approximated by the equation

$$T = T_m + A \sin\left[\frac{2\pi}{365}(t - C)\right]$$

where T_m is the mean annual temperature, A is the maximum temperature deviation from T_m, and C is the phase shift found by counting the number of days from January 1 to the point at which the temperature curve crosses the mean annual temperature line.

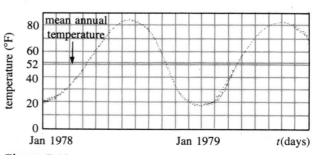

Figure 5.40

Example 4. Write an equation to approximate the temperature variations shown in Figure 5.40.

Solution. From Figure 5.40 we estimate that $T_m = 52°F$, $A = 32°F$, and $C = 102$ days. Figure 5.41 shows both the actual data and the graph of

$$T = 52 + 32 \sin\left[\frac{2\pi}{365}(t - 102)\right]$$

As you can see, there are some specific areas of disagreement, but overall the fit appears to be good.

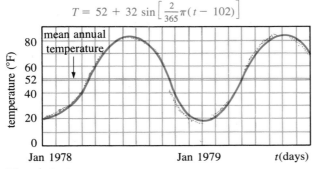

$$T = 52 + 32 \sin\left[\frac{2}{365}\pi(t - 102)\right]$$

Figure 5.41

Biorhythms

An interesting application of the sine function is found in the theory of biorhythms. Briefly, the theory holds that because of certain biological processes, our physical, emotional, and intellectual states of being are periodic and can be represented by sine functions. According to the theory, the cycle of each "feeling" is a constant. Specifically, the periods are 23 days for the physical cycle, 28 days for the emotional cycle, and 33 days for the intellectual cycle. Thus, one's physical state on any day t after birth is given by

$$P = \sin\left(\frac{2\pi}{23}t\right)$$

one's emotional state by

$$E = \sin\left(\frac{2\pi}{28}t\right)$$

and one's intellectual state by

$$I = \sin\left(\frac{2\pi}{33}t\right)$$

Good days are characterized as those for which P, E, and I are positive, and bad days as those for which they are negative. The closer the state is to

+1, the better the day is for that particular phase of your well-being. Your overall state is usually obtained by averaging the three values.

To use the formulas for the biorhythm state, you must compute the number of days from the given birthdate to the present day. One convenient method is to compute the number of days from January 1, 1900 for each date and then subtract the two numbers. Since leap year occurs every four years (but not in 1900), the number of days in the twentieth century for any particular date is given by

$$T = (\text{year of interest} - 1900)365 + (\text{leap years since 1900})$$
$$+ (\text{number of days from Jan. 1 in year of interest})$$

Example 5. Find the biorhythm state of an individual on February 10, 1980 if the person was born on August 8, 1933.

Solution. We first compute the number of days since 1900 for the two days in question. Notice that although 1980 is a leap year, we use 19 and not 20 for the number of leap years since February 2 occurs prior to February 29.

$$T(8/8/33) = (1933 - 1900)365 + 8 + 220 = 12{,}273 \text{ days}$$
$$T(2/10/80) = (1980 - 1900)365 + 19 + 41 = 29{,}260 \text{ days}$$

Thus, the number of days between the two dates is

$$t = 29{,}260 - 12{,}273 = 16{,}987 \text{ days}$$

Using this value in the formulas for P, E, and I, we have

$$P = \sin\left(\frac{2\pi}{23} \cdot 16{,}987\right) = \sin 4640.54 = -0.398$$

$$E = \sin\left(\frac{2\pi}{28} \cdot 16{,}987\right) = \sin 3811.87 = -0.901$$

$$I = \sin\left(\frac{2\pi}{33} \cdot 16{,}987\right) = \sin 3234.32 = -0.999$$

The above values can be determined on a calculator in radian mode. The average of the three numbers is -0.766. All indications are that this person should have stayed in bed on this date.

Exercises for Section 5.5

1. The motion of a spring-mass system similar to the one shown in Figure 5.36 is modeled by the equation $x = 15 \cos 2t$, where x is the displacement in centimeters from the rest position.
   (a) Determine the time required for one oscillation.
   (b) Determine the position of the weight when $t = 1$ sec.

2. The motion in a spring-mass system is described by $y = 2 \cos 3.2t$, where y is the displacement in meters from the rest position and t is the elapsed time in seconds.
   (a) Determine the time required for one oscillation of the mass.
   (b) Determine the location of the weight after 5 sec.

3. Suppose the water wheel illustrated in Figure 5.42 rotates at 6 rpm. Two seconds after a stopwatch is started, point D on the rim of the wheel is at its greatest height.
   (a) Model the distance h of point D from the surface of the water in terms of the time t in seconds on the stopwatch.
   (b) Sketch the graph of the sinusoidal model.
   (c) Find the time when D first emerges from below the water.

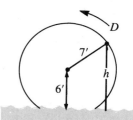

Figure 5.42

4. A space shuttle is fired into a circular orbit from its launch pad in Florida. Ten minutes after it leaves, it reaches its farthest distance north of the Equator, which is 4000 km. A half-cycle later it reaches its farthest distance south of the Equator (also 4000 km).
   (a) Write a sinusoidal model describing the relationship between distance from the Equator, D, and time from launch, t, in minutes. Consider distance south of the Equator to be negative distance.
   (b) How much time will elapse before the shuttle will cross the Equator for the first time?
   (c) How far north of the Equator is the launch site? Assume a 90-min orbit.

5. Tarzan is swinging back and forth on his grapevine, alternately going over land and water. (See Figure 5.43.) Consider values of y, the distance from the river bank, to be positive if Tarzan is over water and negative if he is over land. Write a sinusoidal model to describe Tarzan's distance from the river bank. Assume that Jane is measuring Tarzan's motion and finds that 2 sec after she starts her stopwatch, Tarzan is at one end of his swing, where $y = -23$ ft. When $t = 5$ sec, $y = 17$ ft.
   (a) Sketch a graph of this sinusoidal function.
   (b) Write the sinusoidal model.
   (c) Predict y when $t = 2.8$, 6.3, and 15.
   (d) Where was Tarzan when Jane started her stopwatch?
   (e) Determine the time registered on the stopwatch when Tarzan first reached a point directly over the bank.

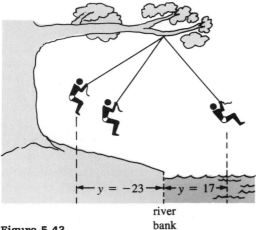

Figure 5.43

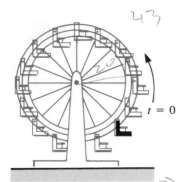

Figure 5.44

6. As you ride a ferris wheel, your distance from the ground varies sinusoidally with time. When the last seat is filled and the ferris wheel begins, your seat is at the position shown in Figure 5.44. It takes you 3 sec to reach the top, which is 43 ft above the ground. The wheel, which travels at 7.5 rpm, has a diameter of 40 ft.
   (a) Sketch a graph of this sinusoid.
   (b) Write an equation for the sinusoid.
   (c) What is the lowest point you reach as the wheel turns?
   (d) How high above the ground are you when $t = 0$? When $t = 9$?
   (e) What is the value of t when, in your first descent, you reach a point 23 ft above the ground?

7. The number of deer in an ecological region is given by $D = 1500 + 400 \sin 0.4t$ and the number of pumas in the region by $P = 500 + 200 \sin (0.4t - 0.8)$, where t is the time in years. Sketch the variation in these two populations on the same set of coordinates.

8. The rabbit population in an ecological region is given by the expression $R = 1000 + 200 \sin 4t$ and the fox population by $F = 100 + 10 \sin (4t - 0.8)$, where t is the time in years. Discuss the variation and sketch the populations on the same coordinate system.

9. Draw the graph of the normal mean temperature variation for a 2-year period if the approximating equation is

$$T = 55 + 38 \sin \left[\frac{2\pi}{365} (t - 100) \right]$$

Assume T is in °F and t is the number of days from January 1. What is the mean annual temperature of the area described by this equation?

10. Write the equation for the temperature variation shown in Figure 5.45.

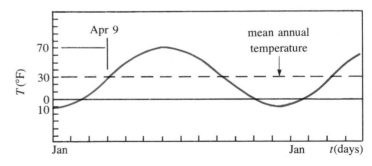

Figure 5.45

11. For several hundred years, astronomers have kept track of the number of solar flares, or sunspots, that occur on the surface of the sun. The number of sunspots counted in a given year varies periodically from a minimum of about 10 per year to a maximum of about 110 per year. Between the maximums that occurred in the years 1750 and 1948, there were 18 complete cycles. (Look in the September 1975 issue of *Scientific American*, page 166, to see how closely the sunspot cycle resembles a sinusoid.)

(a) What is the period of the sunspot cycle?

(b) Assume that the number of sunspots counted in a year varies sinusoidally with the year. Sketch a graph of two sunspot cycles, starting with 1948.

(c) Write the equation expressing the number of sunspots per year (y) in terms of the year (t).

(d) How many sunspots should you expect this year? In the year 2000?

(e) What is the first year after 2000 in which the maximum number of sunspots will occur?

12. A portion of a roller coaster track is to be built in the shape of a sinusoid. (See Figure 5.46.)

(a) The high and low points on the track are separated horizontally by 50 m and vertically by 30 m. The low point is 3 m below the ground. Letting y be the number of meters the track is above the ground and x the number of meters horizontally from the high point, write an equation expressing y in terms of x.

(b) How long is the vertical timber at the high point, the one at $x = 4$ m, and the one at $x = 32$ m?

(c) How long is the horizontal timber that is 25 m above the ground and the one that is 5 m above the ground?

(d) Where does the track first go below the ground?

(e) The vertical timbers are spaced every 2 m, starting at $x = 0$ and ending where the track goes below the ground. What is the total length of timber needed to make the vertical supports? (If you have access to a computer, this exercise should be done by writing a program that prints out the length of each timber and then totals these lengths.)

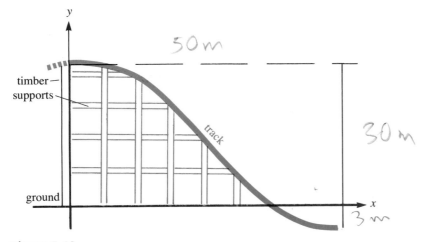

Figure 5.46

13. Find the biorhythm state on Jan. 25, 1983 for each of the given birth dates.

(a) 3/11/1935   (b) 2/27/1963   (c) 5/20/1965

(d) 7/30/1966   (e) 2/7/1968   (f) 11/18/1969

14. Calculate your biorhythm state for today's date.

5.6 Addition of Ordinates

Functions that are written as a sum of more elementary functions, such as

$$y_1 = \sin x + \cos x \qquad \text{and} \qquad y_2 = x + \sin x$$

occur frequently. It can be a very tedious process to graph such functions if you use the method of substituting values of x and determining corresponding ordinates. Sometimes a technique called **addition of ordinates** can be useful in plotting such functions. Suppose $h(x) = f(x) + g(x)$. We sketch the graphs of $f(x)$ and $g(x)$ on the same coordinate system, as in Figure 5.47. Then for particular values of x, such as x_1, we find $h(x_1)$ as the sum of $f(x_1)$ and $g(x_1)$. A vertical line is usually drawn at the point $(x_1, 0)$, and then ordinates $f(x_1)$ and $g(x_1)$ are added by using a set of dividers or markings on the edge of a strip of paper. This process is repeated as often as necessary to get a representation of the desired graph.

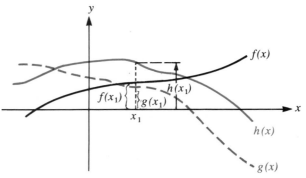

Figure 5.47

Example 1. Use the method of addition of ordinates to sketch the function $y = \sin x + \cos 2x$.

Solution. In Figure 5.48, both $\sin x$ and $\cos 2x$ are sketched along with their sum. The period of the given function is 2π even though that of $\cos 2x$ is π.

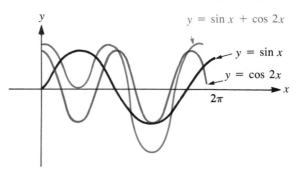

Figure 5.48

Example 2. Sketch the graph of $y = x + \sin x$.

Solution. See Figure 5.49. In this case the basic sine curve oscillates about the curve $y = x$. Note that the given function is *not* periodic and is *not* bounded.

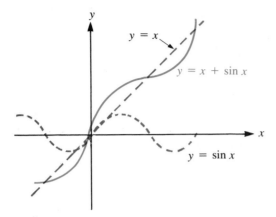

Figure 5.49

Exercises for Section 5.6

In Exercises 1–20, use the method of addition of ordinates to sketch the graph of each function. In each case, tell whether the function is periodic.

1. $y = 3 + \sin x$
2. $y = 1 + 2 \cos x$
3. $y = -1 + \cos x$
4. $y = -0.5 + \sin x$
5. $y = \sin x + 2$
6. $y = x - \sin x$
7. $y = \frac{1}{2}x + \cos x$
8. $y = \cos x - x$
9. $y = 2 - x + \sin x$
10. $y = 1 + x - \cos x$
11. $y = 0.1x^2 + \sin 2x$
12. $y = 0.1x^2 + \cos x$
13. $y = \sin x + \cos x$
14. $y = \sin 2x + 2 \sin x$
15. $y = \sin \frac{1}{2}x - 2 \cos x$
16. $y = \sin \frac{1}{2}x - \sin x$
17. $y = \sin x + \sin (x - \frac{1}{4}\pi)$
18. $y = \cos x + \sin (x - \frac{1}{4}\pi)$
19. $y = 2 \sin \pi x + \sin x$
20. $y = \sin \pi x - \cos 2x$

21. Compare the graphs of the following functions

$$y = \sqrt{2} \sin (x + \tfrac{1}{4}\pi) \quad \text{and} \quad y = \cos x + \sin x$$

22. Compare the graphs of the following functions

$$y = \cos^2 x \quad \text{and} \quad y = \tfrac{1}{2} + \tfrac{1}{2} \cos 2x$$

23. An object oscillating on a spring has a motion described by

$$y(t) = y_0 \cos 2t + \left(\frac{v_0}{2}\right) \sin 2t$$

where y_0 is the initial displacement from equilibrium and v_0 is the initial velocity. Given that $y_0 = 5\,\text{cm}$ and $v_0 = 24\,\text{cm/sec}$, make a graph of the motion for $0 \le t \le 2\pi$.

24. An electronic tuner/amplifier accepts signals of the form $I = I_m \sin 2\pi ft$. Graph the input to the tuner if two waves arrive simultaneously at the terminals, one with $I_m = 3\,\text{mA}$, $f = 1000\,\text{Hz}$, and the other with $I_m = 5\,\text{mA}$, $f = 2000\,\text{Hz}$. The tuner sees this as the sum of the two waves.

25. The current on a transmission line may be considered as the sum of an outgoing and a reflected wave, of the forms $\sin(x + ct)$ and $\sin(x - ct)$, respectively. Graph the sum of these two waveforms for $c = 1$, when $t = 1$ for $0 \le x \le 2\pi$.

5.7 Graphs of the Tangent and Cotangent Functions

The analytic properties of the tangent and cotangent functions are summarized here. Each property affects the nature of the graph in a very important manner.

(1) Both $\tan x$ and $\cot x$ are periodic with period π. Thus, only a one-period interval need be analyzed, such as $-\frac{1}{2}\pi < x < \frac{1}{2}\pi$ or $0 < x < \pi$.

(2) Both $\tan x$ and $\cot x$ are **unbounded**, which means that their values become arbitrarily large. Tan x becomes unbounded near odd multiples of $\frac{1}{2}\pi$, whereas $\cot x$ becomes unbounded near multiples of π. The lines $x = (2n + 1)\pi/2$ are called **vertical asymptotes** of the graph of $y = \tan x$; the lines $x = n\pi$ are vertical asymptotes of the graph of $y = \cot x$.

(3) Tan x is zero for $x = 0, \pm\pi, \pm 2\pi$, and so on. Cot x is zero at $x = \pm\frac{1}{2}\pi$, $\pm\frac{3}{2}\pi$, $\pm\frac{5}{2}\pi$, and so on. The graph crosses the x-axis at these places.

(4) Numerically (ignoring sign), the values of both functions are completely determined in the first quadrant—that is, for $0 < x < \frac{1}{2}\pi$.

Figure 5.50 shows a graph of several periods of the tangent function and of the cotangent function. For purposes of graphing, the x-intercepts and the asymptotes are emphasized. The asymptotes for $\tan x$ are $x = \pm\frac{1}{2}\pi$, $\pm\frac{3}{2}\pi$, $\pm\frac{5}{2}\pi$, and so on. The asymptotes for $\cot x$ are $x = 0, \pm\pi, \pm 2\pi, \pm 3\pi$, and so on.

The graphs of the more general functions $y = A \tan(Bx - C)$ and $y = A \cot(Bx - C)$ are analyzed in a manner similar to that described in Section 5.3.

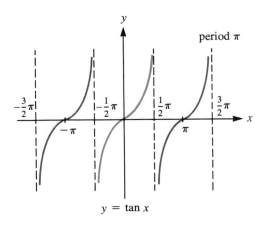

$y = \tan x$

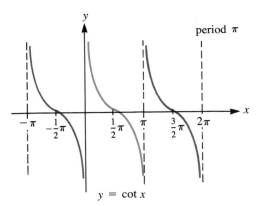

$y = \cot x$

x	0	$\frac{1}{6}\pi$	$\frac{1}{4}\pi$	$\frac{1}{3}\pi$	$\frac{1}{2}\pi$	$\frac{3}{4}\pi$	π
$\tan x$	0	$\sqrt{3}/3$	1	$\sqrt{3}$	undef.	-1	0
$\cot x$	undef.	$\sqrt{3}$	1	$\sqrt{3}/3$	0	-1	undef.

Figure 5.50

In the case of $y = A \tan x$, we do not call A the amplitude because this would imply that the function was bounded. The constant A multiplies each functional value but has no other graphical significance.

The period of $\tan Bx$ is π/B. Thus, if $B > 1$, the period is shorter than that of the basic tangent function; if $B < 1$, the period is larger. The constant C is a phase shift constant and translates the basic function to the right or to the left.

Example 1. Sketch the function $y = \tan (4x - \frac{1}{3}\pi)$.

Solution. The period of this function is $\frac{1}{4}\pi$. The phase shift is located by determining where the argument $4x - \frac{1}{3}\pi$ is equal to zero. Thus, the phase shift is $\frac{1}{12}\pi$. The graph is shown in Figure 5.51.

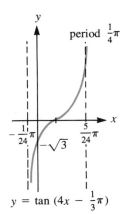

$y = \tan (4x - \frac{1}{3}\pi)$

Figure 5.51

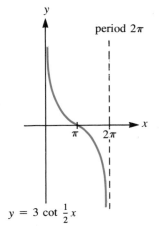

$y = 3 \cot \frac{1}{2} x$

Figure 5.52

Example 2. Draw the graph of $y = 3 \cot \frac{1}{2} x$.

Solution. The period of this graph is $\pi / \frac{1}{2} = 2\pi$, and the phase shift is zero. The graph is shown in Figure 5.52.

5.8 Graphs of the Secant and Cosecant Functions

The graphs of sec x and csc x can be sketched directly from the graphs of cos x and sin x, since they are reciprocals of the respective functions. These functions have several important general properties:

(1) Both functions are unbounded. In fact, since both sin x and cos x are bounded by ± 1, the graphs of csc x and sec x lie above $y = 1$ and below $y = -1$.

(2) Both sec x and csc x are periodic with period 2π.

(3) Sec x is an even function; csc x is an odd function.

(4) Both sec x and csc x are never 0.

(5) Numerically, the functional values are determined for $0 < x < \frac{1}{2}\pi$.

The graphs of both functions are sketched in Figure 5.53. In each case, the reciprocal function is sketched lightly on the same coordinate system to show the relation between the two.

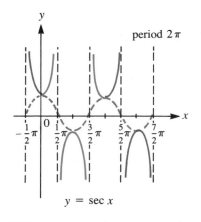

$y = \sec x$

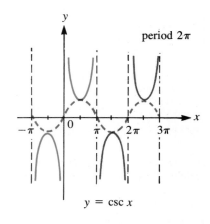

$y = \csc x$

x	0	$\frac{1}{6}\pi$	$\frac{1}{4}\pi$	$\frac{1}{3}\pi$	$\frac{1}{2}\pi$	π	$\frac{3}{2}\pi$	2π
sec x	1	$2/\sqrt{3}$	$\sqrt{2}$	2	undef.	-1	undef.	1
csc x	undef.	2	$\sqrt{2}$	$2/\sqrt{3}$	1	undef.	-1	undef.

Figure 5.53

The manner of sketching the more general functions

$$y = A \sec (Bx - C) \quad \text{and} \quad y = A \csc (Bx - C)$$

is similar to that discussed in Section 5.2. The basic waveforms remain the same, but the constants A and B exert a vertical and horizontal stretching while C effects a horizontal translation.

Example 1. Sketch the graph of $y = 0.3 \sec (x + \frac{1}{4}\pi)$.

Solution. Since 0.3 is the coefficient of the secant, the range is outside the interval $-0.3 < y < 0.3$. The period is 2π, and the phase shift is $\frac{1}{4}\pi$ units to the left. The graph is shown in Figure 5.54.

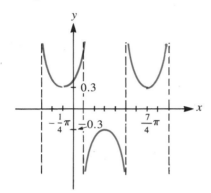

Figure 5.54

Exercises for Sections 5.7 and 5.8

In Exercises 1–8, find the period of each of the functions.

1. $\cot \frac{1}{2}x$ 2. $\csc 3x$ 3. $\sec \pi x$ 4. $\tan 3\pi x$

5. $\tan \frac{1}{2}\pi x$ 6. $\sec \frac{1}{3}x$ 7. $\cot \frac{5}{6}x$ 8. $\sec \frac{1}{\pi} x$

In Exercises 9–17, sketch the graph of each of the given functions over at least two periods. Give the period, phase shift, and the asymptotes.

9. $y = \tan 2x$ 10. $y = \tan (x + \frac{1}{2}\pi)$

11. $y = \cot (\frac{1}{4}\pi - x)$ 12. $y = 2 \sec (x - \frac{1}{2}\pi)$

13. $y = \tan (2x + \frac{1}{3}\pi)$ 14. $y = 2 \csc 2x$

15. $y = \csc (2x - 3\pi)$ 16. $y = \sec (x + \frac{1}{3}\pi)$

17. $y = -\tan (x - \frac{1}{4}\pi)$

18. Does $\tan x$ exist at its asymptote?

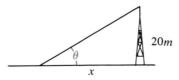

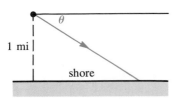

Figure 5.55

19. How are the graphs of tan x and cot x related?

20. How are the graphs of sec x and csc x related?

21. How are the zeros of the tangent function and the asymptotes of the cotangent function related?

22. How are the zeros of the sine function and the asymptotes of the cosecant function related?

23. How are the graphs of $y = \tan x$ and $y = \tan(-x)$ related?

24. How are the graphs of $y = \tan x$ and $y = -\tan x$ related?

25. Show that the distance x to the base of the antenna shown in Figure 5.55 is given by $x = 20 \cot \theta$. Make a sketch of x as a function of θ.

26. The force in a cable is given by $T = T_x \sec \theta$, where T_x is a measurable horizontal component of T. If T_x is kept constant at 10 lb, sketch the graph of T as a function of θ for $0 \le \theta \le 60°$.

27. To determine the coefficient of sliding friction, μ, physicists perform a simple experiment in which the angle of inclination θ of an inclined plane is increased until the block starts to slide. Then $\mu = \tan \theta$. Graph μ as a function of θ for $0 \le \theta \le 80°$.

28. A revolving light 1 mi from a straight shoreline makes 3 revolutions per minute. See Figure 5.56. If the light is initially directed parallel to the shore, show that the distance, x, is given by $x = \cot \frac{1}{10}\pi t$, where t is the time in seconds, $0 \le t \le 5$ sec. Make a sketch of x as a function of t for this time interval.

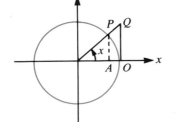

Figure 5.56

5.9 A Fundamental Inequality

The inequality

$$\sin x < x < \tan x$$

is fundamental to trigonometric analysis because it relates the arc length on the unit circle to two of the trigonometric functions. To demonstrate the validity of this inequality, we construct a unit circle as shown in Figure 5.57. From the figure, it is obvious that the length of AP is less than the length of the arc OP, which in turn is less than OQ:

$$\overline{AP} < \overline{OP} < \overline{OQ}$$

Since the circle has radius 1, $\sin x = \sin \theta = \overline{AP}$ and $\tan x = \tan \theta = \overline{OQ}$, and hence

$$\sin x < x < \tan x$$

Figure 5.57

Figure 5.58 shows the three functions, $y = \sin x$, $y = x$, and $y = \tan x$, and at the same time exhibits the fact that the inequality is true only for $0 < x < \frac{1}{2}\pi$.

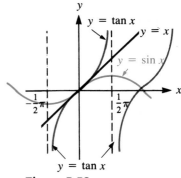

The ratio $(\sin x)/x$ is important in calculus. In the exercises for Section 5.1, you were asked to compute the values of this ratio for a few values of x near zero. You should have found that this ratio is just slightly less than 1 when x is a very small number. Using the fundamental inequality, we can show this a bit more rigorously. Divide the members of the fundamental inequality by $\sin x$ to obtain

$$1 < \frac{x}{\sin x} < \frac{1}{\cos x}$$

(we are assuming x is positive). Inverting and reversing the sense of the inequalities, we obtain

$$\cos x < \frac{\sin x}{x} < 1$$

Figure 5.58

Since $\cos x$ is near 1 when x is small and since the ratio is trapped between two functions close to 1, it follows that it also has values close to 1. (A similar argument for negative x leads to the same conclusion.)

The function $(\sin x)/x$ is undefined for $x = 0$, but if you look at a sketch of its graph (Figure 5.59), it looks as though its value "should be" 1.

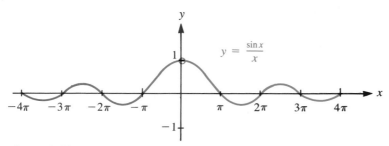

Figure 5.59

One of the side results of the fact that $(\sin x)/x$ is close to 1 for x close to zero is that for small values of x, $\sin x \approx x$. Thus, in some applications, if the absolute value of x is small, $\sin x$ is replaced by x.

Exercises for Section 5.9

1. By carefully examining the graphs of $y = 2x/\pi$ and $y = \sin x$, convince yourself that $\dfrac{2x}{\pi} \le \sin x \le 1$ for $0 < x < \dfrac{\pi}{2}$.

2. What are the zeros of $\dfrac{\sin x}{x}$?

3. Sketch the function $f(x) = \dfrac{\sin 2x}{x}$ for $x \neq 0$. What do you think the values of the function are approaching when x is near 0?

Key Topics for Chapter 5

Define and/or discuss each of the following.

Periodicity

Graphs of Trigonometric Functions

Bounded and Unbounded Trigonometric Functions

Even and Odd Functions

Cycle

Amplitude

Phase Shift

Harmonic Motion

Predator-Prey Problem

Addition of Ordinates

Asymptotes

Fundamental Inequality

Review Exercises for Chapter 5

1. Let $f(x) = 2 \sin (3x - 1)$. Compute
   (a) $f(0)$ (b) $f(\frac{1}{3})$ (c) $f(\frac{2}{3}\pi)$

2. Determine where the function in Exercise 1 is equal to 0.

3. Determine where the function in Exercise 1 is equal to 1.

In Exercises 4–10, sketch the graph of each function and give its period and phase shift. Also give, where applicable, amplitude, mean value (avg), asymptotes, and intercepts.

4. $y = -\sin (x + \frac{1}{6}\pi)$

5. $y = 2 + \sin (x - \frac{1}{3}\pi)$

6. $y = \tan (\pi x + \pi) - 1$

7. $y = 4 \cot (\frac{1}{2}x + \frac{1}{8}\pi)$

8. $y = 2 \cos (2 - x) + \frac{1}{2}$

9. $y = \sec \frac{1}{3}(\pi - x) + 2$

10. $y = \csc (2x + 1)$

11. Sketch the graph of the harmonic motion described by $x = 9 \cos 3t$ and give the amplitude and period.

12. Use the method of addition of ordinates to sketch $y = x + \cos x$.

13. Locate the asymptotes for the graph of $y = 2 \tan (x + 3)$.

14. Completely describe the impact of the constants A, B, C, and D on the graph of $y = D + A \sin (Bx + C)$.

15. A weight at the end of a spring oscillates according to the formula

$$y(t) = 10 \cos (3t - \tfrac{1}{2}\pi)$$

Make a sketch of the motion of the weight.

16. The current in a certain coil is given by

$$i(t) = 10 \sin (120\pi t - \tfrac{1}{6}\pi)$$

Make a sketch of current vs. time.

17. A block attached to a spring oscillates with decreasing amplitude according to the following formula:

$$x(t) = \begin{cases} \tfrac{5}{2} \cos \tfrac{3}{2}t + \tfrac{1}{2}, & 0 \le t \le \tfrac{2}{3}\pi \\ \tfrac{3}{2} \cos \tfrac{3}{2}t - \tfrac{1}{2}, & \tfrac{2}{3}\pi \le t \le \tfrac{4}{3}\pi \\ \tfrac{1}{2} \cos \tfrac{3}{2}t + \tfrac{1}{2}, & \tfrac{4}{3}\pi \le t \le 2\pi \\ 0, & 2\pi \le t \end{cases}$$

Make a sketch of the motion of the block.

18. The current in a particular circuit is given by $i = 0.5 \sin 1$. Find the value of i.

19. The horizontal displacement of a simple oscillator is found to be

$$x = A \cos 2\pi f t \text{ cm}$$

What is x if $A = 1.4$, $f = 0.1$ and $t = 4$?

20. Compare the values of $\sin x$ for $x = 1$ radian, 1 degree, and the real number 1.

21. For which values of x is $\sin x = \tan x = 0$?

22. Why are the trigonometric functions called the circular functions?

23. A piston is connected to the rim of a wheel as shown in Figure 5.60. The radius of the wheel is 2 ft, and the length of the connecting rod ST is 5 ft. The wheel rotates counterclockwise at the rate of 1 revolution per second. Find a formula for the position of the point S, t seconds after it has coordinates (2, 0). Find the position of the point S when $t = \tfrac{1}{2}, \tfrac{3}{4}$, and 2.

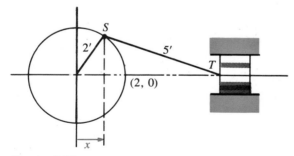

Figure 5.60

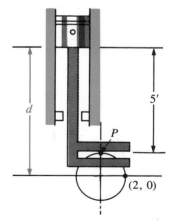

Figure 5.61

24. One end of a shaft is fastened to a piston that moves vertically. The other end is connected to the rim of the wheel by means of prongs, as shown in Figure 5.61. If the radius of the wheel is 2 ft and the shaft is 5 ft long, find a formula for the distance d ft between the bottom of the piston and the x-axis, t seconds after P is at (2, 0). Assume the wheel rotates at 2 revolutions per second.

6

Identities, Equations, and Inequalities

6.1 Fundamental Trigonometric Relations

Any combination of trigonometric functions such as $3 \sin x + \cos x$ or $\sec^2 x + \tan^2 x + 2 \sin x$ is called a **trigonometric expression**. One of the important skills you will learn in this chapter is how to simplify or alter the form of trigonometric expressions using certain fundamental trigonometric relations.

There are eight *fundamental* relations, or identities, that you must know if you are to work the problems in the remainder of this book efficiently. You are already familiar with most of these relations, but they are listed here for completeness. The fundamental relations fall into three groups: the reciprocal relations, the quotient relations, and the Pythagorean relations.

The Reciprocal Relations

$$\sin \theta = \frac{1}{\csc \theta} \tag{6.1}$$

$$\cos \theta = \frac{1}{\sec \theta} \tag{6.2}$$

$$\tan \theta = \frac{1}{\cot \theta} \tag{6.3}$$

We establish Equation (6.1) by observing that for any angle θ in standard position and (x, y) on the terminal side with length r,

$$\sin \theta = \frac{y}{r} = \frac{1}{r/y} = \frac{1}{\csc \theta}$$

The other two relations are established in a similar manner.

The Quotient Relations

$$\tan \theta = \frac{\sin \theta}{\cos \theta} \tag{6.4}$$

$$\cot \theta = \frac{\cos \theta}{\sin \theta} \tag{6.5}$$

To establish Equation (6.4), we note that

$$\cos \theta \tan \theta = \left(\frac{x}{r}\right)\left(\frac{y}{x}\right) = \frac{y}{r} = \sin \theta$$

Therefore,

$$\tan \theta = \frac{\sin \theta}{\cos \theta}$$

The Pythagorean Relations

$$\sin^2 \theta + \cos^2 \theta = 1 \tag{6.6}$$
$$\tan^2 \theta + 1 = \sec^2 \theta \tag{6.7}$$
$$\cot^2 \theta + 1 = \csc^2 \theta \tag{6.8}$$

We prove Equation (6.6) by dividing $x^2 + y^2 = r^2$ by r^2 to get

$$\frac{x^2}{r^2} + \frac{y^2}{r^2} = 1$$

Then, since

$$\sin \theta = \frac{y}{r} \qquad \text{and} \qquad \cos \theta = \frac{x}{r}$$

we have

$$\cos^2 \theta + \sin^2 \theta = 1$$

Note that Equation (6.7) is derived from Equation (6.6). Dividing both sides of Equation (6.6) by $\cos^2 \theta$, we get

$$\frac{\sin^2 \theta}{\cos^2 \theta} + 1 = \frac{1}{\cos^2 \theta}$$

Then, using the fact that $\dfrac{\sin \theta}{\cos \theta} = \tan \theta$ and $\dfrac{1}{\cos \theta} = \sec \theta$, we arrive at

$$\tan^2 \theta + 1 = \sec^2 \theta$$

Similarly, Equation (6.8) is derived from Equation (6.6) by first dividing both sides by $\sin^2 \theta$ and then applying Equations (6.1) and (6.5).

These eight **fundamental identities** of trigonometry are valid for all values of the argument for which the functions in the expression have meaning. As before, the variable (often the letter x is chosen instead of θ) may be regarded as either a real number or an angle, the interpretation depending on the context.

Using the fundamental identities, you can manipulate (sometimes ingeniously) trigonometric expressions into alternative forms.

Example 1. Write the following expression as a single trigonometric term:

$$\frac{\tan x \csc^2 x}{1 + \tan^2 x}$$

Solution. By Equation (6.7), the denominator may be written as $\sec^2 x$. Thus,

$$\frac{\tan x \csc^2 x}{1 + \tan^2 x} = \frac{\tan x \csc^2 x}{\sec^2 x}$$

We now express $\tan x$, $\csc x$, and $\sec x$ in terms of the sine and cosine functions:

$$\frac{\tan x \csc^2 x}{1 + \tan^2 x} = \frac{\dfrac{\sin x}{\cos x} \cdot \dfrac{1}{\sin^2 x}}{\dfrac{1}{\cos^2 x}}$$

$$= \frac{\cos^2 x \sin x}{\sin^2 x \cos x} \qquad \text{Inverting } \frac{1}{\cos^2 x} \text{ and multiplying}$$

$$= \frac{\cos x}{\sin x} \qquad \text{Cancellation law}$$

$$= \cot x \qquad \frac{\cos x}{\sin x} = \cot x$$

Therefore, we have shown that $\dfrac{\tan x \csc^2 x}{1 + \tan^2 x} = \cot x$.

As the preceding example shows, a large part of the process is algebraic. The steps used in the example are not the only way to simplify the expression. For example, we could have initially expressed the complete expression in terms of the sine and cosine functions. However, writing the entire expression in terms of sine and cosine functions is not necessarily the shortest or easiest method.

Example 2. Simplify the expression $(\sec x + \tan x)(1 - \sin x)$.

Solution. We write each of the functions in terms of the sine and cosine functions:

$$(\sec x + \tan x)(1 - \sin x) = \left(\frac{1}{\cos x} + \frac{\sin x}{\cos x}\right)(1 - \sin x)$$

$$= \frac{(1 + \sin x)(1 - \sin x)}{\cos x} \qquad \text{Adding fractions}$$

$$= \frac{(1 - \sin^2 x)}{\cos x} \qquad (a + b)(a - b) = a^2 - b^2$$

$$= \frac{\cos^2 x}{\cos x} \qquad 1 - \sin^2 x = \cos^2 x$$

$$= \cos x \qquad \text{Cancellation law}$$

Therefore, we have shown that $(\sec x + \tan x)(1 - \sin x) = \cos x$.

Example 3. Expand and simplify the expression $(\sin x + \cos x)^2$.

Solution. Note that this is *not* the same expression as $\sin^2 x + \cos^2 x$. By squaring the expression, we obtain

$$(\sin x + \cos x)^2 = \sin^2 x + 2 \sin x \cos x + \cos^2 x \qquad \text{Expanding the binomial}$$

$$= 1 + 2 \sin x \cos x \qquad \sin^2 x + \cos^2 x = 1$$

Example 4. Simplify the expression $\sin^4 x - \cos^4 x + \cos^2 x$.

Solution. We write the expression in a form involving only the cosine function. To make this simplification, we note that $\sin^4 x = (\sin^2 x)^2 = (1 - \cos^2 x)^2$. Thus,

$$\sin^4 x - \cos^4 x + \cos^2 x = (1 - \cos^2 x)^2 - \cos^4 x + \cos^2 x$$

$$= 1 - 2 \cos^2 x + \cos^4 x - \cos^4 x + \cos^2 x$$

$$= 1 - \cos^2 x$$

$$= \sin^2 x$$

Therefore, we have shown that $\sin^4 x - \cos^4 x + \cos^2 x = \sin^2 x$.

Certain algebraic expressions encountered in calculus are often transformed into trigonometric expressions in which, after simplification, "hard to handle" terms such as radicals disappear.

Example 5. Using the substitution $x = 2 \sin \theta$, simplify the expression $\sqrt{4 - x^2}$ and determine an interval for the variable θ that corresponds to $0 \le x \le 2$ in a one-to-one manner. What is $\tan \theta$?

Solution. Substituting $x = 2 \sin \theta$ into the radical, we have

$$\sqrt{4 - x^2} = \sqrt{4 - 4 \sin^2 \theta}$$

$$= \sqrt{4(1 - \sin^2 \theta)}$$

$$= |2 \cos \theta| = 2|\cos \theta|$$

When $x = 0, \theta = 0$, and when $x = 2, \theta = \frac{1}{2}\pi$, so the interval $0 \le x \le 2$ corresponds to $0 \le \theta \le \frac{1}{2}\pi$. On this interval $\cos \theta \ge 0$, so $|\cos \theta| = \cos \theta$. Hence,

$$\sqrt{4 - x^2} = 2 \cos \theta \quad \text{for } 0 \le x \le 2 \text{ and } 0 \le \theta \le \frac{1}{2}\pi$$

Since $\sin \theta = x/2$, the right triangle in Figure 6.1 shows the relations necessary to establish that

$$\tan \theta = \frac{x}{\sqrt{4 - x^2}}$$

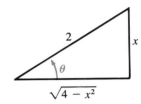

Figure 6.1

Exercises for Section 6.1

In Exercises 1–20, reduce each expression to a single trigonometric function.

1. $\cos \theta + \tan \theta \sin \theta$

2. $\csc \theta - \cot \theta \cos \theta$

3. $(\tan x + \cot x) \sin x$

4. $\dfrac{1 + \cos x}{1 + \sec x}$

5. $\dfrac{(\tan x)(1 + \cot^2 x)}{1 + \tan^2 x}$

6. $\sec x - \sin x \tan x$

7. $\cos x \csc x$

8. $\cos x(\tan x + \cot x)$

9. $(\cos^2 x - 1)(\tan^2 x + 1)$

10. $\dfrac{\sec^2 x - 1}{\sec^2 x}$

11. $\dfrac{\sec x - \cos x}{\tan x}$

12. $\dfrac{1 + \tan^2 x}{\tan^2 x}$

13. $(\sin^2 x + \cos^2 x)^3$

14. $\dfrac{1 + \sec x}{\tan x + \sin x}$

15. $(\csc x - \cot x)^4(\csc x + \cot x)^4$

16. $\dfrac{\sec x}{\tan x + \cot x}$

17. $(\tan x)(\sin x + \cot x \cos x)$

18. $1 + \dfrac{\tan^2 x}{1 + \sec x}$

19. $\dfrac{\tan x \sin x}{\sec^2 x - 1}$

20. $(\cos x)(1 + \tan^2 x)$

In Exercises 21–26, use substitutions to reduce each given expression to one involving only trigonometric functions. Assume $0 \le \theta < \frac{1}{2}\pi$.

21. $\sqrt{a^2 + x^2}$, let $x = a \tan \theta$. What is $\sin \theta$?

22. $\sqrt{36 + 16x^2}$, let $x = \frac{3}{2} \tan \theta$. What is $\sin \theta$?

23. $\dfrac{\sqrt{x^2 - 4}}{x}$, let $x = 2 \sec \theta$.

24. $x^2\sqrt{4 + 9x^2}$, let $x = \frac{2}{3} \tan \theta$.

25. $\sqrt{3 - 5x^2}$, let $x = \sqrt{\frac{3}{5}} \sin \theta$.

26. $\sqrt{2 + x^2}$, let $x = \sqrt{2} \tan \theta$.

27. In calculus class, a dispute arises over an answer to a problem. One group says that the answer is $\tan x$, and the other group says that the answer is

$$\frac{\cos x - \cos^3 x}{\sin x - \sin^3 x}$$

Can you show that both groups are correct?

28. The polar equation of a parabola can be written as

$$r = \frac{k}{1 - \cos \theta}$$

Show that this is equivalent to

$$r = k \csc \theta \, (\csc \theta + \cot \theta)$$

29. Working a problem in statics, you get an answer of $\sec x + \tan x$. The back of the book has $(\sec x - \tan x)^{-1}$. Does the book have a printing error?

6.2 Trigonometric Identities

When both sides of an equation are equal for all values of the variable for which the equation is defined, the equation is called an **identity**. The eight fundamental relations (6.1)–(6.8) are trigonometric identities. In addition to the fundamental identities, there are many other trigonometric identities that arise in applications of mathematics. In most cases you are required to verify, or **prove**, that the given relation is an identity. There are several techniques that can be used to prove an identity, but the most common one involves using other known identities to transform one side of the equation into precisely the same form as the other. In this section you will learn to use the Fundamental Identities (6.1)–(6.8) to prove other identities. Although there is no general approach to proving identities, you may find it desirable to write the given expressions in terms of sines and cosines only. Then you can often see the manipulations necessary to complete the verification.

Comment: It is incorrect to verify an identity by beginning with the assumption that it *is* an identity. Thus, the expressions on either side of the equality are not equated initially. Operations valid for conditional equations (such as transposing) are not valid.

Example   Verify the identity $\cot x + \tan x = \csc x \sec x$.

Solution.   Here we express the left-hand side in terms of sines and cosines. Thus,

$$\cot x + \tan x = \frac{\cos x}{\sin x} + \frac{\sin x}{\cos x} \qquad \text{Changing to sine and cosine}$$

$$= \frac{\cos^2 x + \sin^2 x}{\sin x \cos x} \qquad \text{Adding fractions}$$

$$= \frac{1}{\sin x \cos x} \qquad \cos^2 x + \sin^2 x = 1$$

$$= \csc x \sec x \qquad \frac{1}{\sin x} = \csc x; \; \frac{1}{\cos x} = \sec x$$

Therefore, we have shown that $\cot x + \tan x = \csc x \sec x$.

Example 2.   Verify the identity

$$\frac{\sec^4 x - \tan^4 x}{\sec^2 x} = 1 + \sin^2 x$$

Solution.   We begin by factoring the numerator as the difference of two squares:

$$\frac{\sec^4 x - \tan^4 x}{\sec^2 x} = \frac{(\sec^2 x - \tan^2 x)(\sec^2 x + \tan^2 x)}{\sec^2 x} \qquad \text{Factoring } \sec^4 x - \tan^4 x$$

$$= \frac{\sec^2 x + \tan^2 x}{\sec^2 x} \qquad \text{Replacing } \sec^2 x - \tan^2 x \text{ with } 1$$

$$= \frac{\dfrac{1}{\cos^2 x} + \dfrac{\sin^2 x}{\cos^2 x}}{\dfrac{1}{\cos^2 x}} \qquad \text{Changing to } \sin x \text{ and } \cos x$$

$$= 1 + \sin^2 x \qquad \begin{array}{l}\text{Inverting the denominator} \\ \text{and multiplying}\end{array}$$

Therefore, we have verified that $\dfrac{\sec^4 x - \tan^4 x}{\sec^2 x} = 1 + \sin^2 x$.

Example 3.   Show that $\sin \theta(\csc \theta - \sin \theta) = \cos^2 \theta$ is an identity.

Solution.   Here the most expedient approach is to expand the left-hand side. Thus,

$$\sin \theta(\csc \theta - \sin \theta) = \sin \theta \csc \theta - \sin^2 \theta \qquad \text{Expanding}$$

$$= \sin \theta \frac{1}{\sin \theta} - \sin^2 \theta \qquad \csc \theta = \frac{1}{\sin \theta}$$

$$= 1 - \sin^2 \theta \qquad \text{Cancellation law}$$

$$= \cos^2 \theta \qquad 1 - \sin^2 \theta = \cos^2 \theta$$

Therefore, we have shown that $\sin \theta(\csc \theta - \sin \theta) = \cos^2 \theta$.

Example 4. Verify the identity

$$\frac{\cos x}{1 - \sin x} = \frac{1 + \sin x}{\cos x}$$

Solution. Here we start on the left-hand side. One way to get $1 + \sin x$ into the numerator of the left side is to multiply the left side by $\dfrac{1 + \sin x}{1 + \sin x}$, as shown below.

$$\frac{\cos x}{1 - \sin x} = \frac{\cos x}{1 - \sin x} \cdot \frac{1 + \sin x}{1 + \sin x} \qquad \text{Multiplying by } \frac{1 + \sin x}{1 + \sin x}$$

$$= \frac{\cos x(1 + \sin x)}{1 - \sin^2 x} \qquad \text{Multiplication}$$

$$= \frac{\cos x(1 + \sin x)}{\cos^2 x} \qquad 1 - \sin^2 x = \cos^2 x$$

$$= \frac{1 + \sin x}{\cos x} \qquad \text{Cancellation law}$$

Therefore, we have proved that $\dfrac{\cos x}{1 - \sin x} = \dfrac{1 + \sin x}{\cos x}$

Example 5. Verify the identity

$$(\csc x + \cot x)^2 = \frac{1 + \cos x}{1 - \cos x}$$

Solution. We start on the left-hand side by squaring the binomial:

$$(\csc x + \cot x)^2 = \csc^2 x + 2 \csc x \cot x + \cot^2 x \qquad \text{Expanding}$$

$$= \frac{1}{\sin^2 x} + \frac{2 \cos x}{\sin^2 x} + \frac{\cos^2 x}{\sin^2 x} \qquad \text{Changing to } \sin x \text{ and } \cos x$$

$$= \frac{1 + 2 \cos x + \cos^2 x}{\sin^2 x} \qquad \text{Adding fractions}$$

$$= \frac{(1 + \cos x)^2}{\sin^2 x} \qquad \text{Factoring the numerator}$$

$$= \frac{(1 + \cos x)^2}{1 - \cos^2 x} \qquad \sin^2 x = 1 - \cos^2 x$$

$$= \frac{(1 + \cos x)^2}{(1 + \cos x)(1 - \cos x)} \qquad \text{Factoring the denominator}$$

$$= \frac{1 + \cos x}{1 - \cos x} \qquad \text{Cancellation law}$$

Therefore, we have proved that $(\csc x + \cot x)^2 = \dfrac{1 + \cos x}{1 - \cos x}$.

Sometimes you can verify an identity by manipulating the left-hand side and the right-hand side into forms that are precisely the same.

Example 6. Verify the identity

$$\cos^2 x \tan^2 x + 1 = \sec^2 x + \sin^2 x - \sin^2 x \sec^2 x$$

Solution. We transform the two sides of the given expression into precisely the same form.

(1) The left-hand side becomes

$$\cos^2 x \tan^2 x + 1 = \cos^2 x \left(\frac{\sin^2 x}{\cos^2 x} \right) + 1 \qquad \text{Changing to sine and cosine}$$

$$= \sin^2 x + 1 \qquad \text{Cancellation law}$$

(2) The right-hand side may be transformed as follows:

$$\sec^2 x + \sin^2 x - \sin^2 x \sec^2 x = \frac{1}{\cos^2 x} + \sin^2 x - \frac{\sin^2 x}{\cos^2 x} \qquad \begin{array}{l}\text{Changing to sine}\\ \text{and cosine}\end{array}$$

$$= \frac{1 - \sin^2 x}{\cos^2 x} + \sin^2 x \qquad \text{Combining fractions}$$

$$= \frac{\cos^2 x}{\cos^2 x} + \sin^2 x \qquad 1 - \sin^2 x = \cos^2 x$$

$$= 1 + \sin^2 x \qquad \text{Cancellation law}$$

Since the right- and left-hand sides of the identity have been transformed into the same expression, the identity is verified. Therefore, we have verified that $\cos^2 x \tan^2 x + 1 = \sec^2 x + \sin^2 x - \sin^2 x \sec^2 x$.

To show that a relation is not an identity, we only have to show that the two expressions are unequal for a particular value of x.

Example 7. Show that the expression $\sin x = \sqrt{\sin^2 x}$ is not an identity.

Solution. Note that there are many values of x for which the two expressions $\sin x$ and $\sqrt{\sin^2 x}$ *are* equal. But consider $x = -\pi/4$. The value of $\sin(-\pi/4)$ is $-\sqrt{2}/2$, whereas

$$\sqrt{\sin^2 \left(\frac{-\pi}{4} \right)} = \sqrt{\left(\frac{-\sqrt{2}}{2} \right)^2} = \sqrt{\frac{1}{2}} = \frac{\sqrt{2}}{2}$$

Hence, the two expressions are not equal for this value of x (other values of x could have been used), and the given expression is not an identity.

Following are some guidelines for verifying trigonometric identities.

■ Know your fundamental identities well. For example, you should know not only $\sin^2 x + \cos^2 x = 1$ but also all the variations of this identity, such as $\sin^2 x = 1 - \cos^2 x$. Quick recall of the fundamental identities is the single most important ability you can develop.

■ Start with the most complicated side. If both appear fairly complicated, you might try manipulating both sides *independently* until they are in the same form.

■ Sometimes it helps to change everything into sines and cosines or, in some instances, into the same function.

■ Perform algebraic simplifications whenever possible, but watch for two (or more) possibilities. For example,

$$\frac{1}{\sin x} - \sin x = \frac{1 - \sin^2 x}{\sin x} = \frac{\cos^2 x}{\sin x}$$

or

$$\frac{1}{\sin x} - \sin x = \frac{1 - \sin^2 x}{\sin x} = \frac{(1 - \sin x)(1 + \sin x)}{\sin x}$$

■ Be guided in your simplification process by what comes next, or by what result you wish to achieve. Consider the expressions

$$\frac{1 - \sin^2 x}{1 + \sin x} \quad \text{and} \quad \frac{1 - \sin^2 x}{\cos x}$$

both of which have the same numerator. With the first expression, the logical next step is to factor the numerator and simplify as follows:

$$\frac{(1 - \sin x)(1 + \sin x)}{1 + \sin x} = 1 - \sin x$$

With the second expression, the best thing to do is to recall the identity $1 - \sin^2 x = \cos^2 x$ and simplify as follows:

$$\frac{1 - \sin^2 x}{\cos x} = \frac{\cos^2 x}{\cos x} = \cos x$$

Exercises for Section 6.2

In Exercises 1–82, verify the identities.

1. $\sin x \cot x = \cos x$

2. $\cos x \tan x = \sin x$

3. $\sec x \cot x = \csc x$

4. $(1 + \tan^2 x) \sin^2 x = \tan^2 x$

5. $\sin^2 x(1 + \cot^2 x) = 1$

6. $\cot^2 x - \cos^2 x = \cot^2 x \cos^2 x$

7. $\csc x - \sin x = \cot x \cos x$

8. $\sec^2 x \csc^2 x = \sec^2 x + \csc^2 x$

9. $(\sin^2 x - 1)(\cot^2 x + 1) = 1 - \csc^2 x$

10. $\dfrac{\sin^2 x + \cos^2 x}{\cos^2 x} = \sec^2 x$

11. $\dfrac{2 + \sec x}{\csc x} - 2 \sin x = \tan x$

12. $\dfrac{\sin^4 x - \cos^4 x}{\sin x - \cos x} = \sin x + \cos x$

13. $\dfrac{\sin x}{1 - \cos x} = \csc x + \cot x$

14. $\dfrac{\tan x - 1}{\tan x + 1} = \dfrac{1 - \cot x}{1 + \cot x}$

15. $\dfrac{\cot x + 1}{\cot x - 1} = -\dfrac{\tan x + 1}{\tan x - 1}$

16. $\dfrac{\cos x}{\sec x} + \dfrac{\sin x}{\csc x} = \sec^2 x - \tan^2 x$

17. $\dfrac{1 + \sec x}{\sin x + \tan x} = \csc x$

18. $\sec^2 x - \csc^2 x = \tan^2 x - \cot^2 x$

19. $\dfrac{1 - \sin x}{1 + \sin x} = (\sec x - \tan x)^2$

20. $\cos^2 x - \sin^2 x = 2 \cos^2 x - 1$

21. $(\sin^2 x + \cos^2 x)^4 = 1$

22. $\dfrac{\csc^2 x - \cot^2 x}{\sec^2 x} = \cos^2 x$

23. $\dfrac{\tan x + \cot x}{\tan x - \cot x} = \dfrac{\sec^2 x}{\tan^2 x - 1}$

24. $\sin^2 x \sec^2 x + 1 = \sec^2 x$

25. $\dfrac{\sin x}{\csc x(1 + \cot^2 x)} = \sin^4 x$

26. $\sec^2 x - (\cos^2 x + \tan^2 x) = \sin^2 x$

27. $1 - \tan^4 x = 2 \sec^2 x - \sec^4 x$

28. $\sec x - \cos x = \sin x \tan x$

29. $(\cot x + \csc x)^2 = \dfrac{1 + \cos x}{1 - \cos x}$

30. $\sin^2 x(\csc^2 x - 1) = \cos^2 x$

31. $\sec x \csc x - 2 \cos x \csc x = \tan x - \cot x$

32. $\sec^4 x + \tan^4 x = 1 + 2 \sec^2 x \tan^2 x$

33. $\tan^2 x - \cot^2 x = \sec^2 x - \csc^2 x$

34. $\dfrac{1 - \tan^2 x}{1 - \cot^2 x} = 1 - \sec^2 x$

35. $(1 - \sin^2 x)(1 + \tan^2 x) = 1$

36. $\dfrac{\tan x + \sin x}{\tan x - \sin x} = \dfrac{\sec x + 1}{\sec x - 1}$

37. $\dfrac{\cos x + \tan x}{\sin x \cos x} = \csc x + \sec^2 x$

38. $\cos^2 x \tan x = \dfrac{2 \sin x}{\cos x + \sec x + \sin^2 x \sec x}$

39. $(\sin x - \cos x)^2 = 1 - 2 \sin x \cos x$

40. $\dfrac{\cos x}{\cos x - \sin x} = \dfrac{1}{1 - \tan x}$

41. $\cos^2 x - \sin x \tan x = \cos x \cot x \sin x - \tan x \sin x$

42. $\dfrac{\tan^2 x}{\sin^4 x} = \dfrac{1 + \tan^2 x}{1 - \cos^2 x}$

43. $\tan x - \cot x = -\dfrac{\cos x - \sin^2 x \sec x}{\sin x}$

44. $1 - \sin x = \dfrac{\cot x - \cos x}{\cot x}$

45. $\csc^4 x + \cot^4 x = 1 + 2 \csc^2 x \cot^2 x$

46. $\dfrac{\sec^2 x + 2 \tan x}{1 + \tan x} = 1 + \tan x$

47. $(1 + \cos x)^2 = \sin^2 x \,\dfrac{\sec x + 1}{\sec x - 1}$

48. $(\sec x - \tan x)^2 = \dfrac{1 - \sin x}{1 + \sin x}$

49. $(\sec x + \tan x)^2 = \dfrac{\sec x + \tan x}{\sec x - \tan x}$

50. $(\csc x - \cot x)^2 = \dfrac{\csc x - \cot x}{\csc x + \cot x}$

51. $\dfrac{\csc x}{\csc x - \tan x} = \dfrac{\cos x}{\cos x - \sin^2 x}$

52. $(\tan x - 1) \cos x = \sin x - \cos x$

53. $\dfrac{1}{1 - \sin x} - \dfrac{1}{1 + \sin x} = 2 \tan x \sec x$

54. $\dfrac{\tan x - \csc x}{\tan x + \csc x} = \dfrac{\sin^2 x - \cos x}{\sin^2 x + \cos x}$

55. $\sec^4 x - \tan^4 x = \dfrac{1 + \sin^2 x}{\cos^2 x}$

56. $\dfrac{\tan x}{\sec x - \cos x} = \csc x$

57. $(\csc x - \cot x)(\sec x + 1) = \tan x$

58. $\dfrac{\cos^2 x}{1 + \sin x} = 1 - \sin x$

59. $(1 + \sin x)(\sec x - \tan x) = \cos x$

60. $\cos^4 x - \sin^4 x = 1 - 2 \sin^2 x$

61. $\sec x \csc x - 2 \cos x \csc x = \tan x - \cot x$

62. $\dfrac{\sin x}{\sin x + \cos x} = \dfrac{\tan x}{1 + \tan x}$

63. $\dfrac{\sin x}{1 + \cos x} + \dfrac{1 + \cos x}{\sin x} = 2 \csc x$

64. $2 \sin^4 x - 3 \sin^2 x + 1 = \cos^2 x(1 - 2 \sin^2 x)$

65. $\dfrac{\csc x}{\tan x + \cot x} = \cos x$

66. $\dfrac{1 - \sin x}{1 + \sin x} = \left(\dfrac{\cos x}{1 + \sin x}\right)^2$

67. $\dfrac{1 + \cos x}{1 - \cos x} = (\csc x + \cot x)^2$

68. $\dfrac{\sec^3 x - \cos^3 x}{\sec x - \cos x} = 1 + \cos^2 x + \sec^2 x$

69. $\dfrac{\cos^2 x}{1 - \sin x + \cos^2 x} = \dfrac{1 + \sin x}{2 + \sin x}$

70. $(1 + \tan x)^2 = \sec^2 x(1 + 2 \cos x \sin x)$

71. $(1 + \cot x)^2 = \csc^2 x(1 + 2 \cos x \sin x)$

72. $(\sin x + \cos x)^2 = \dfrac{\sec x \csc x + 2}{\sec x \csc x}$

73. $(\cos x + \sin x + \tan x)^2 = \sec^2 x(1 + 2 \sin^2 x \cos x) + 2 \sin x(1 + \cos x)$

74. $\dfrac{1 + \cos x}{2 - \cos x} = \dfrac{\sin^2 x}{2 - 3 \cos x + \cos^2 x}$

75. $(1 + \sin^2 x)^4 = 16 - 32 \cos^2 x + 24 \cos^4 x - 8 \cos^6 x + \cos^8 x$

76. $\dfrac{\tan x - \cot x}{\sec^2 x - \csc^2 x} = \sin x \cos x$

77. $\dfrac{\tan x - \tan y}{\cot x - \cot y} = \dfrac{1 - \tan x \tan y}{1 - \cot x \cot y}$

78. $\dfrac{\cos x \cos y - \sin x \sin y}{\cos x \sin y + \cos y \sin x} = \dfrac{1 - \tan x \tan y}{\tan x + \tan y}$

79. $\tan x + \cot x = \sec x \csc x$

80. $\dfrac{\tan x}{1 - \cot x} + \dfrac{\cot x}{1 - \tan x} - 1 = \sec x \csc x$

81. $(2 \cos x - \sin x)^2 + (2 \sin x + \cos x)^2 = 5$

82. $(a \cos x - b \sin x)^2 + (a \sin x + b \cos x)^2 = a^2 + b^2$

In Exercises 83–94, show that the given expressions are not identities.

83. $\cos t = \sqrt{\cos^2 t}$

84. $1 = \tan (\cot x)$

85. $1 = \sec(\cos x)$

86. $\sin x = (1 - \cos x)^2$

87. $\cos^2 x = \dfrac{1 - \sin x}{2}$

88. $\sin x + \cos x = \sqrt{\sin^2 x + \cos^2 x}$

89. $\sin \frac{1}{2}x = \frac{1}{2} \sin x$

90. $\tan 2x = 2 \tan x$

91. $\sin(x + \pi) = \sin x$

92. $\cos(x + \pi) = \cos x - 1$

93. $\cos x^2 = \cos^2 x$

94. $\sin x^2 = \cos(1 - x^2)$

In Exercises 95–104, determine which of the expressions are identities.

95. $(\cos x - \sin x)(\cos x + \sin x) = 2 \cos^2 x - 1$

96. $\sin x \sec x = \tan x$

97. $\cos x = \cot x$

98. $1 - \cot x = \cot x \tan x - \cot x$

99. $1 - \dfrac{2}{\sec^2 x} = \sin^2 x - \cos^2 x$

100. $\cos x + 1 = \sin x$

101. $\dfrac{\cos x}{1 - \sin x} = \dfrac{1 + \sin x}{\cos x}$

102. $\sin x \cot x \tan^2 x = \sec x - \sin x \cot x$

103. $\sin x \tan x + \cos x = \sec x$

104. $\sin x \tan^2 x = \sin x$

105. A student finds that two calculus books give different formulas for the derivative of $\sec x$. The first book gives $\sec x \tan x$ and the other one gives $\sin x/\cos^2 x$. Show that the two forms are equivalent.

106. In calculating the slope of the path of a projectile, an aerospace engineer is confronted with the expression

$$\frac{-2v_0 \sin \theta + v_0 \sin \theta}{v_0 \cos \theta}$$

where v_0 is the muzzle velocity and θ is the initial elevation angle. Show that this expression is equal to $-\tan \theta$.

107. When you find the point of intersection of

$$r = \frac{1}{1 - \cos \theta} \quad \text{and} \quad r = \frac{3}{1 + \cos \theta}$$

the equality

$$\frac{1 + \cos \theta}{1 - \cos \theta} = 3$$

must be simplified. Show that the left-hand side may be written as $\dfrac{(1 + \cos \theta)^2}{\sin^2 \theta}$.

6.3 Trigonometric Equations

A **trigonometric equation** is any statement involving a conditional equality of two trigonometric expressions. A **solution** to the trigonometric equation is a value of the variable (within the domain of the function) that makes the statement true. The **solution set** is the set of all values of the variable that are solutions. Solving a trigonometric equation means finding the solution set for some indicated domain. If no domain is specifically mentioned, the domain is assumed to be all values of the independent variable for which the terms of the equation have meaning.

To solve trigonometric equations, we proceed in a series of steps until we reach a point at which an explicit determination of the solution set can be made. Usually, some specific knowledge about certain values of the trigonometric functions is necessary to make this determination.

We say that two trigonometric equations are **equivalent** if they have the same solution sets. Any operation on a given equation is **allowable** if the consequence of the operation is an equivalent equation. The permissible operations are (1) adding or subtracting the same expression to both sides of an equality and (2) multiplying or dividing both sides by the same nonzero expression.

Generally trigonometric equations have infinitely many solutions, but we are often content to list only those roots over some fundamental interval, or period, of the functions. All other roots can be obtained from these by simply adding multiples of the period. Unless otherwise stated, the fundamental interval is chosen to be $\phi \leq x < \phi + p$, where ϕ is the phase shift and p is the period of the trigonometric function.

Example 1. Solve the equation $\cos x = \frac{1}{2}$ on $0 \leq x < 2\pi$.

Solution. The period of $\cos x$ is 2π. Since $\cos \frac{1}{3}\pi = \frac{1}{2}$, the only solutions to this equation on the interval $0 \leq x < 2\pi$ are $x = \frac{1}{3}\pi$ and $\frac{5}{3}\pi$. (The complete solution set is composed of the values that can be written in the form $\frac{1}{3}\pi + 2n\pi$ and $\frac{5}{3}\pi + 2n\pi$, where n is an integer.) Figure 6.2 illustrates the nature of the solution set as the points of intersection of the curve $y = \cos x$ with the line $y = \frac{1}{2}$.

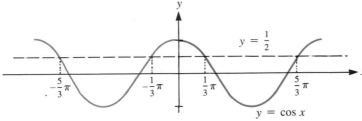

Figure 6.2

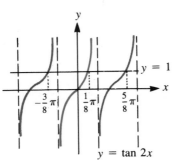

Figure 6.3

Example 2. Solve the trigonometric equation $\tan 2x = 1$ on $-\frac{1}{4}\pi < x < \frac{1}{4}\pi$.

Solution. Since the period of this function is $\frac{1}{2}\pi$, it is sufficient to examine for roots on the interval $-\frac{1}{4}\pi \leq x < \frac{1}{4}\pi$. The value of θ at which $\tan \theta = 1$ is $\theta = \frac{1}{4}\pi$. Hence $\tan 2x = 1$ has the solution $x = \frac{1}{8}\pi$. Figure 6.3 shows a graphical interpretation of the solution set as the intersection of the curve $y = \tan 2x$ with the line $y = 1$. (The complete solution is given by $\frac{1}{8}\pi \pm \frac{1}{2}n\pi$, where n is an integer.)

Example 3. Find the solution to the equation $|\sin x| = \frac{1}{2}$ on $0 \leq x < \pi$.

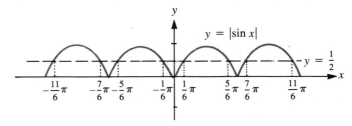

Figure 6.4

Solution. Figure 6.4 shows the intersection points of the curve $y = |\sin x|$ with $y = \frac{1}{2}$. Since $|\sin x|$ is periodic with period π, we need only find those values of x on the interval $0 \leq x < \pi$ for which the equation is true. From our knowledge of the sine function, we know that these values are $x = \frac{1}{6}\pi$ and $x = \frac{5}{6}\pi$. Notice that all solutions are given by $x = \frac{1}{6}\pi + n\pi$ and $\frac{5}{6}\pi + n\pi$, where n is an integer.

Trigonometric equations that are quadratic in one of the functions can be factored into a product of linear factors. The total solution set is found by finding the solution to each of the resulting linear equations.

Example 4. Solve the equation $2 \cos^2 \theta + 3 \cos \theta + 1 = 0$ on $0 \leq \theta < 2\pi$.

Solution. This is a quadratic equation in $\cos \theta$ and may be factored as

$$(2 \cos \theta + 1)(\cos \theta + 1) = 0$$

Equating each factor to zero and solving for θ, we get

$$
\begin{array}{c|c}
2 \cos \theta + 1 = 0 & \cos \theta + 1 = 0 \\
\cos \theta = -\frac{1}{2} & \cos \theta = -1 \\
\theta = \frac{2}{3}\pi, \frac{4}{3}\pi & \theta = \pi
\end{array}
$$

Hence the solutions are $\theta = \frac{2}{3}\pi$, π, and $\frac{4}{3}\pi$.

If more than one trigonometric function occurs in the equation, use trigonometric identities to write an equivalent equation involving only one function.

Example 5. Solve the equation $2 \cos^2 x - \sin x - 1 = 0$ on $0 \leq x < 2\pi$.

Solution. Since $\cos^2 x = 1 - \sin^2 x$,

$$2(1 - \sin^2 x) - \sin x - 1 = 0$$
$$2 - 2\sin^2 x - \sin x - 1 = 0$$
$$2\sin^2 x + \sin x - 1 = 0$$

Factoring,

$$(2\sin x - 1)(\sin x + 1) = 0$$

Equating each factor to zero and solving for x, we have

$2\sin x - 1 = 0$	$\sin x + 1 = 0$
$\sin x = \frac{1}{2}$	$\sin x = -1$
$x = \frac{1}{6}\pi, \frac{5}{6}\pi$	$x = \frac{3}{2}\pi$

Hence, the solutions are $x = \frac{1}{6}\pi, \frac{5}{6}\pi$, and $\frac{3}{2}\pi$.

The following is normally a good procedure for solving trigonometric equations.

(1) Gather the entire expression on one side of the equality.

(2) Use the fundamental identities to express the conditional equality in terms of one function, or, failing this, as a product of two expressions, each involving one function.

(3) Use some algebraic technique, such as substitution into the quadratic formula or techniques of factoring, to write the expression as a product of linear factors.

(4) Determine the zeros, if any, of each of the linear factors. The solution set consists of all zeros of these linear factors.

Example 6. Solve the equation $2\tan\theta\sec\theta - \tan\theta = 0$ on $0° \le \theta < 360°$.

Solution. The given equation can be factored as

$$\tan\theta (2\sec\theta - 1) = 0$$

Equating each factor to zero, we get

$\tan\theta = 0$	$2\sec\theta - 1 = 0$
$\theta = 0°$ and $180°$	$\sec\theta = \frac{1}{2}$
	No solution possible since $\sec\theta \ge 1$.

Thus, the solutions are $\theta = 0°$ and $180°$.

> **Warning:** Squaring both sides of an equality is not an allowable operation since it does not necessarily yield an equivalent equation. In practice you need not restrict yourself to allowable operations, but when you use nonallowable operations to solve an equation, you must check each apparent solution for validity. Of course, it is *always* good practice to check your work.

Example 7. Solve the equation $\sin x + \cos x = 1$ on $[0, 2\pi)$.

Solution. Squaring both sides of this equation, we get

$$\sin^2 x + 2 \sin x \cos x + \cos^2 x = 1$$

Since $\sin^2 x + \cos^2 x = 1$,

$$\sin x \cos x = 0$$

The solution to this equation consists of the values of x for which $\sin x = 0$ (that is, $x = 0$ and π), and the values of x for which $\cos x = 0$ (that is, $x = \frac{1}{2}\pi$ and $\frac{3}{2}\pi$). Hence, the possible solutions are

$$x = 0, \tfrac{1}{2}\pi, \pi, \text{ and } \tfrac{3}{2}\pi.$$

Since "squaring" does not yield an equivalent equation, you must check these values to determine if they are solutions to the original equation. It is easy to show that only $x = 0$ and $x = \frac{1}{2}\pi$ are valid solutions.

Example 8.   Solve the equation $\sin^2 x + 3 \sin x - 2 = 0$ on $0 \le x \le 2\pi$.

Solution.   Since the given quadratic does not factor, we use the quadratic formula to obtain

$$\sin x = \frac{-3 \pm \sqrt{3^2 - 4(1)(-2)}}{2(1)} = \frac{-3 \pm \sqrt{17}}{2}$$

From this, we can write

$$\sin x = \frac{-3 + \sqrt{17}}{2} = 0.5616 \quad \text{and} \quad \sin x = \frac{-3 - \sqrt{17}}{2} = -3.5616$$

Since $\sin x$ cannot be less than -1, $\sin x = -3.5616$ has no solution. Using a calculator in the radian mode to solve $\sin x = 0.5616$, we find that $x = 0.596$. Since $\sin x$ is also positive in the second quadrant, another solution is $\pi - 0.596 = 2.546$. Therefore, the desired solutions are $x = 0.596$ and 2.546.

Exercises for Section 6.3

In Exercises 1–10, solve each equation over a fundamental interval of the function. Make a sketch showing the solution set as the intersection of a line with the graph of some trigonometric function.

1. $\sin x = \frac{1}{2}$

2. $\cos 2x = \frac{1}{2}\sqrt{2}$

3. $\tan x = \sqrt{3}$

4. $\cos x = 1$

5. $\sin x = \frac{1}{2}\sqrt{3}$

6. $\cos x = -\frac{1}{2}$

7. $\sin 2x = -\frac{1}{2}$

8. $\tan \frac{1}{2}x = 1$

9. $|\cos x| = \frac{1}{2}\sqrt{2}$

10. $|\sin x| = \frac{1}{2}\sqrt{3}$

In Exercises 11–37, solve each trigonometric equation over the interval $[0, 2\pi)$ unless another interval is indicated.

11. $2 \sin x + 1 = 0$

12. $\sin 2x + 1 = 0$, $[0, \pi)$

13. $\cos 3x = 1$, $[0, \frac{2}{3}\pi)$

14. $\tan 2x + 1 = 0$, $[0, \frac{1}{2}\pi)$

15. $\cos^2 x + 2 \cos x + 1 = 0$

16. $\tan^2 x - 1 = 0$, $[0, \pi)$

17. $2 \sin^2 x = \sin x$

18. $\sec^2 2x = 1$, $[0, \pi)$

19. $\sec^2 x + 1 = 0$

20. $\cos^2 x = 2$

21. $\cos x = \sin x$

22. $2 \sec x \tan x + \sec^2 x = 0$

23. $\sec^2 x - 2 = \tan^2 x$

24. $4 \sin^2 x - 1 = 0$

25. $2 \cos^2 x - \sin x = 1$

26. $2 \sec x + 4 = 0$

27. $\sin^2 x - 2 \sin x + 1 = 0$

28. $\cot^2 x - 5 \cot x + 4 = 0$, $[0, \pi)$

29. $\tan^2 x - \tan x = 0$

30. $\cos 2x + \sin 2x = 0$, $[0, \pi)$

31. $\sin x \tan^2 x = \sin x$

32. $\cos x + 2 \sin^2 x = 1$

33. $\tan x + \sec x = 1$

34. $\tan x + \cot x = \sec x \csc x$

35. $\cos x + 1 = \sin x$

36. $2 \tan x - \sec^2 x = 0$, $[0, \frac{1}{2}\pi)$

37. $\csc^5 x - 4 \csc x = 0$

In Exercises 38–51, solve each equation for all values of θ over the interval $[0°, 360°)$.

38. $4 \sin^2 \theta = 3$

39. $2 \cos^2 \theta = 1$

40. $\tan^2 \theta - 3 = 0$

41. $\csc^2 \theta = 1$

42. $(2 \cos \theta - \sqrt{3})(\sqrt{2} \sin \theta + 1) = 0$

43. $(\sin \theta - 1)(2 \cos \theta + \sqrt{3}) = 0$

44. $\csc \theta = 1 + \cot \theta$

45. $2 \sin \theta \tan \theta + \sqrt{3} \tan \theta = 0$

46. $2 \cos^3 \theta = \cos \theta$

47. $\tan^2 \theta - \tan \theta = 0$

48. $2 \sin^2 \theta = 21 \sin \theta + 11$

49. $2 \cos^2 \theta + 7 \sin \theta = 5$

50. $5 \tan^2 \theta - 11 \sec \theta + 7 = 0$

51. $\sin \theta + \sqrt{3} \cos \theta = 0$

In Exercises 52–60, solve each equation for x, $0 \le x < 2\pi$. Note: For these problems, the solutions are not necessarily multiples of π.

52. $\tan^2 x = 3.2$

53. $\sin^2 x + 2 \cos^2 x = 1.7$

54. $\cos x + 2 \sin x \tan x = 1$

55. $3 \sin^2 x + \sin x = 0$

56. $\tan^2 x + 2 \sec^2 x = 1$

57. $2 \sin x - \cot x \cos x = 1$

58. $\csc^2 x + 2 \cot^2 x - 1 = 0$ 59. $\sin x - \csc x + 1 = 0$

60. $2 \tan^2 x + 3 \sec x - 5 = 0$

61. To solve a vibrations problem, a student must find the values of x for which $\sqrt{3} \cos \pi x + \sin \pi x = 0$. What are those values?

62. The undamped motion of an object at the end of a spring is governed by the formula $y = \cos t + \sqrt{3} \sin t$. For which values of t is $y = 0$?

63. In radio engineering, the carrier wave is described by $V \sin (\omega_c t + \phi)$, where V is the amplitude and ϕ is the phase angle of the carrier. If V is varied, the wave is said to be amplitude modulated. If $V \neq 0$, for what values of t is the carrier wave equal to 0?

64. The approximate solution to Bessel's equation, which is used in solving vibration problems, is given by

$$\frac{A}{\sqrt{x}} \cos (x + \beta)$$

Find the values of x for which this function is zero.

6.4 Parametric Equations

A convenient way to define the locus* of a point in the plane is by using two equations, one for x and one for y in terms of some variable, say t. Thus, two functions of t

$$x = f(t) \quad \text{and} \quad y = g(t)$$

determine the location of the set of points. As t varies, the point describes a curve in the plane. The equations are called **parametric equations** of the curve; the variable t is called the **parameter**. Each value of t gives a pair of values x and y that represents a point on the curve. For any interval of the parameter, a smooth curve can be drawn connecting these points. Such a curve is said to be defined **parametrically** in terms of the parameter t. (The parameter t may be thought of as representing time.)

Example 1. Determine the values of (x, y) for values of t between 0 and 2π if $x = 2 \cos t$, $y = 2 \sin t$. Then sketch the curve given by these points.

Solution. The following table shows the values for $0 < t < \pi$. The values for $\pi < t < 2\pi$ are the same for x but opposite in sign for y.

* The **locus** of a point in the plane is the path taken by the point in satisfying a given condition, such as a mathematical formula.

t	0	$\frac{1}{6}\pi$	$\frac{1}{4}\pi$	$\frac{1}{3}\pi$	$\frac{1}{2}\pi$	$\frac{2}{3}\pi$	$\frac{3}{4}\pi$	$\frac{5}{6}\pi$	π
x	2	$\sqrt{3}$	$\sqrt{2}$	1	0	-1	$-\sqrt{2}$	$-\sqrt{3}$	-2
y	0	1	$\sqrt{2}$	$\sqrt{3}$	2	$\sqrt{3}$	$\sqrt{2}$	1	0

Figure 6.5 shows the curve. It is a circle of radius 2 centered at the origin.

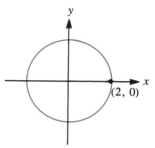

Figure 6.5

A major advantage of the parametric method is that curves are **oriented** by the parameter. For example, consider the curve described parametrically by $x = 2 \sin t$, $y = 3 \cos t$. (The graph is shown in Figure 6.6.) Note that when $t = 0$, the point is at $(0, 3)$. The point goes around the curve clockwise as t increases, completing a revolution every 2π units.

Sometimes a curve defined parametrically may also be expressed in the form of an equation involving only x and y. The basic idea is to *eliminate the parameter t* from the two given parametric equations. Two approaches are possible.

- Use trigonometric identities so that certain algebraic combinations of x and y will cause the parameter to disappear.
- Solve for t in terms of either x or y and substitute into the remaining equation.

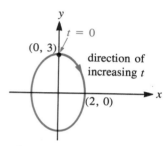

Figure 6.6

Example 2. Eliminate the parameter from the following parametrically defined curves:

(a) $x = 2 \cos t$, $y = 2 \sin t$ (b) $x = 1 + 2t$, $y = 2 - t$

Solution

(a) Square both sides of equations $x = 2 \cos t$, $y = 2 \sin t$ to obtain $x^2 = 4 \cos^2 t$, $y^2 = 4 \sin^2 t$. Adding these two equations gives

$$x^2 + y^2 = 4 \cos^2 t + 4 \sin^2 t$$
$$= 4 (\cos^2 t + \sin^2 t)$$
$$= 4$$

We note that $x^2 + y^2 = 4$ is the equation of a circle centered at the origin with a radius of 2. Hence, $x = 2 \cos t$ and $y = 2 \sin t$ are parametric equations of a circle.

(b) For the equations $x = 1 + 2t$, $y = 2 - t$, solve the second equation for t and substitute into the first one. Thus, $t = 2 - y$ and

$$x = 1 + 2t = 1 + 2(2 - y) = 1 + 4 - 2y$$
$$= 5 - 2y$$

Rearranging, we have

$$x + 2y = 5$$

which is the equation of a straight line. Hence, $x = 1 + 2t$, $y = 2 - t$ are parametric equations of a straight line.

When you are eliminating the parameter, be careful that the process of

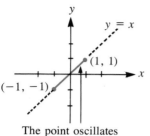

The point oscillates
back and forth on
this line segment.

Figure 6.7

elimination does not include (or exclude) more points than the original parametric equations give.

Example 3. Sketch the graph of the curve described parametrically by $x = \sin t$, $y = \sin t$.

Solution. The parameter is eliminated because x and y represent the same function. Hence, the nonparametric relation is $y = x$. The graph of $y = x$ is the line that splits the first and third quadrants. (See Figure 6.7.) However, by the nature of the parametric equations, the values of x and y are limited between 1 and -1. Thus, the point moves on the line $y = x$, but it oscillates between the points $(1, 1)$ and $(-1, -1)$.

Example 4. Eliminate t from the parametric equations

$$x = a + b \sec t \qquad \text{and} \qquad y = c + d \tan t$$

where a, b, c, and d are positive constants.

Solution. Rewrite the two equations as

$$\frac{x - a}{b} = \sec t \qquad \text{and} \qquad \frac{y - c}{d} = \tan t$$

Squaring and subtracting, we have

$$\left(\frac{x - a}{b}\right)^2 - \left(\frac{y - c}{d}\right)^2 = 1$$

This is the equation of a hyperbola; thus $x = a + b \sec t$, $y = c + d \tan t$ are parametric equations of a hyperbola.

When the parameters are sinusoids, the wave forms can be displayed on an oscilloscope. Voltages corresponding to the x and y functions are imposed on the horizontal and vertical input terminals of the scope, respectively. When the parametric equations are used in this manner, the resulting figure is called a **Lissajous figure**.

Example 5. Plot the Lissajous figure corresponding to input voltages of $x = \sin 2t$, $y = \cos t$.

Solution. Elimination of the parameter is not convenient in this case, so we generate a table of values for $0 \le t \le 2\pi$.

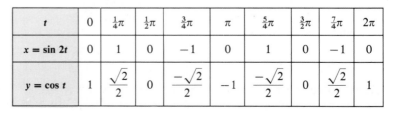

t	0	$\frac{1}{4}\pi$	$\frac{1}{2}\pi$	$\frac{3}{4}\pi$	π	$\frac{5}{4}\pi$	$\frac{3}{2}\pi$	$\frac{7}{4}\pi$	2π
$x = \sin 2t$	0	1	0	-1	0	1	0	-1	0
$y = \cos t$	1	$\frac{\sqrt{2}}{2}$	0	$\frac{-\sqrt{2}}{2}$	-1	$\frac{-\sqrt{2}}{2}$	0	$\frac{\sqrt{2}}{2}$	1

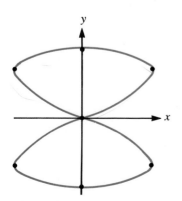

Figure 6.8

Using these points, we obtain Figure 6.8.

Exercises for Section 6.4

In Exercises 1–10, sketch the curve described by each parametric equation. In each case, indicate the orientation of the curve.

1. $x = \sin t, \ y = \cos t$

2. $x = 2 \cos t, \ y = \sin t$

3. $x = \cos^2 t, \ y = 1$

4. $x = 3t, \ y = 2$

5. $x = t^2, \ y = t^3$

6. $x = \cos t, \ y = \sin 2t$

7. $x = \sin t, \ y = \cos 2t$

8. $x = \sin t + \cos t, \ y = \sin t - \cos t$

9. $x = 2 \sin t + \cos t, \ y = 2 \cos t - \sin t$

10. $x = \sin t, \ y = \sin 2t$

In Exercises 11–20, eliminate t from each parametric equation to obtain one equation in x and y. Sketch the curve, being careful to observe any limitations imposed by the parameterization.

11. $x = \sin t, \ y = \cos t$

12. $x = 2 \cos t, \ y = \sin t$

13. $x = t, \ y = 1 + t$

14. $x = -\cos t, \ y = \sin t$

15. $x = \cos t, \ y = \cos t$

16. $x = 1 - t^2, \ y = t$

17. $x = \sin^2 t, \ y = t$

18. $x = \cos^2 t, \ y = \cos^2 t$

19. $x = t - 3, \ y = t^2 + 1$

20. $x = 2 \sin t + \cos t, \ y = 2 \cos t - \sin t$

21. The approximate path of the earth about the sun is given by $x = aR \cos 2\pi t$, $y = bR \sin 2\pi t$, where t is the time in years, R is the radius of the earth, and b and a are constants that are very nearly equal. Sketch this path for $a = 1$ and $b = 1.1$.

22. A piston is connected to the rim of a wheel as shown in Figure 6.9. At a time t seconds after it has coordinates $(2, 0)$, the point S has coordinates given by $x = 2 \cos 2\pi t, \ y = 2 \sin 2\pi t$. Find the position of the point S when $t = \frac{1}{2}, \frac{3}{4}$, and 2. Eliminate the parameter to find an equation relating x and y.

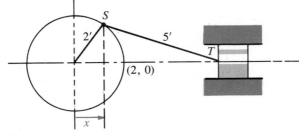

Figure 6.9

23. Lissajous figures are sometimes described by saying that the plotting is at right angles. The right angle plot idea comes from the fact that the voltages are impressed at right angles on the scope. Sketch the Lissajous figure for the equations $x = \sin t, \ y = \cos (t - \frac{1}{4}\pi)$.

24. Sketch the Lissajous figure for $x = 2 \sin 3t$, $y = \cos t$.

25. Sketch the Lissajous figure for $x = 2 \cos 3t$, $y = \sin 2t$.

26. The curve traced by a point on the circumference of a wheel of a bicycle tire is given by the set of parametric equations $x = t - \sin t$, $y = 1 - \cos t$. Sketch this curve. (The curve is called a **cycloid**.)

6.5 Graphical Solutions of Trigonometric Equations

Equations containing a mixture of trigonometric and other functions may be quite difficult to solve by analytic methods. A graphical analysis will usually yield at least an approximation to the roots and will often give helpful information even when a problem can be solved analytically.

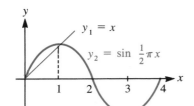

Figure 6.10

Example 1. Graphically solve the equation $x = \sin \frac{1}{2} \pi x$.

Solution. The functions $y_1 = x$ and $y_2 = \sin \frac{1}{2}\pi x$ are sketched in Figure 6.10. Since both functions are odd (that is, their graphs are symmetric with respect to the origin), the graph is drawn for positive values of x only. The solution set to the equation is the set of x-coordinates of the points of intersection of the two curves. The figure shows that the values are $x = 0$ and $x = 1$; hence, by the symmetry of the graph, the solution set is $\{-1, 0, 1\}$.

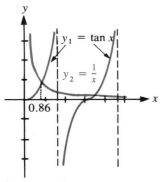

Figure 6.11

Example 2. Graphically solve the equation $x \tan x = 1$.

Solution. We write the equation in the form $\tan x = 1/x$ and then graph the two functions $y_1 = \tan x$ and $y_2 = 1/x$, as shown in Figure 6.11. As in the previous example, we may, without loss of generality, sketch only the part of the graphs for $x \geq 0$. Figure 6.11 shows that there are infinitely many solutions, of which the first positive one is approximately 0.86.

6.6 Trigonometric Inequalities

A **trigonometric inequality** is any conditional statement of inequality involving trigonometric expressions. Like the solution set of a trigonometric equation, the solution set of a trigonometric inequality is defined as the set of values for which the conditional statement is true. Other related terminology is consistent with that for trigonometric equations. The allowable operations are slightly more restricted for inequalities in that multiplication or division of both sides is permitted only if the expressions are positive. Multiplication of both sides by a negative quantity results in a reversal of the inequality.

Aside from some of the very basic kinds of inequalities such as $\sin x < 1$ and $\cos x > 0$, most trigonometric inequalities are best solved by some combination of graphical and analytical methods.

Example 1. Solve the inequality $\sin x > \cos x$.

Solution. In Figure 6.12, both the sine and the cosine function are sketched. The points of intersection occur at the real numbers that can be written in the form $\frac{1}{4}\pi + n\pi$. The graph shows that the sine function is greater than the cosine function over those intervals for which the left-hand end point is given for even values of n. In Figure 6.12, the solution set for one period is shown as the interval $\frac{1}{4}\pi < x < \frac{5}{4}\pi$.

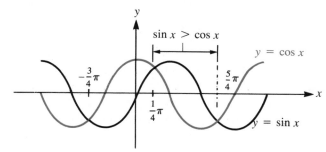

Figure 6.12

Inequalities that are not strictly trigonometric but that include other functions are also best analyzed graphically.

Example 2. Solve the inequality $x > \cos x$.

Solution. From Figure 6.13, we see that $x = \cos x$ at approximately $x = 0.74$. Hence, from the graph, the solution is $x > 0.74$.

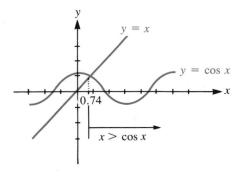

Figure 6.13

Exercises for Sections 6.5 and 6.6

In Exercises 1–10, use graphical methods to approximate the solutions to each equation on one fundamental period.

1. $x = \sin x$
2. $x = \sin 2x$
3. $x \sin x = 1$
4. $x = \tan x$
5. $x = \cos x$
6. $x^2 = \cos 2\pi x$
7. $x^2 - \tan x = 0$
8. $\sin x = \cos x$
9. $\tan x = \cos x$
10. $\tan x = \sin x$

In Exercises 11–20, solve each trigonometric inequality on one fundamental period.

11. $\sin x \geq \frac{1}{2}$
12. $\frac{1}{2} > \cos x$
13. $2 \cos^2 x \leq 1$
14. $\tan^2 x \leq 3$
15. $\sec x > \csc x$
16. $x > \sin x$
17. $x > \sin 2\pi x$
18. $x \sin x > 1$
19. $|\sin x| < |\cos x|$
20. $\sec^2 x < 1$

21. Two ramps are built, one in the shape of the straight line $y = 2x$ and the other in the shape of the curve $y = \tan x$. For which values of x is the straight line above the trigonometric ramp?

22. To solve a problem involving phase modulation, a communications engineer must find the zeros of

$$\sin t(\tfrac{1}{2}t - \sin t) = 0$$

Find the zeros of this function corresponding to $\frac{1}{2}t - \sin t = 0$.

Key Topics for Chapter 6

Define and/or discuss each of the following.

Fundamental Identities
Verifying and Proving Identities
Trigonometric Equations
Parametric Equations

Graphical Solutions to Trigonometric
   Equations
Trigonometric Inequalities

Review Exercises for Chapter 6

In Exercises 1–10, prove each of the given identities.

1. $1 + \sec x = \csc x(\tan x + \sin x)$

2. $\cos^2 x = \dfrac{\csc^2 x - 1}{\csc^2 x}$

3. $\sin x = \dfrac{1 - \cos^2 x}{\sin x}$

4. $\cos x = \sec x - \sin x \tan x$

5. $\cos x + \sin x = \dfrac{2\cos^2 x - 1}{\cos x - \sin x}$

6. $\sec x + \tan x = \dfrac{1}{\sec x - \tan x}$

7. $1 - \sin x = \dfrac{\cos x}{\sec x + \tan x}$

8. $\dfrac{\csc x + 1}{\cot x} = \dfrac{\cot x}{\csc x - 1}$

9. $\sec x - \cos x = \dfrac{\tan^2 x}{\sec x}$

10. $2 \sin^2 x = \sin^4 x - \cos^4 x + 1$

In Exercises 11–15, solve each trigonometric equation over one fundamental interval.

11. $\sin x = -\sqrt{3}/2$

12. $\cos 2x = 0.6789$

13. $\sin x + 2 \cos^2 x = 1$

14. $\sin^2 x - \sin x = 0$

15. $\sec^5 x - 4 \sec x = 0$

In Exercises 16–20, eliminate the parameter and sketch.

16. $x = 3 \sin t,\ y = \cos t$

17. $x = \tan t,\ y = 2 \sec t$

18. $x = 1 + \sin t,\ y = 2 \cos t$

19. $x = \tan t,\ y = \tan t,\ 0 \le t \le \pi$

20. $x = 1,\ y = \sin t$

21. Using graphical methods, find approximate solutions to $x^2 = \sin x$ on $(-\pi, \pi)$.

22. Approximate solutions to $x \cos x = 1$ on $[0, 2\pi]$.

23. Solve the inequality $\cos x > \sin x$ on $[0, 2\pi]$.

24. Solve the inequality $\sec x > 2$ on $[-\pi, \pi]$.

25. Plot the Lissajous figure given by $x = \cos t,\ y = 2 \cos t$.

26. Plot the Lissajous figure given by $x = 2 \sin t,\ y = \cos 3t$.

27. In analyzing conditions corresponding to overthrust faulting, a geologist is confronted with the expression

$$\tfrac{1}{2}(\sigma_x - \sigma_z) = [\tfrac{1}{2}(\sigma_x + \sigma_z) + \tau_0 \cot \phi] \sin \phi$$

Show that this simplifies to

$$\sigma_x = a + b\sigma_z$$

where

$$a = 2\tau_0\sqrt{b} \quad \text{and} \quad b = \dfrac{1 + \sin \phi}{1 - \sin \phi}$$

28. The values of time t for which the function $y(t) = 6 \cos t - 8 \sin t$ is equal to 0 are times at which a mass at the end of a spring attains its extreme positions. Find these times.

29. In simplifying a calculus problem, a student substitutes $x = 2 \tan \theta$ in $x\sqrt{x^2 + 4}$. Carry out this substitution and simplify.

30. Using the substitution $x = 3 \sin \theta$, reduce the expression $\sqrt{9 - x^2}$ to one with one trigonometric function. What is $\tan \theta$?

31. A civil engineer designs two ramps for a parking garage, the curves of which approximate those of $y = x^2$ and $y = \sin 3x$, respectively, for small values of x. Determine graphically where these curves meet.

32. The function $y = \begin{cases} \cos x & 0 \le x \le x_1 \\ \frac{3}{4} - \frac{3}{8}x, & x_1 \le x \le 2 \end{cases}$ can be used to approximate a temperature distribution on a wire. Sketch a graph to find the value of x_1 such that $\cos x_1 = \frac{3}{4} - \frac{3}{8}x_1$.

33. Voltages of $3 \sin t$ and $2 \cos t$ are applied to the horizontal and vertical terminals of an oscilloscope, respectively. Make a sketch of the waveform that appears on the screen.

7
Composite Angle Identities

7.1 The Cosine of the Difference or Sum of Two Angles

The cosine of the difference of A and B is denoted by $\cos (A - B)$. Beginning trigonometry students tend to think that $\cos (A - B)$ should equal $\cos A - \cos B$. However, as the next example shows, these two expressions are not equivalent.

Example 1. Show that $\cos (A - B) = \cos A - \cos B$ is not an identity.

Solution. Suppose we let $A = \frac{1}{2}\pi$ and $B = \frac{1}{6}\pi$. Then

$$\cos (A - B) = \cos (\tfrac{1}{2}\pi - \tfrac{1}{6}\pi) = \cos \tfrac{1}{3}\pi = \tfrac{1}{2}$$

But

$$\cos A - \cos B = \cos \tfrac{1}{2}\pi - \cos \tfrac{1}{6}\pi = 0 - \tfrac{1}{2}\sqrt{3}$$

Since the two expressions yield different values for the chosen numbers, we conclude that $\cos (A - B) = \cos A - \cos B$ is not an identity.

The purpose of this section is to show precisely which formulas the trigonometric functions of sums and differences *do* obey. The formulas will be derived from the viewpoint of angles, but they are also valid for real numbers. The most basic of the so-called addition formulas is the formula for $\cos (A - B)$. This formula is derived directly from the definitions of the trigonometric functions.

Let A and B represent angles in standard position superimposed on a circle of radius 1. Figure 7.1(a) is a picture of the general situation.

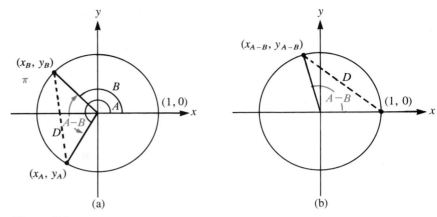

Figure 7.1

The terminal side of A intersects the unit circle at the point $(x_A, y_A) = (\cos A, \sin A)$. Similarly, the terminal side of B intersects the circle at $(x_B, y_B) = (\cos B, \sin B)$. The distance D between these two points is given by

$$D^2 = (x_A - x_B)^2 + (y_A - y_B)^2$$
$$= (\cos A - \cos B)^2 + (\sin A - \sin B)^2$$
$$= (\cos^2 A - 2 \cos A \cos B + \cos^2 B)$$
$$+ (\sin^2 A - 2 \sin A \sin B + \sin^2 B)$$

Since $\cos^2 A + \sin^2 A = 1$ and $\cos^2 B + \sin^2 B = 1$, we have

$$D^2 = 2(1 - \cos A \cos B - \sin A \sin B)$$

Now we rotate the angle $A - B$ until it is in standard position, as shown in Figure 7.1(b). The coordinates of the point of intersection of the terminal side of the angle $A - B$ and the unit circle are

$$(x_{A-B}, y_{A-B}) = [\cos (A - B), \sin (A - B)]$$

D is now the distance connecting $(1, 0)$ to this point. Using the distance formula, we have

$$D^2 = (x_{A-B} - 1)^2 + (y_{A-B})^2$$
$$= [\cos (A - B) - 1]^2 + \sin^2 (A - B)$$
$$= \cos^2 (A - B) - 2 \cos (A - B) + 1 + \sin^2 (A - B)$$

Since $\cos^2 (A - B) + \sin^2 (A - B) = 1$, we can write D^2 as

$$D^2 = 2[1 - \cos (A - B)]$$

Equating this result to the first expression we derived for D^2, we have

$$1 - \cos A \cos B - \sin A \sin B = 1 - \cos (A - B)$$

Simplifying this expression, we have the fundamental formula

$$\cos (A - B) = \cos A \cos B + \sin A \sin B \qquad (7.1)$$

Formula (7.1) was derived under the conditions that $A > B$ and that A and B are between 0 and 2π. However, since

$$\cos (A - B) = \cos (B - A) = \cos (A - B + 2n\pi)$$

it follows that the formula is general. Since it is true for all angles A and B and, consequently, for all real numbers, it is an **identity**.

The principal use of this formula is to derive other important relations. However, it can also be used to obtain the value of the cosine function at a particular angle (or real number) if that angle can be expressed as the difference of two angles for which the exact value of the cosine is known.

Example 2. Without the use of tables, find the exact value of $\cos \frac{1}{12}\pi$.

Solution. Since $\frac{1}{12}\pi = \frac{1}{3}\pi - \frac{1}{4}\pi$,

$$\begin{aligned}
\cos \tfrac{1}{12}\pi &= \cos (\tfrac{1}{3}\pi - \tfrac{1}{4}\pi) \\
&= \cos \tfrac{1}{3}\pi \cos \tfrac{1}{4}\pi + \sin \tfrac{1}{3}\pi \sin \tfrac{1}{4}\pi \\
&= \frac{1}{2}\left(\frac{\sqrt{2}}{2}\right) + \left(\frac{\sqrt{3}}{2}\right)\left(\frac{\sqrt{2}}{2}\right) \\
&= \frac{\sqrt{2} + \sqrt{6}}{4}
\end{aligned}$$

If we replace B by $-B$ in Formula (7.1), we obtain

$$\cos (A - (-B)) = \cos A \cos (-B) + \sin A \sin (-B)$$

Since the cosine function is even, $\cos (-B) = \cos B$; since the sine function is odd, $\sin (-B) = -\sin B$. Hence,

$$\cos (A + B) = \cos A \cos B - \sin A \sin B \qquad (7.2)$$

Example 3
(a) $-\sin 2\alpha \sin 3\alpha + \cos 2\alpha \cos 3\alpha = \cos (2\alpha + 3\alpha) = \cos 5\alpha$
(b) $7 \cos (2x + 4y) = 7 \cos 2x \cos 4y - 7 \sin 2x \sin 4y$

Example 4. Find the exact value of $\cos 75°$.

Solution. Since $75° = 30° + 45°$,

$$\begin{aligned}
\cos 75° &= \cos (30° + 45°) \\
&= \cos 30° \cos 45° - \sin 30° \sin 45° \\
&= \frac{\sqrt{3}}{2} \cdot \frac{\sqrt{2}}{2} - \frac{1}{2} \cdot \frac{\sqrt{2}}{2} \\
&= \frac{\sqrt{6} - \sqrt{2}}{4}
\end{aligned}$$

Example 5. Find the value of $\cos(A - B)$ given that $\sin A = \frac{3}{5}$ in quadrant II and $\tan B = \frac{1}{2}$ in quadrant I.

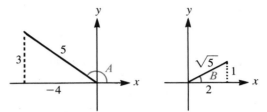

Figure 7.2

Solution. From Figure 7.2, we see that $\cos A = \dfrac{-4}{5}$, $\sin B = \dfrac{1}{\sqrt{5}}$, and $\cos B = \dfrac{2}{\sqrt{5}}$. Hence,

$$\cos(A - B) = \cos A \cos B + \sin A \sin B$$
$$= \left(-\frac{4}{5}\right)\left(\frac{2}{\sqrt{5}}\right) + \left(\frac{3}{5}\right)\left(\frac{1}{\sqrt{5}}\right)$$
$$= -\frac{5}{5\sqrt{5}} = -\frac{\sqrt{5}}{5}$$

In certain applied problems, we encounter trigonometric functions of the form $c_1 \cos Bx + c_2 \sin Bx$. It is easy to see that this function is periodic with a period of $2\pi/B$; however, it is not easy to recognize the amplitude and phase shift of the oscillation of the function in its present form. To obtain these properties, as well as the period, we make use of the identity

$$c_1 \cos Bx + c_2 \sin Bx = A \cos(Bx - C)$$

where $A = \sqrt{c_1^2 + c_2^2}$ and $\tan C = c_2/c_1$.

To verify this identity, we note that by Equation (7.2)

$$A \cos(Bx - C) = A \cos C \cos Bx + A \sin C \sin Bx$$

Therefore,

$$c_1 \cos Bx + c_2 \sin Bx = A \cos C \cos Bx + A \sin C \sin Bx$$

if and only if $c_1 = A \cos C$ and $c_2 = A \sin C$. Squaring c_1 and c_2 and adding, we get

$$c_1^2 + c_2^2 = A^2 \cos^2 C + A^2 \sin^2 C = A^2(\cos^2 C + \sin^2 C)$$

or

$$A = \sqrt{c_1^2 + c_2^2}$$

Also, the ratio of c_2 to c_1 yields

$$\frac{c_2}{c_1} = \frac{A \sin C}{A \cos C} = \tan C$$

It should be noted that C is not just *any* angle for which $\tan C = c_2/c_1$. The angle must be chosen so that its terminal side passes through the point (c_1, c_2).

Example 6. Express $f(x) = \sin x + \cos x$ as a cosine function and sketch.

Solution. Using the preceding formulas, we see that

$$f(x) = A \cos (x - C)$$

where $A = \sqrt{1^2 + 1^2} = \sqrt{2}$ and $\tan C = 1$. Since $c_1 = c_2 = 1$ and $(1, 1)$ is in the first quadrant, we let $C = \frac{1}{4}\pi$. Therefore,

$$f(x) = \sqrt{2} \cos (x - \tfrac{1}{4}\pi)$$

This is a function with amplitude $\sqrt{2}$, period 2π, and phase shift $\frac{1}{4}\pi$. The graph is shown in Figure 7.3, along with the graphs of $\sin x$ and $\cos x$.

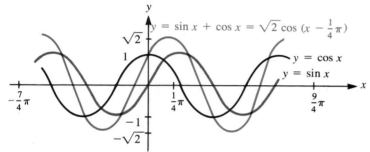

Figure 7.3

Exercises for Section 7.1

1. Show that $\sin (x + y) \neq \sin x + \sin y$. (Let $x = \frac{1}{3}\pi$ and $y = \frac{1}{6}\pi$.)

2. Show that $\tan (x + y) \neq \tan x + \tan y$. (Let $x = \frac{1}{3}\pi$ and $y = \frac{1}{6}\pi$.)

3. Show that $\sin 2x \neq 2 \sin x$. (Let $x = \frac{1}{4}\pi$.)

4. Show that $\cos 2x \neq 2 \cos x$. (Let $x = \frac{1}{4}\pi$.)

5. Show that $\tan 2x \neq 2 \tan x$. (Let $x = \frac{1}{4}\pi$.)

In Exercises 6–10, find the *exact* value of each trigonometric function.

6. $\cos 105°$ 　　7. $\cos \frac{1}{12}\pi$ 　　8. $\cos \frac{11}{12}\pi$ 　　9. $\cos 195°$ 　　10. $\cos 345°$

Use Equation (7.1) to show that the expressions in Exercises 11–13 are true.

11. $\cos (\pi - \theta) = -\cos \theta$ 　　　　　　12. $\cos (\frac{1}{2}\pi - \theta) = \sin \theta$

13. $\cos (\frac{1}{2}\pi + \theta) = -\sin \theta$

14. Using Equation (7.1), give a proof that the cosine is an even function.

In Exercises 15–19, verify each of the given identities.

15. $\cos\left(\frac{1}{3}\pi - x\right) = \dfrac{\cos x + \sqrt{3}\sin x}{2}$

16. $\cos\left(\frac{1}{4}\pi + \theta\right) = \dfrac{\cos\theta - \sin\theta}{\sqrt{2}}$

17. $\cos\left(\frac{3}{2}\pi + x\right) = \sin x$

18. $\cos(x + y)\cos(x - y) = \cos^2 x - \sin^2 y$

19. $\cos(x + y) + \cos(x - y) = 2\cos x \cos y$

In Exercises 20–22, reduce each of the given expressions to a single term.

20. $\cos 2x \cos 3x + \sin 2x \sin 3x$

21. $\cos 7x \cos x - \sin 7x \sin x$

22. $\cos\frac{1}{6}x \cos\frac{5}{6}x - \sin\frac{1}{6}x \sin\frac{5}{6}x$

In Exercises 23–26, find the value of $\cos(A + B)$ for each of the given conditions.

23. $\cos A = \frac{1}{3}$, $\sin B = -\frac{1}{2}$, A in quadrant I, B in quadrant IV

24. $\cos A = \frac{3}{5}$, $\tan B = \frac{12}{5}$, both A and B acute

25. $\tan A = \frac{24}{7}$, $\sec B = \frac{5}{3}$, A in quadrant III, B in quadrant I

26. $\sin A = \frac{1}{4}$, $\cos B = \frac{1}{2}$, both A and B in quadrant I

In Exercises 27–28, let A and B both be positive acute angles.

27. Find $\cos A$ if $\cos(A + B) = \frac{5}{6}$ and $\sin B = \frac{1}{3}$.

28. Find $\cos A$ if $\cos(A - B) = \frac{3}{4}$ and $\cos B = \frac{2}{3}$.

In Exercises 29–32, graph each of the given functions and give the amplitude and phase shift.

29. $f(x) = \cos x - \sin x$

30. $f(x) = 2\cos x + 2\sin x$

31. $f(x) = \cos 2x + \sqrt{3}\sin 2x$

32. $f(x) = -\cos 2x + \sqrt{3}\sin 2x$

In Exercises 33–38, show that $\cos(A + B) \neq \cos A + \cos B$. Then verify that $\cos(A + B) = \cos A \cos B - \sin A \sin B$. Use a calculator when necessary.

33. $A = 0$, $B = 0$

34. $A = 30°$, $B = 30°$

35. $A = 0.987$, $B = 0.111$

36. $A = -0.912$, $B = 0.912$

37. $A = 23.1$, $B = 14.14$

38. $A = 1.57$, $B = -1.57$

39. The output voltage of an a.c. generator is approximated by

$$v = 156\cos\left(2\pi f t - \tfrac{1}{3}\pi\right)$$

where $f = 60$ Hz. Express this output as the sum of a sine and a cosine.

40. A transverse wave on a stretched string is represented by the sum of the two waves

$$y = A \cos \frac{2\pi}{\lambda} (x \pm vt)$$

where v and λ are constants. Write this expression as the sum of a sine and a cosine wave where $t = 1$.

41. A student knows that $\tan \phi = \frac{2}{3}$ and $\tan \theta = \frac{3}{2}$, both in quadrant I. Find $\cos (\theta - \phi)$.

42. If a mass is attached to a spring, its motion may be approximated by

$$y = A \cos \omega t + \frac{F}{\omega Z} \sin \omega t$$

Write this expression as one cosine function.

43. A vibrating string has a vertical deflection that can be approximated by

$$y = 3 (\cos 2\pi x \cos ct + \sin 2\pi x \sin ct)$$

Write this expression as one cosine function.

44. In the theory of phase modulation, one is confronted with the sum

$$\cos k\theta \cos (x \sin \theta) + \sin k\theta \sin (x \sin \theta)$$

Show that this sum is equal to $\cos (k\theta - x \sin \theta)$.

7.2 Other Addition Formulas

From the formula for the cosine of the difference, it is easy to establish that

$$\cos (\tfrac{1}{2}\pi - \theta) = \sin \theta \quad \text{and} \quad \sin(\tfrac{1}{2}\pi - \theta) = \cos \theta \qquad (7.3)$$

Note that this is a general statement of the identity relating cofunctions of complementary angles.

Example 1
(a) $\cos (-10°) = \cos (90° - 100°) = \sin 100°$
(b) $\sin (\tfrac{5}{6}\pi) = \sin [\tfrac{1}{2}\pi - (-\tfrac{1}{3}\pi)] = \cos (-\tfrac{1}{3}\pi) = \cos \tfrac{1}{3}\pi$

The fundamental Pythagorean relationship is $\sin^2 x + \cos^2 x = 1$. A similar formula involving *two* numbers x and y is true if x and y have a sum of $\tfrac{1}{2}\pi$.

Example 2. If x and y are complementary numbers (that is, their sum is $\frac{1}{2}\pi$), show that $\sin^2 x + \sin^2 y = 1$.

Solution. Since $x + y = \frac{1}{2}\pi$, $y = \frac{1}{2}\pi - x$. Therefore, $\sin y = \sin\left(\frac{1}{2}\pi - x\right) = \cos x$. Hence, from the Pythagorean relation,

$$\sin^2 x + \cos^2 x = 1$$

Assuming $x + y = \frac{1}{2}\pi$, we have, using $\sin y = \cos x$,

$$\sin^2 x + \sin^2 y = 1$$

If we let $\theta = A + B$ in Equation (7.3), we can write

$$\sin(A + B) = \cos\left[\frac{1}{2}\pi - (A + B)\right]$$
$$= \cos\left[\left(\frac{1}{2}\pi - A\right) - B\right]$$

Using Equation (7.1) for the cosine of a difference, we obtain

$$\sin(A + B) = \cos\left(\frac{1}{2}\pi - A\right)\cos B + \sin\left(\frac{1}{2}\pi - A\right)\sin B$$

Then applying Equation (7.3) again yields

$$\sin(A + B) = \sin A \cos B + \cos A \sin B \qquad (7.4)$$

Similarly, if $\theta = A - B$,

$$\sin(A - B) = \sin A \cos(-B) + \cos A \sin(-B)$$

Since the cosine function is even and the sine function is odd, this becomes

$$\sin(A - B) = \sin A \cos B - \cos A \sin B \qquad (7.5)$$

Example 3. Find the exact value of $\sin\left(\frac{1}{12}\pi\right)$.

Solution. Since $\frac{1}{12}\pi = \frac{1}{3}\pi - \frac{1}{4}\pi$,

$$\sin\left(\frac{1}{12}\pi\right) = \sin\left(\frac{1}{3}\pi\right)\cos\left(\frac{1}{4}\pi\right) - \cos\left(\frac{1}{3}\pi\right)\sin\left(\frac{1}{4}\pi\right)$$

$$= \frac{\sqrt{3}}{2}\frac{\sqrt{2}}{2} - \frac{1}{2}\frac{\sqrt{2}}{2}$$

$$= \frac{1}{4}\left(\sqrt{6} - \sqrt{2}\right)$$

Example 4. Show that $\sin x + \cos x = \sqrt{2}\sin\left(x + \frac{1}{4}\pi\right)$.

Solution. If $\sin x + \cos x = A \sin(x + C) = A \sin x \cos C + A \cos x \sin C$, then $A \cos C = 1$ and $A \sin C = 1$. Squaring these two equations and adding the results, we get

$$A^2 \cos^2 C + A^2 \sin^2 C = 1^2 + 1^2 = 2$$

or $A = \sqrt{2}$.

Also,

$$\frac{A \sin C}{A \cos C} = \frac{1}{1}$$

so $\tan C = 1$.

We choose C to be a first-quadrant angle, since its terminal side must pass through $(1, 1)$. Thus, $C = \frac{1}{4}\pi$, and hence

$$\sin x + \cos x = \sqrt{2} \sin (x + \tfrac{1}{4}\pi)$$

The sketch of this function is precisely that given in Figure 7.3. Can you verify this?

Sum and difference formulas for the tangent follow directly from those for the sine and cosine.

$$\tan (A + B) = \frac{\sin (A + B)}{\cos (A + B)}$$

$$= \frac{\sin A \cos B + \cos A \sin B}{\cos A \cos B - \sin A \sin B}$$

Now we divide both the numerator and denominator by $\cos A \cos B$:

$$\tan (A + B) = \frac{\dfrac{\sin A \cos B}{\cos A \cos B} + \dfrac{\cos A \sin B}{\cos A \cos B}}{\dfrac{\cos A \cos B}{\cos A \cos B} - \dfrac{\sin A \sin B}{\cos A \cos B}}$$

Simplifying, we get

$$\tan (A + B) = \frac{\tan A + \tan B}{1 - \tan A \tan B} \tag{7.6}$$

Similarly,

$$\tan (A - B) = \frac{\tan A - \tan B}{1 + \tan A \tan B} \tag{7.7}$$

Example 5. Verify that $\tan (\theta + \pi) = \tan \theta$.

Solution. Using Equation (7.6) and the fact that $\tan \pi = 0$, we have

$$\tan (\theta + \pi) = \frac{\tan \theta + \tan \pi}{1 - \tan \theta \tan \pi} = \tan \theta$$

Example 6. Reduce $\dfrac{\tan (x + y) - \tan (x - y)}{1 + \tan (x + y) \tan (x - y)}$ to a single term.

Solution. We recognize this expression as the right-hand side of Equation (7.7), with $A = x + y$ and $B = x - y$. Thus,

$$\frac{\tan (x + y) - \tan (x - y)}{1 + \tan (x + y) \tan (x - y)} = \tan [(x + y) - (x - y)] = \tan 2y$$

The sum and difference formulas that we have derived can be summarized as follows:

$$\sin (A \pm B) = \sin A \cos B \pm \cos A \sin B$$

$$\cos (A \pm B) = \cos A \cos B \mp \sin A \sin B$$

$$\tan (A \pm B) = \frac{\tan A \pm \tan B}{1 \mp \tan A \tan B}$$

By convention, when the symbols $\pm$ and $\mp$ are used in the same formula, the top signs go together and the bottom signs go together.

Exercises for Section 7.2

In Exercises 1–5, find the exact value of each expression.

1. $\sin \left(\frac{5}{12}\pi\right)$ 2. $\tan 15°$ 3. $\sin \left(\frac{7}{12}\pi\right)$

4. $\sin (345°)$ 5. $\cot \left(\frac{5}{12}\pi\right)$

In Exercises 6–13, verify each identity.

6. $\sin \left(A + \frac{1}{4}\pi\right) = \dfrac{\sqrt{2}}{2} (\sin A + \cos A)$ 7. $\tan \left(A + \frac{1}{2}\pi\right) = -\cot A$

8. $\tan \left(A + \frac{1}{4}\pi\right) = \dfrac{1 + \tan A}{1 - \tan A}$ 9. $\cot (A + B) = \dfrac{\cot A \cot B - 1}{\cot A + \cot B}$

10. $\dfrac{\sin (A + B)}{\sin (A - B)} = \dfrac{\tan A + \tan B}{\tan A - \tan B}$

11. $\sin (A + B) \sin (A - B) = \sin^2 A - \sin^2 B$

12. $\sin (A + B) + \sin (A - B) = 2 \sin A \cos B$

13. $\tan A + \tan B = \dfrac{\sin (A + B)}{\cos A \cos B}$

In Exercises 14–17, reduce each expression to a single term.

14. $\sin 2x \cos 3x + \sin 3x \cos 2x$ 15. $\dfrac{\tan 3x - \tan 2x}{1 + \tan 3x \tan 2x}$

16. $\sin \frac{1}{3}x \cos \frac{2}{3}x + \sin \frac{2}{3}x \cos \frac{1}{3}x$

17. $\dfrac{\tan (x + y) + \tan z}{1 - \tan (x + y) \tan z}$

Find the values of $\sin (A + B)$ and $\tan (A + B)$ if

18. $\sin A = \frac{3}{5}$, $\cos B = \frac{4}{5}$, both A and B in quadrant I

19. $\tan A = -\frac{7}{24}$, $\tan B = \frac{5}{12}$, A in quadrant II, B in quadrant III

20. $\cos A = \frac{1}{3}$, $\cos B = -\frac{1}{3}$, A in quadrant IV, B in quadrant III

In Exercises 21–24, express each function as a sine function with a phase shift and sketch.

21. $\sin 2x + \cos 2x$

22. $\cos x$

23. $\sqrt{3} \sin \pi x + \cos \pi x$

24. $7 \sin 2x - 24 \cos 2x$

In Exercises 25–30, use a calculator to show that in each case the left-hand side is equal to the right-hand side.

25. $\sin (27° + 96°) = \sin 27° \cos 96° + \cos 27° \sin 96°$

26. $\cos (12.6° + 8.7°) = \cos 12.6° \cos 8.7° - \sin 12.6° \sin 8.7°$

27. $\cos (1.1 - 0.3) = \cos 1.1 \cos 0.3 + \sin 1.1 \sin 0.3$

28. $\sin (2.8 - 1.6) = \sin 2.8 \cos 1.6 - \cos 2.8 \sin 1.6$

29. $\tan (0.4 + 0.3) = \dfrac{\tan 0.4 + \tan 0.3}{1 - \tan 0.4 \tan 0.3}$

30. $\tan (2.9 - 1.2) = \dfrac{\tan 2.9 - \tan 1.2}{1 + \tan 2.9 \tan 1.2}$

31. In Figure 7.4, the component of the force in the direction 30° above the horizontal is given by $100 \cos (45° - 30°)$. Find the exact value of this expression without using a calculator.

32. The angle α in Figure 7.5 subtends the picture hanging on the wall. This angle can be changed by raising or lowering the picture or moving the origin of the angle closer to the wall. Find $\tan \alpha$ given that $\tan \theta = 2$.

33. The output of a 60-volt signal generator is

$$v(t) = 60 \sin (2\pi ft + \tfrac{1}{3}\pi)$$

Express $v(t)$ as a sum of sine and cosine terms.

34. In the analysis of luminous intensity from a point source, one meets an expression similar to $2.1 \cos \alpha t - 3.2 \sin \alpha t$. Write this expression as one sine function.

35. In the discussion of the angle β between two curves, one meets the expression $\tan (\theta_2 - \theta_1)$, where θ_2 and θ_1 are the angles with respect to the horizontal of each of the curves. Express $\tan \beta = \tan (\theta_2 - \theta_1)$ in terms of $m_2 = \tan \theta_2$ and $m_1 = \tan \theta_1$.

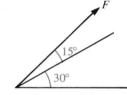

Figure 7.4

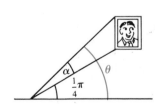

Figure 7.5

7.3 Double- and Half-Angle Formulas

In the two previous sections, we have been primarily interested in expanding trigonometric functions whose arguments are $A \pm B$. Now we will derive formulas for functions of $2A$ and $\frac{1}{2}A$. If A represents an angle, the formulas are called the **double-** and **half-angle formulas**, respectively.

The double-angle formulas are easily proved by choosing $B = A$ in the formulas for the sum of two angles. Thus,

$$\sin 2A = \sin (A + A)$$
$$= \sin A \cos A + \sin A \cos A$$

or

$$\sin 2A = 2 \sin A \cos A \qquad (7.8)$$

and

$$\cos 2A = \cos (A + A)$$
$$= \cos A \cos A - \sin A \sin A$$

or

$$\cos 2A = \cos^2 A - \sin^2 A \qquad (7.9)$$

Using the Pythagorean relation $\sin^2 A + \cos^2 A = 1$, we can express this last formula in the equivalent forms

$$\cos 2A = 2 \cos^2 A - 1 \qquad (7.9a)$$

and

$$\cos 2A = 1 - 2 \sin^2 A \qquad (7.9b)$$

Similarly,

$$\tan 2A = \frac{\tan A + \tan A}{1 - \tan A \tan A}$$

$$\tan 2A = \frac{2 \tan A}{1 - \tan^2 A} \qquad (7.10)$$

Example 1. Find $\sin 2A$ if $\sin A = \frac{1}{3}$ and A is in quadrant II.

Solution. Since $\sin A = \frac{1}{3}$, we have from Figure 7.6 that $\cos A = -\frac{1}{3}\sqrt{8}$, and hence

$$\sin 2A = 2 \sin A \cos A$$

$$= 2\left(\frac{1}{3}\right)\left(\frac{-\sqrt{8}}{3}\right) = \frac{-2\sqrt{8}}{9}$$

Note that $\sin 2A \neq 2 \sin A$.

Example 2. Sketch the graph of $y = \sin x \cos x$. Where does the maximum value of this function occur?

Solution. By multiplying and dividing the right-hand side of this function by 2, we obtain

$$y = \frac{2 \sin x \cos x}{2} = \frac{1}{2} \sin 2x$$

Thus, the graph of this function is a sine wave with amplitude $\frac{1}{2}$ and period π. The maximum value of $\frac{1}{2}$ occurs at $x = \frac{1}{4}\pi + n\pi$. (See Figure 7.7.)

The half-angle formulas are directly related to the formulas for the cosine of a double angle. Since

$$\cos 2A = 2 \cos^2 A - 1$$

we have, upon solving for $\cos A$,

$$\cos A = \pm \sqrt{\frac{1 + \cos 2A}{2}}$$

The positive sign is used when angle A is in either quadrant I or quadrant IV since $\cos A$ is positive in those quadrants. The negative sign is used when angle A is in either quadrant II or quadrant III.

Similarly, using the formula $\cos 2A = 1 - 2 \sin^2 A$ and solving for $\sin A$, we have

$$\sin A = \pm \sqrt{\frac{1 - \cos 2A}{2}}$$

Letting $A = \frac{1}{2}x$ in both of these formulas, we have

$$\cos \tfrac{1}{2}x = \pm \sqrt{\frac{1 + \cos x}{2}} \qquad (7.11)$$

and

$$\sin \tfrac{1}{2}x = \pm \sqrt{\frac{1 - \cos x}{2}} \qquad (7.12)$$

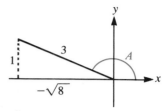

Figure 7.6

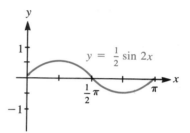

$y = \frac{1}{2} \sin 2x$

Figure 7.7

In Equations (7.11) and (7.12), whether the plus or the minus sign is used depends on the quadrant in which the angle $\frac{1}{2}x$ lies.

To get the formula for $\tan \frac{1}{2}x$, we write

$$\left|\tan \tfrac{1}{2}x\right| = \frac{\left|\sin \tfrac{1}{2}x\right|}{\left|\cos \tfrac{1}{2}x\right|}$$

$$= \frac{\sqrt{(1 - \cos x)/2}}{\sqrt{(1 + \cos x)/2}}$$

$$= \sqrt{\frac{1 - \cos x}{1 + \cos x}}$$

Multiplying the numerator and denominator of this expression by $(1 + \cos x)$, we get

$$\left|\tan \tfrac{1}{2}x\right| = \sqrt{\frac{1 - \cos^2 x}{(1 + \cos x)^2}}$$

$$= \sqrt{\frac{\sin^2 x}{(1 + \cos x)^2}}$$

$$= \frac{\left|\sin x\right|}{\left|1 + \cos x\right|}$$

The expression $1 + \cos x$ is nonnegative because $\cos x$ is never less than -1. To show that $\tan \frac{1}{2}x$ and $\sin x$ have the same sign for all x for which $\tan \frac{1}{2}x$ is defined, we note that

$$\tan \tfrac{1}{2}x \sin x = \tan \tfrac{1}{2}x(2 \sin \tfrac{1}{2}x \cos \tfrac{1}{2}x)$$

$$= 2 \sin^2 \tfrac{1}{2}x$$

Since $\sin^2 \frac{1}{2}x$ is always positive, $\tan \frac{1}{2}x$ and $\sin x$ must have the same sign. Thus we can drop the absolute value signs and write

$$\tan \tfrac{1}{2}x = \frac{\sin x}{1 + \cos x} \tag{7.13}$$

Example 3. If $\tan \theta = -\frac{4}{3}$ and $-\frac{1}{2}\pi < \theta < 0$, find $\sin \frac{1}{2}\theta$ and $\cos \frac{1}{2}\theta$.

Solution. Figure 7.8 shows the angle θ in standard position. From this figure we can see that $\cos \theta = \frac{3}{5}$, and hence

$$\sin \tfrac{1}{2}\theta = -\sqrt{\frac{1 - (\frac{3}{5})}{2}} = -\frac{\sqrt{5}}{5}$$

(The minus sign is chosen because $\sin \frac{1}{2}\theta$ is negative for θ in the fourth quadrant.) Similarly,

$$\cos \tfrac{1}{2}\theta = \sqrt{\frac{1 + (\frac{3}{5})}{2}} = \sqrt{\frac{4}{5}} = \frac{2\sqrt{5}}{5}$$

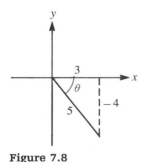

Figure 7.8

Example 4. Find an expression for $\sin 3\theta$ in terms of $\sin \theta$.

Solution

$\sin 3\theta = \sin (2\theta + \theta)$

$\qquad = \sin 2\theta \cos \theta + \sin \theta \cos 2\theta$ Using $\sin (A + B) =$ $\sin A \cos B + \sin B \cos A$

$\qquad = 2 \sin \theta \cos^2 \theta + \sin \theta(1 - 2 \sin^2 \theta)$ Using $\sin 2A = 2 \sin A \cos A$ and $\cos 2A = 1 - 2 \sin^2 A$

$\qquad = 2 \sin \theta(1 - \sin^2 \theta) + \sin \theta(1 - 2 \sin^2 \theta)$ Using $\cos^2 A = 1 - \sin^2 A$

$\qquad = 3 \sin \theta - 4 \sin^3 \theta$ Expanding and collecting like terms

This proves that $\sin 3\theta = 3 \sin \theta - 4 \sin^3 \theta$.

Example 5. Solve the equation $\cos 2x = \cos x$ on the interval $0 \le x \le 2\pi$.

Solution. We use the identity for $\cos 2x$ to transform the equation into one involving $\cos x$:

$$2 \cos^2 x - 1 = \cos x$$

Subtracting $\cos x$ from both sides gives

$$2 \cos^2 x - \cos x - 1 = 0$$

Factoring, we get

$$(2 \cos x + 1)(\cos x - 1) = 0$$

We now have two separate equations

$$2 \cos x + 1 = 0 \qquad \text{and} \qquad \cos x - 1 = 0$$

The solution to the first equation on $0 \le x \le 2\pi$ is $x = \frac{2}{3}\pi$ and $\frac{4}{3}\pi$; the solution to the second is $x = 0$ and 2π. Hence, the complete solution is 0, $\frac{2}{3}\pi$, $\frac{4}{3}\pi$, and 2π.

Example 6. If $\cot 2\theta = -\frac{7}{24}$ and θ is in quadrant I, find $\sin \theta$ and $\cos \theta$.

Solution. A sketch such as the one shown in Figure 7.9 is helpful for this kind of problem. From this figure we can see immediately that $\cos 2\theta = -\frac{7}{25}$. Hence,

$$\sin \theta = \left(\frac{1 - \cos 2\theta}{2}\right)^{1/2}$$

$$= \left(\frac{1 + \frac{7}{25}}{2}\right)^{1/2}$$

$$= \left(\frac{25 + 7}{50}\right)^{1/2}$$

$$= \frac{4}{5}$$

Similarly, $\cos \theta = \frac{3}{5}$.

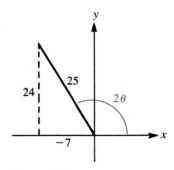

Figure 7.9

A summary of the double- and half-angle formulas follows.

$$\sin 2A = 2 \sin A \cos A$$

$$\cos 2A = \cos^2 A - \sin^2 A$$

$$= 2 \cos^2 A - 1$$

$$= 1 - 2 \sin^2 A$$

$$\tan 2A = \frac{2 \tan A}{1 - \tan^2 A}$$

$$\sin \tfrac{1}{2}A = \pm \sqrt{\frac{1 - \cos A}{2}}$$

$$\cos \tfrac{1}{2}A = \pm \sqrt{\frac{1 + \cos A}{2}}$$

$$\tan \tfrac{1}{2}A = \frac{\sin A}{1 + \cos A}$$

Exercises for Section 7.3

In Exercises 1–6, express each of the expressions as a single trigonometric function.

1. $2 \sin 3x \cos 3x$

2. $6 \sin \tfrac{1}{2}x \cos \tfrac{1}{2}x$

3. $\sin^2 4x - \cos^2 4x$

4. $4 \sin^2 x \cos^2 x$

5. $\dfrac{2 \tan \frac{1}{6}x}{1 - \tan^2 \frac{1}{6}x}$

6. $\dfrac{\sin 6x}{1 + \cos 6x}$

7. Sketch the graph of the function $f(x) = \sin 2x \cos 2x$. What is the maximum value of the function? Where does it occur?

8. Sketch the graph of the two functions $f(x) = \cos^2 x - \sin^2 x$ and $g(x) = \cos^2 x + \sin^2 x$. Tell what the maximum values are for each.

9. Sketch the graph of the function $f(x) = \sec x \csc x$. Where is this function undefined?

10. Sketch the graph of the function

$$f(x) = \frac{2 \tan x}{1 + \tan^2 x}$$

What is the maximum value of this function? What is the period?

11. Sketch the graph of the function $f(x) = \tan x + \cot x$. Where is this function undefined? What is the period?

12. Sketch the graph of the function $f(x) = \cot x - \tan x$. What are the zeros of this function? Is it bounded or unbounded? What is the period?

Find sin 2A, cos 2A, and tan 2A, given that

13. $\sin A = \frac{3}{5}$, A in quadrant I   **14.** $\cos A = -\frac{12}{13}$, A in quadrant III

15. $\tan A = \frac{7}{24}$, A in quadrant III   **16.** $\sec A = -\frac{13}{5}$, A in quadrant II

In Exercises 17–21, use the double-angle identities to determine the exact value of each expression.

17. $\sin \frac{1}{8}\pi$ **18.** $\cos \frac{5}{8}\pi$

19. $\tan 157.5°$ **20.** $\sin 67.5°$

21. $\cos \frac{1}{12}\pi$

In Exercises 22–42, verify each of the identities.

22. $(\sin x + \cos x)^2 = 1 + \sin 2x$

23. $\cos 3x = 4 \cos^3 x - 3 \cos x$

24. $\sin 4x = 4 \cos x \sin x(1 - 2 \sin^2 x)$

25. $\tan x + \cot x = 2 \csc 2x$

26. $\cos 4x = 8 \cos^4 x - 8 \cos^2 x + 1$

27. $\cot^2 \frac{1}{2} x = \dfrac{\sec x + 1}{\sec x - 1}$

28. $\cos^4 x = \frac{3}{8} + \frac{1}{2} \cos 2x + \frac{1}{8} \cos 4x$

29. $\sec^2 x = \dfrac{4 \sin^2 x}{\sin^2 2x}$

30. $\tan 2x = \dfrac{2}{\cot x - \tan x}$ **31.** $2 \cot 2x \cos x = \csc x - 2 \sin x$

32. $\dfrac{1 - \tan x}{1 + \tan x} = \dfrac{1 - \sin 2x}{\cos 2x}$ **33.** $\tan^2 x + \cos 2x = 1 - \cos 2x \tan^2 x$

34. $\dfrac{2 \tan x}{\tan 2x} = 1 - \tan^2 x$ **35.** $\dfrac{\cos 3x}{\sec x} - \dfrac{\sin x}{\csc 3x} = \cos^2 2x - \sin^2 2x$

36. $\dfrac{\csc x - \sec x}{\csc x + \sec x} = \dfrac{\cos 2x}{1 + \sin 2x}$ **37.** $1 + \cot x \cot 3x = \dfrac{2 \cos 2x}{\cos 2x - \cos 4x}$

38. $-4 \sin^3 x + 3 \sin x = \sin 3x$ **39.** $\sin 6x \tan 3x = 2 \sin^2 3x$

40. $\dfrac{\cos^3 x + \sin^3 x}{2 - \sin 2x} = \dfrac{1}{2} (\sin x + \cos x)$

41. $\cos 4x \sec^2 2x = 1 - \tan^2 2x$

42. $16 \cos^5 x = 5 \cos 3x + 10 \cos x + \cos 5x$

In Exercises 43–45, find the indicated functional value. (Assume 2θ is in quadrant I.)

43. $\tan \theta$ if $\sin 2\theta = \frac{5}{13}$ **44.** $\sin \theta$ if $\sin 2\theta = \frac{3}{5}$

45. $\cos \theta$ if $\cos 2\theta = \frac{24}{25}$

In Exercises 46–54, solve each equation on the interval $0 \leq x \leq 2\pi$.

46. $\sin 2x = \sin x$

47. $\sin x = \cos x$

48. $\sin 2x \sin x + \cos x = 0$

49. $\tan 2x = \tan x$

50. $\cos x - \sin 2x = 0$

51. $\sin 2x + \cos 2x = 0$

52. $\sin 2x - 2 \cos x + \sin x - 1 = 0$

53. $2(\sin^2 2x - \cos^2 2x) = 1$

54. $\sin 2x \cos x - \frac{1}{2} \sin 3x = \frac{1}{2} \sin x$

In Exercises 55–59, show that each expression reduces to 1.

55. $(\sin x + \cos x)^2 - \sin 2x$

56. $\dfrac{\sin 2x \sin x}{2 \cos x} + \cos^2 x$

57. $\sec^4 x - \tan^4 x - 2 \tan^2 x$

58. $\left[\dfrac{\sin 2x}{\sin x} - \dfrac{\cos 2x}{\cos x} \right] \sec x - \tan^2 x$

59. $\cos 2x + \sin 2x \tan x$

In Exercises 60–64, use a calculator to show that the left-hand side of each equation is equal to the right-hand side.

60. $\sin 2(1.05) = 2 \sin 1.05 \cos 1.05$

61. $\cos^2 0.47 = \frac{1}{2}[1 + \cos 2(0.47)]$

62. $\sin^2 1.3 = \frac{1}{2}[1 - \cos 2(1.3)]$

63. $\cos 2(0.22) = \cos^2 0.22 - \sin^2 0.22$

64. $\tan \dfrac{1}{2}(3.0) = \dfrac{\sin 3.0}{1 + \cos 3.0}$

65. The angle of rotation, θ, of coordinate axes is often given in terms of $\tan 2\theta$. Find $\cos \theta$ and $\sin \theta$ (θ acute) when $\tan 2\theta = 2$.

66. Find the cosine and sine of the acute rotation angle θ in Exercise 65 when $\tan 2\theta = -2$.

67. The horizontal displacement of a projectile fired at an angle θ with initial velocity v_0 is given by

$$R = 32v_0^2 \sin \theta \cos \theta$$

Express R as a function of 2θ. (See Figure 7.10.)

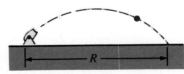

Figure 7.10

68. In the analysis of the electron microscope, one is confronted with the expression

$$\sin \frac{\alpha}{2} = \frac{\lambda}{a}$$

Find $\sin \alpha$.

Sum Formulas

Sometimes you will need to write a sum of sines and cosines as a product. A scheme based on the addition formulas follows.

Example 1. Factor $\sin 7x + \sin 3x$.

Solution. We write

$$\sin 7x + \sin 3x = \sin (5x + 2x) + \sin (5x - 2x)$$
$$= \sin 5x \cos 2x + \sin 2x \cos 5x + \sin 5x \cos 2x - \sin 2x \cos 5x$$
$$= 2 \sin 5x \cos 2x$$

The method illustrated in Example 1 is called the **average angle method**. The procedure for the general case is as follows:

$$\sin A + \sin B = \sin \left(\frac{A + B}{2} + \frac{A - B}{2} \right) + \sin \left(\frac{A + B}{2} - \frac{A - B}{2} \right)$$
$$= \sin \frac{A + B}{2} \cos \frac{A - B}{2} + \cos \frac{A + B}{2} \sin \frac{A - B}{2}$$
$$+ \sin \frac{A + B}{2} \cos \frac{A - B}{2} - \cos \frac{A + B}{2} \sin \frac{A - B}{2}$$

Combining terms, we conclude that

$$\sin A + \sin B = 2 \sin \frac{A + B}{2} \cos \frac{A - B}{2} \qquad (7.14)$$

The following formulas are derived analogously. If you understand the preceding technique, you will not have to memorize them.

$$\sin A - \sin B = 2 \cos \frac{A + B}{2} \sin \frac{A - B}{2} \qquad (7.15)$$

$$\cos A + \cos B = 2 \cos \frac{A + B}{2} \cos \frac{A - B}{2} \qquad (7.16)$$

$$\cos A - \cos B = -2 \sin \frac{A + B}{2} \sin \frac{A - B}{2} \qquad (7.17)$$

Example 2. The difference quotient of a function f is denoted Δf and is defined by

$$\Delta f = \frac{f(x + h) - f(x)}{h}$$

Find the difference quotient for the sine function and express it as the product of a sine and a cosine.

Solution. First compute $\sin (x + h) - \sin x$ using Equation (7.15):

$$\sin (x + h) - \sin x = 2 \cos \frac{(x + h) + x}{2} \sin \frac{(x + h) - x}{2}$$

$$= 2 \sin \frac{h}{2} \cos \frac{2x + h}{2}$$

Thus, the difference quotient for the sine function is

$$\frac{\sin (h/2) \cos [x + (h/2)]}{h/2}$$

Example 3. Prove the identity

$$\frac{\sin 6x - \sin 4x}{\cos 6x + \cos 4x} = \tan x$$

Solution

$$\frac{\sin 6x - \sin 4x}{\cos 6x + \cos 4x} = \frac{2 \cos \frac{1}{2}(6x + 4x) \sin \frac{1}{2}(6x - 4x)}{2 \cos \frac{1}{2}(6x + 4x) \cos \frac{1}{2}(6x - 4x)}$$

$$= \frac{\cos 5x \sin x}{\cos 5x \cos x}$$

$$= \tan x$$

Example 4. Solve the equation $\sin 3x + \sin x = 0$ on the interval $0 \le x \le \pi$.

Solution. To put the given equation in factored form, use Equation (7.14) with $A = 3x$ and $B = x$:

$$\sin 3x + \sin x = 2 \sin 2x \cos x$$

The given equation is then

$$\sin 3x + \sin x = 2 \sin 2x \cos x = 0$$

Now, since $\sin 2A = 2 \sin A \cos A$, we can write

$$2(2 \sin x \cos x) \cos x = 0$$

$$4 \sin x \cos^2 x = 0$$

Equating each factor to zero, we have

$$\sin x = 0 \qquad \Big| \qquad \cos^2 x = 0$$

$$x = 0, \pi \qquad \Big| \qquad x = \tfrac{1}{2}\pi$$

The desired solution set is $x = 0, \tfrac{1}{2}\pi, \pi$.

Example 5. In the analysis of some types of harmonic motion, the governing equation is $y(t) = A(\cos \omega t - \cos \omega_0 t)$, where the difference between ω and ω_0 is considered to be very small. Make a sketch of the graph of this function.

Solution. Using Equation (7.17), we write

$$y(t) = 2A \sin \frac{\omega_0 - \omega}{2} t \sin \frac{\omega_0 + \omega}{2} t$$

If ω is close to ω_0, the resultant oscillation can be interpreted as having a frequency close to $\omega_0/2\pi$ (and close to $\omega/2\pi$) with variable amplitude given by

$$2A \sin \frac{\omega_0 - \omega}{2} t$$

which fluctuates with frequency $(\omega - \omega_0)/\pi$. Oscillations of this type are called **beats**. (See Figure 7.11.)

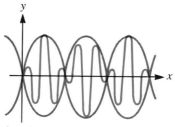

Figure 7.11

Product Formulas

Sometimes you may need to express products of trigonometric functions as sums of trigonometric functions. If we add the formulas for $\sin (A + B)$ and for $\sin (A - B)$, we obtain

$$\sin A \cos B = \tfrac{1}{2}[\sin (A + B) + \sin (A - B)] \qquad (7.18)$$

Subtracting $\sin (A - B)$ from $\sin (A + B)$ and simplifying, we have

$$\cos A \sin B = \tfrac{1}{2}[\sin (A + B) - \sin (A - B)] \qquad (7.19)$$

In like manner, by first adding and then subtracting the formulas for $\cos (A + B)$ and $\cos (A - B)$, we obtain

$$\cos A \cos B = \tfrac{1}{2}[\cos (A + B) + \cos (A - B)] \qquad (7.20)$$

and

$$\sin A \sin B = \tfrac{1}{2}[\cos (A - B) - \cos (A + B)] \qquad (7.21)$$

Example 6. Express $\sin mx \cos nx$ as a sum of functions.

Solution. Using Equation (7.18) with $A = mx$ and $B = nx$, we have

$$\sin mx \cos nx = \tfrac{1}{2}[\sin (mx + nx) + \sin (mx - nx)]$$
$$= \tfrac{1}{2}[\sin (m + n)x + \sin (m - n)x]$$

Example 7. Prove that

$$\frac{\sin(x+y) + \sin(x-y)}{\sin(x+y) - \sin(x-y)} = \tan x \cot y$$

Solution

$$\frac{\sin(x+y) + \sin(x-y)}{\sin(x+y) - \sin(x-y)} = \frac{\frac{1}{2}[\sin(x+y) + \sin(x-y)]}{\frac{1}{2}[\sin(x+y) - \sin(x-y)]}$$

$$= \frac{\sin x \cos y}{\cos x \sin y}$$

$$= \tan x \cot y$$

Exercises for Section 7.4

In Exercises 1–7, express each sum or difference as a product.

1. $\sin 3\theta + \sin \theta$

2. $\cos 3\alpha - \cos 8\alpha$

3. $\sin 8x + \sin 2x$

4. $\sin \frac{1}{2}x - \sin \frac{1}{4}x$

5. $\cos 50° - \cos 30°$

6. $\sin \frac{3}{4}\pi - \sin \frac{1}{4}\pi$

7. $\sin \frac{3}{4} - \sin \frac{1}{4}$

In Exercises 8–12, express each product as a sum or difference.

8. $\sin 3x \cos x$

9. $\cos x \sin \frac{1}{2}x$

10. $\cos \frac{1}{3}\pi \sin \frac{2}{3}\pi$

11. $\cos 6x \cos 2x$

12. $\sin \frac{1}{4}\pi \sin \frac{1}{12}\pi$

In Exercises 13–18, verify each identity.

13. $\dfrac{\sin x + \sin y}{\cos x + \cos y} = \tan \dfrac{1}{2}(x+y)$

14. $\dfrac{\sin x + \sin y}{\sin x - \sin y} = \dfrac{\tan \frac{1}{2}(x+y)}{\tan \frac{1}{2}(x-y)}$

15. $\dfrac{\cos 3x + \cos x}{\sin 3x + \sin x} = \cot 2x$

16. $\cos 7x + \cos 5x + 2 \cos x \cos 2x = 4 \cos 4x \cos 2x \cos x$

17. $\dfrac{\sin 2x + \sin 2y}{\cos 2x + \cos 2y} = \tan(x+y)$

18. $\dfrac{\sin 9x - \sin 5x}{\sin 14x} = \dfrac{\sin 2x}{\sin 7x}$

19. Find the difference quotient for $\cos x$. (See Example 2.)

In Exercises 20–23, solve each of the given equations on $0 \le x \le \pi$.

20. $\sin 3x + \sin 5x = 0$

21. $\sin x - \sin 5x = 0$

22. $\cos 3x - \cos x = 0$ **23.** $\cos 2x - \cos 3x = 0$

24. Let $f(x) = \sin (2x + 1) + \sin (2x - 1)$. Make a sketch of the graph of the function and give the period and the amplitude.

25. Let $f(x) = \cos (3x + 1) + \cos (3x - 1)$. Make a sketch of the graph of this function and give the period and the amplitude.

26. Make a sketch of the graph of the function $f(x) = \cos 99x - \cos 101x$.

In Exercises 27–30, use a calculator to show that the left-hand side of each equation is equal to the right-hand side.

27. $\sin 0.3 + \sin 1.2 = 2 \sin \frac{1}{2}(0.3 + 1.2) \cos \frac{1}{2}(0.3 - 1.2)$

28. $\cos 2.05 + \cos 0.72 = 2 \cos \frac{1}{2}(2.05 + 0.72) \cos \frac{1}{2}(2.05 - 0.72)$

29. $\cos 200° - \cos 76° = -2 \sin \frac{1}{2}(200° + 76°) \sin \frac{1}{2}(200° - 76°)$

30. $\sin 15° \cos 29° = \frac{1}{2}[\sin (15° + 29°) + \sin (15° - 29°)]$

31. The Fourier analysis of the vibrating string for a particular initial condition yields an expression whose first two terms are

$$\cos \tfrac{1}{2}x + \cos \tfrac{3}{2}x$$

Using trigonometric identities, express this sum as a product.

32. In the analysis of a half-wave rectified sine wave, a communications engineer encounters the product

$$\sin \tfrac{1}{3}\pi x \sin \tfrac{1}{3}n\pi x$$

Show that this product can be written as a difference of cosines.

Key Topics for Chapter 7

Define and/or discuss each of the following.

Cosine of Sum and Difference   Half-Angle Formulas
Sine of Sum and Difference   Sum and Difference Formulas
Tangent of Sum and Difference   Product Formulas
Double-Angle Formulas

Review Exercises for Chapter 7

1. Find $\sin (A + B)$ if $\cos A = \frac{3}{5}$, $\sin B = \frac{2}{3}$, A in quadrant I, B in quadrant II.

2. Find $\cos (A + B)$ if $\sin A = \frac{1}{4}$, $\tan B = 1$, A and B in quadrant I.

3. Find $\tan 2A$ if $\cos A = -\frac{4}{5}$ in quadrant II.

4. Find $\sin 2A$ if $\tan A = \frac{1}{2}$ in quadrant III.

In Exercises 5-20, verify each identity.

5. $\cos(x - \frac{1}{3}\pi) = \frac{1}{2}(\cos x + \sqrt{3} \sin x)$

6. $\sin(2x + 1) = \sin 2x \cos 1 + \cos 2x \sin 1$

7. $\tan\left(2x + \frac{1}{4}\pi\right) = \dfrac{1 + \tan 2x}{1 - \tan 2x}$

8. $2 \cot 2x = \cot x - \tan x$

9. $\csc x = \frac{1}{2} \csc \frac{1}{2}x \sec \frac{1}{2}x$

10. $\dfrac{\cos 2x}{\cos x} = \cos x - \tan x \sin x$

11. $\cos(1 + h) - \cos 1 = -2 \sin \frac{1}{2}(2 + h) \sin \frac{1}{2}h$

12. $\sin 4x - \sin 2x = 2 \sin x \cos 3x$

13. $\dfrac{\cos 2x - \cos 4x}{\sin 2x + \sin 4x} = \tan x$

14. $\dfrac{\cos 3t}{\cos t} = 1 - 4 \sin^2 t$

15. $\dfrac{\sin 3\theta}{\sin \theta} = 3 - 4 \sin^2 \theta$

16. $\dfrac{\sin x + \sin 3x}{\cos x + \cos 3x} = \tan 2x$

17. $\sin 4a \sin 2a + \sin^2 a = \frac{1}{2}(1 - \cos 6a)$

18. $\dfrac{\sin^3 x - \cos^3 x}{\sin x - \cos x} = 1 + \frac{1}{2} \sin 2x$

19. $\cos 4x = 4 \cos 2x - 3 + 8 \sin^4 x$

20. $\sin \dfrac{1}{2} A = \dfrac{\sec A - 1}{2 \sin \frac{1}{2}A \sec A}$

In Exercises 21-25, write each of the functions as a cosine and indicate the amplitude, the period, and the phase shift. Sketch the graph.

21. $f(x) = 5 \cos 2x - 12 \sin 2x$ 22. $f(x) = 3 \sin x - 4 \cos x$

23. $f(x) = \cos 3x - \sin 3x$ 24. $f(x) = \sin 3x + \cos 3x$

25. $f(x) = \sqrt{3} \sin x + \cos x$

In Exercises 26-29, find the exact value for $\sin \theta$. Do not use a calculator.

26. $\sin 2\theta = \frac{1}{4}, 0 \le \theta \le \frac{1}{2}\pi$ 27. $\sin 2\theta = -\frac{1}{3}, -\frac{1}{2}\pi \le \theta \le 0$

28. $\tan 2\theta = -0.7, 0 \le \theta \le \frac{1}{2}\pi$ 29. $\sin \frac{1}{2}\theta = \frac{4}{5}, \frac{1}{2}\pi \le \theta \le \pi$

30. In the study of Mohr's stress circle in geology, we must solve the two equations

$$\sigma_x \cos \alpha = \tau \sin \alpha + \sigma \cos \alpha$$
$$\sigma_z \sin \alpha = \tau \cos \alpha + \sigma \sin \alpha$$

Show that, when solving for σ and τ and simplifying, you obtain

$$\sigma = \tfrac{1}{2}(\sigma_x + \sigma_z) + \tfrac{1}{2}(\sigma_x - \sigma_z)\cos 2\alpha$$
$$\tau = \tfrac{1}{2}(\sigma_x - \sigma_z)\sin 2\alpha$$

31. Rotation of the coordinate axes can be used to simplify some expressions in analytic geometry and calculus. Given that the rotation angle is $\tan 2\theta = -\tfrac{4}{3}$, find $\sin \theta$ and $\cos \theta$. Find the exact values; do not use your calculator. (Assume θ is acute.)

32. The voltage across the output terminals of a signal generator is given by $V = 3 \cos 2\pi ft - 4 \sin 2\pi ft$. Express V as one cosine function and sketch its graph, given that $f = 60$ Hz. Then express V as one sine function.

33. The expression $\sin x + \sin 2x$ represents an approximation to the solution of a problem in harmonic analysis. Write this sum as a product of functions.

8

The Inverse Trigonometric Functions

8.1 Relations, Functions, and Inverses

Functions and Relations

The concept of a function, as explained in Chapter 1, is based on the pairing of numbers. Let X represent the set of values assigned to x, and let Y represent the corresponding set of values assigned to y. Then, a rule of correspondence f that assigns to each x in X exactly one element y in Y is called a **function**. The set X is called the **domain** of the function, and the set of ys is called the **range** of the function. The general pairing indicated by f is written

$$y = f(x)$$

and is depicted as shown in Figure 8.1. The figure emphasizes the idea that the pairings are ordered; that is, each y is obtained from an x. The figure also shows that the functional pairing is unique; that is, only one value of y is paired with each x.

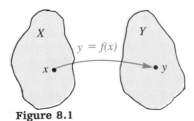

Figure 8.1

Functions are frequently expressed by a formula such as $y = 3x^2$, but other means may be used. For instance, a set of ordered pairs $\{(x, y)\}$ is a function if no two distinct pairs have the same first element. This follows from the fact that a function assigns to each x in X a unique y in Y. Thus the set $\{(2, 3), (-1, 4), (0, -5), (3, 4)\}$ is a function, since each first element is paired with a unique second element. The following illustrations clarify the important points of the definition of a function.

- A formula such as $y = \pm\sqrt{x}$ does *not* define y as a function of x, since it assigns two values to each nonzero value of x. For example, $+2$ and -2 are both images of $x = 4$. However, the formulas $y = \sqrt{x}$ and $y = -\sqrt{x}$ taken separately do define functions.

- The expression $y = 8$ defines a function, since y has the value 8 for every value of x. The definition does not require that y have a different value for each x, but only that the pairing be unique.

- The expression $x = 5$ is not a function, because many values of y correspond to $x = 5$. For example, $(5, -1), (5, 0)$, and $(5, 3)$ are some of the ordered pairs that satisfy the expression $x = 5$.

- The set $\{(-3, 4), (2, 5), (2, -6), (9, 7)\}$ is not a function, because the distinct pairs $(2, 5)$ and $(2, -6)$ have the same first element. Therefore, two different values of y are assigned to the same x.

Comment: Rules of correspondence that permit more than one value of y to be paired with a value of x are called **relations**. Hence $y = \pm\sqrt{x}$, $x = 5$, and $\{(-3, 4), (2, 5), (2, -6), (9, 7)\}$ are relations. Note that every function is a relation but a relation is not necessarily a function.

The domain of a function can be quite arbitrary. For example, we could limit the domain of $y = 3x^2$ to $x = 0, 1, 2$. In this case, the range elements are 0, 3, 12, and the functional pairings are $\{(0, 0), (1, 3), (2, 12)\}$. *If the domain is not specified, we assume that it consists of all real numbers for which the rule of correspondence will yield a real number.* Examples 2 and 3 show functions with restricted domains.

Example 1. The equation $y = x^2 + 5$ defines a function, since each value of x determines only one value of y. The domain consists of all real numbers, and the range consists only of those real numbers greater than or equal to 5 (since the smallest possible value of x^2 is 0).

Example 2. Find the domain and range of the function $y = \sqrt{x}$.

Solution. If we substitute a negative real number for x in $y = \sqrt{x}$, we do not get a real number for y. However, each nonnegative real number substituted for x yields a nonnegative real number for y. Therefore, both the domain and the range of this function consist of all nonnegative real numbers.

Example 3. Find the domain and range of the function $y = \dfrac{4}{x-3}$.

Solution. Since division by zero is not allowed, we must exclude 3 as a domain element; we conclude that the domain consists of all real numbers except 3.

To find the range of this function, we solve for x and note any restrictions on y. Thus,

$$y = \frac{4}{x-3}$$

$$xy - 3y = 4 \qquad \text{Multiplying both sides by } x - 3$$

$$xy = 3y + 4 \qquad \text{Adding } 3y \text{ to both sides}$$

$$x = \frac{3y + 4}{y} \qquad \text{Dividing both sides by } y$$

The only limitation on y is that it cannot equal 0. Therefore, the range consists of all real numbers except 0.

Inverse Functions

An element in the range of a function may correspond to more than one element in its domain. For example, if $f(x) = x^2$, then both 2 and -2 have the same range element 4. This concept is diagramed in Figure 8.2.

If every element in the domain is assigned a different element in the range, the function is called a **one-to-one function**.

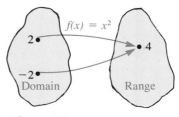

> **Definition 8.1:** A function f is one-to-one if, for a and b in the domain of f, $f(a) = f(b)$ implies that $a = b$.

Figure 8.2

The function

$$y = x^2$$

is *not* one-to-one, because $f(2) = 4 = f(-2)$, but $2 \neq -2$.

Recall from Section 1.2 that if f is a function a vertical line will intersect its graph in at most one point. A one-to-one function is one for which *both* horizontal and vertical lines intersect the graph in, at most, one point. In Figure 8.3, the first function is one-to-one, and the second is not.

If f is a one-to-one function, we know that each element in the range corresponds to a unique element in the domain. Under this condition it is possible to define a function f^{-1} from a given function f merely by interchanging the elements in the functional pairing (x, y). The domain of the function f^{-1} is the range of f, and the range of f^{-1} is the domain of f.

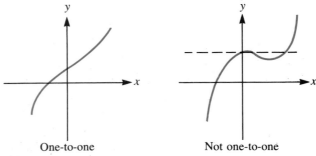

One-to-one   Not one-to-one

Figure 8.3

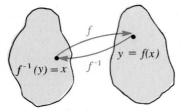

Figure 8.4

Thus, f^{-1} is called the **inverse function** of f. The relationship between f and f^{-1} is shown in Figure 8.4.

Warning: The notation f^{-1} does not mean f to a power of -1; that is,

$$f^{-1}(x) \neq \frac{1}{f(x)}$$

Example 4.   The function $f = \{(2, 1), (-3, 2), (0, 5)\}$ has the inverse $f^{-1} = \{(1, 2), (2, -3), (5, 0)\}$.

The algebraic procedure for finding the inverse function of a function given by the formula of $y = f(x)$ is summarized as follows:

(1) Interchange the x and y variables.
(2) Solve the new equation for the y variable. The resulting expression is $f^{-1}(x)$.

Example 5.   If $f(x) = 2x + 4$, find $f^{-1}(x)$.

Solution.   Let $y = 2x + 4$, and then interchange x and y to get

$$x = 2y + 4$$

Solving this equation for y, we have

$$2y = x - 4$$
$$y = \tfrac{1}{2}x - 2$$

Substituting $f^{-1}(x)$ for y, the inverse is

$$f^{-1}(x) = \tfrac{1}{2}x - 2$$

Example 6.   Let $f(x) = \dfrac{2x + 1}{x + 3}$. Find the inverse of f.

Solution.   Let $y = \dfrac{2x + 1}{x + 3}$. Then interchange the x and y variables to get

$$x = \frac{2y + 1}{y + 3}$$

Solving for y, we have

$$xy + 3x = 2y + 1$$

$$xy - 2y = 1 - 3x$$

$$y(x - 2) = 1 - 3x$$

$$y = \frac{1 - 3x}{x - 2}$$

Hence, the inverse of f is

$$f^{-1}(x) = \frac{1 - 3x}{x - 2}$$

Notice that the domain of the inverse function in Example 6 is all real numbers except $x = 2$. Restrictions of this type should always be specified when inverse functions are computed. Further, the domain of f is the range of f^{-1}, and the range of f is the domain of f^{-1}. Thus the function and its inverse are obtained by interchanging the domain and range and keeping the same correspondence.

Scientific calculators have an inv *or* arc *button, but this button can only be used to perform a few selected preprogramed functions and not the inverse operation in general.*

A function that is not one-to-one does not have an inverse function. We sometimes restrict the domain of such a function to a specified interval in order to make a one-to-one function, thereby allowing the possibility of an inverse function.

Example 7. The function $y = x^2 - 1$ is not a one-to-one function because a horizontal line will intersect its graph in two places for all $y > -1$. See Figure 8.5(a). However, if we restrict the domain to the interval $x \geq 0$, the function is one-to-one. We can compute the inverse function on this interval by interchanging x and y in the original function to get $x = y^2 - 1$. Solving for y yields the inverse function $y = \sqrt{x + 1}$. Figure 8.5(b) shows the graph of the inverse function as a dashed line.

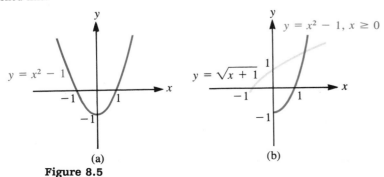

(a)
(b)
Figure 8.5

The graphs of f and f^{-1} have an interesting relationship. As the next two examples show, the graphs of f and f^{-1} are mirror reflections of each other in the line bisecting the first and third quadrants.

Example 8. Draw the graphs of $f = \{(2, 1), (-3, 2), (0, 5)\}$ and $f^{-1} = \{(1, 2), (2, -3), (5, 0)\}$ on the same coordinate axes.

Solution. The second set of ordered pairs is the inverse of the first. The graphs of f and f^{-1} are shown in Figure 8.6. You can see that the points in the graphs are mirror reflections in the solid line.

Example 9. Draw the graphs of $y = 2x + 4$ and $y = \frac{1}{2}x - 2$ on the same coordinate axes.

Solution. The inverse nature of these functions was established in Example 5. Several solution pairs for each equation are given in the respective tables, and the graphs are plotted in Figure 8.7. Notice that the graphs are mirror reflections in the dashed line.

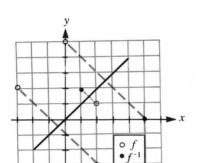

Figure 8.6

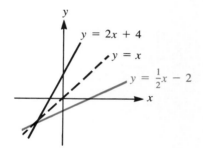

Figure 8.7

$y = 2x + 4$

x	-6	-4	-2	0	2
y	-8	-4	0	4	8

$y = \frac{1}{2}x - 2$

x	-8	-4	0	4	8
y	-6	-4	-2	0	2

Exercises for Section 8.1

1. If $f(x) = 2x + 3$, for which x is $f(x) = 6$?

2. If $f(x) = \dfrac{4}{x}$, for which x is $f(x) = 3$?

3. If $v = 10 + 4t$, for which t is $v = 20$?

4. If $f = \{(3, 0), (2, 6), (1, 5)\}$, for which x is $f(x) = 5$? For which x is $f(x) = 0$?

5. If $f = \{(0, 1), (2, 5), (3, 8), (5, 6)\}$, for which x is $f(x) = 8$? For which x is $f(x) = 3$?

In Exercises 6–21, determine the inverse function if one exists.

6. $\{(2, 6), (3, 5), (0, 4)\}$

7. $\{(3, 7), (5, 9), (7, 3), (9, 5)\}$

8. $\{(-1, 2), (2, 3), (6, -2)\}$

9. $\{(1, 2), (2, 2)\}$

10. $\{(3, 2), (5, 4), (7, 2), (9, 8)\}$

11. $\{(-2, 3), (-1, 4), (0, 0)\}$

12. $\{(0, 1), (1, 0)\}$

13. $\{(0, 3), (1, 5), (2, 3), (6, 7)\}$

14. $f(x) = 3x + 2$

15. $f(x) = x - 3$

16. $f(x) = \frac{1}{2}x + 5$

17. $f(x) = \dfrac{1}{x + 1}$

18. $f(x) = x^2 + 2x$

19. $f(x) = 2 - 3x^2$

20. $f(x) = \dfrac{2x - 1}{3x + 5}$

21. $f(x) = \dfrac{x - 1}{x + 1}$

22. The function $f(x) = x^2 - 4x$ is not one-to-one and hence does not have an inverse. Show how to define an inverse on a restricted domain.

Which of the functions whose graphs are shown in Exercises 23–26 have inverses? Sketch the graph of the inverse where one exists.

23.

24.

25.

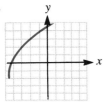

26.

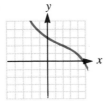

In Exercises 27–34, determine graphically which of the pairs of functions are inverses.

27. $y = -2x, \ y = -\frac{1}{2}x$

28. $y = 3x, \ y = \frac{1}{3}x$

29. $y = x + 1, \ y = x - 2$

30. $y = 6 - 3x, \ y = -\frac{1}{3}x + 2$

31. $y = 5(x + 1), \ y = \frac{1}{5}x - 1$

32. $y = \frac{1}{2}x + 1, \ y = 2x - 1$

33. $y = x^3, \ y = \sqrt[3]{x}$

34. $y = x^2, \ y = \sqrt{x}$

8.2 The Inverse Trigonometric Functions

You learned in Section 5.1 that $y = \sin x$ can be treated as a function whose domain is the set of all real numbers and whose range is the interval $[-1, 1]$. For instance, $y = \sin \frac{1}{6}\pi = \frac{1}{2}$. Sometimes, instead of computing the value of the function for a given domain element, we wish to determine the domain element that corresponds to a given functional value. For example, we might want the value of x for which $\sin x = \frac{1}{2}$. Unfortunately, since $\sin x$ is not a one-to-one function, $x = \frac{1}{6}\pi$ is not the only answer. In fact, there are infinitely many answers, some of which are shown in Figure 8.8.

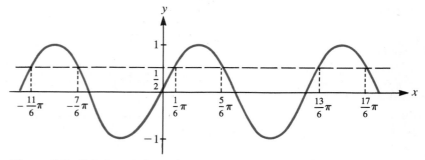

Figure 8.8 Solutions to $\sin x = \frac{1}{2}$

The fact that $-\frac{7}{6}\pi$, $\frac{1}{6}\pi$, $\frac{5}{6}\pi$, $\frac{13}{6}\pi$, etc. all satisfy $\sin x = \frac{1}{2}$ causes certain difficulties. For example, if you were asked to design a calculator to solve $\sin x = \frac{1}{2}$, what value would you have the calculator display as the solution? To avoid ambiguity, it is customary to restrict the domain of $\sin x$ to the interval $[-\frac{1}{2}\pi, \frac{1}{2}\pi]$ so that $\sin x$ is a one-to-one function. With this added restriction, $x = \frac{1}{6}\pi$ is the only value for which $\sin x = \frac{1}{2}$.

The situation described for the sine function is true for each of the trigonometric functions. Hence, we restrict the domain so that each of the six functions has one and only one domain value for each range value. The interval to which the domain of the trigonometric function is restricted in order to make the function one-to-one is called the **principal value interval** of the function. The principal value intervals for each of the six trigonometric functions are given in Table 8.1.

Table 8.1 Table of Principal Value Intervals of the Trigonometric Functions

$\sin x$	$-\frac{1}{2}\pi \leq x \leq \frac{1}{2}\pi$	$\cot x$	$0 < x < \pi$
$\cos x$	$0 \leq x \leq \pi$	$\csc x$	$-\frac{1}{2}\pi \leq x \leq \frac{1}{2}\pi, x \neq 0$
$\tan x$	$-\frac{1}{2}\pi < x < \frac{1}{2}\pi$	$\sec x$	$0 \leq x \leq \pi, x \neq \frac{1}{2}\pi$

Example 1. Find the principal value of x for which (a) $\cos x = -\frac{1}{2}$ and (b) $\sin x = -\frac{1}{2}$.

Solution. Figure 8.9(a) shows the cosine function intersecting the line $y = -\frac{1}{2}$, and Figure 8.9(b) shows the sine function intersecting the line $y = -\frac{1}{2}$. In each case, the principal value interval is printed in color so that you can see that only one value of x on that interval satisfies the equation. Thus, the principal value of x for which $\cos x = -\frac{1}{2}$ is $x = \frac{2}{3}\pi$. The principal value of x for which $\sin x = -\frac{1}{2}$ is $x = -\frac{1}{6}\pi$.

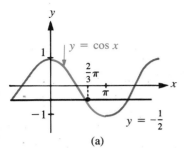
(a)

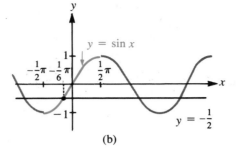
(b)

Figure 8.9

Since $\sin x$ is a one-to-one function on its principal value interval, it has an inverse function on that interval.

Definition 8.2: The inverse sine function, denoted by **Arcsin** or **Sin^{-1}**, is defined by

$$y = \text{Arcsin } x \quad \text{if and only if} \quad \sin y = x$$

where $-1 \le x \le 1$ and $-\tfrac{1}{2}\pi \le y \le \tfrac{1}{2}\pi$.

The notation $y = \text{Arcsin } x$ or $y = \text{Sin}^{-1} x$ means "y is the arclength whose sine is x" and is usually read "the inverse sine of x." We will use Arcsin x and Sin^{-1} x interchangeably throughout this chapter.

Warning:

$$y = \text{Sin}^{-1} x \text{ does NOT mean } y = \frac{1}{\sin x}.$$

Example 2. Evaluate (a) $y = \text{Arcsin } \dfrac{\sqrt{3}}{2}$ and (b) $y = \text{Sin}^{-1} (\cos \tfrac{1}{3}\pi)$.

Solution

(a) Finding y such that $y = \text{Arcsin } \sqrt{3}/2$ is the same as finding the principal value of the angle y for which $\sin y = \sqrt{3}/2$. Recalling the special-angle values, we see that $y = \text{Arcsin } \sqrt{3}/2 = \tfrac{1}{3}\pi$. The equation $y = \text{Arcsin } \sqrt{3}/2$ is equivalent to $\sin y = \sqrt{3}/2$ if y is restricted to its principal value interval.

(b) Since $\cos \tfrac{1}{3}\pi = \tfrac{1}{2}$, we wish to find y such that $\sin y = \tfrac{1}{2}$ for $-\tfrac{1}{2}\pi \le y \le \tfrac{1}{2}\pi$. Hence,

$$y = \text{Sin}^{-1} (\cos \tfrac{1}{3}\pi) = \text{Sin}^{-1} (\tfrac{1}{2}) = \tfrac{1}{6}\pi$$

The definition of each of the other five inverse trigonometric functions parallels that of the inverse sine function. Table 8.2 lists the domain and principal value interval for each of the inverse trigonometric functions.

Table 8.2   The Inverse Trigonometric Functions

Function	Domain	Principal Value Interval
$y = \text{Arcsin } x$ $(x = \sin y)$	$-1 \le x \le 1$	$-\frac{1}{2}\pi \le \text{Arcsin } x \le \frac{1}{2}\pi$
$y = \text{Arccos } x$ $(x = \cos y)$	$-1 \le x \le 1$	$0 \le \text{Arccos } x \le \pi$
$y = \text{Arctan } x$ $(x = \tan y)$	$-\infty < x < \infty$	$-\frac{1}{2}\pi < \text{Arctan } x < \frac{1}{2}\pi$
$y = \text{Arccot } x$ $(x = \cot y)$	$-\infty < x < \infty$	$0 < \text{Arccot } x < \pi$
$y = \text{Arcsec } x$ $(x = \sec y)$	$x \le -1 \text{ or } x \ge 1$	$0 \le \text{Arcsec } x \le \pi, \text{Arcsec } x \ne \frac{1}{2}\pi$
$y = \text{Arccsc } x$ $(x = \csc y)$	$x \le -1 \text{ or } x \ge 1$	$-\frac{1}{2}\pi \le \text{Arccsc } x \le \frac{1}{2}\pi, \text{Arccsc } x \ne 0$

Study Table 8.2 carefully so that you will know what limitations apply to the various inverse trigonometric functions. The information in this table is fundamental to the examples and exercises that follow.

Most scientific calculators are capable of displaying the values of the inverse trigonometric functions, although not necessarily directly. For example, one model requires that you insert the number and then press first the button `inv` *and then the desired trigonometric function. Some other models have separate buttons for* Sin^{-1}, Cos^{-1}, *and* Tan^{-1}. *At this stage, if you haven't already, you should familiarize yourself with how your specific calculator functions.*

Example 3.   Evaluate the following without using a calculator.

(a) $\text{Cos}^{-1}(0.5)$ (b) $\text{Cos}^{-1}(-0.5)$ (c) $\text{Arcsin }(-2)$

Solution

(a) Let $y = \text{Cos}^{-1}(0.5)$. Then

$$0.5 = \cos y \quad \text{where } 0 \le y \le \pi$$

The angle or number whose cosine is 0.5 is $\frac{1}{3}\pi$. Thus, $\text{Cos}^{-1}(0.5) = \frac{1}{3}\pi$.

(b) Let $y = \text{Cos}^{-1}(-0.5)$. Then

$$-0.5 = \cos y \quad \text{where } 0 \le y \le \pi$$

The angle or number whose cosine is -0.5 is $\frac{2}{3}\pi$. Thus, $\text{Cos}^{-1}(-0.5) = \frac{2}{3}\pi$.

(c) Let $y = \text{Arcsin }(-2)$. Then y is undefined because -2 is not in the domain of Arcsin x.

Example 4.   Find Arctan (-1) and Arccot (-1).

Solution.   Let $y = \text{Arctan }(-1)$ and $u = \text{Arccot }(-1)$. Then

$$-1 = \tan y, \quad \text{where } -\frac{1}{2}\pi < y < \frac{1}{2}\pi$$

and

$$-1 = \cot u, \quad \text{where } 0 < u < \pi$$

Thus,

$$y = \text{Arctan} (-1) = -\tfrac{1}{4}\pi \qquad \text{and} \qquad u = \text{Arccot} (-1) = \tfrac{3}{4}\pi$$

> **Comment:** Since the tangent and cotangent are reciprocal functions, you might have expected that Arctan (-1) and Arccot (-1) would yield the same value. The previous example shows the necessity of adhering strictly to the definitions of the inverse functions, giving close attention to the principal value interval.

The following examples concern trigonometric functions of some inverse trigonometric functions. In these circumstances it helps to show the functional value by drawing a right triangle, always keeping track of the principal value interval.

Example 5. Find $\sin (\text{Cos}^{-1} \tfrac{1}{2})$.

Solution. We first let $\theta = \text{Cos}^{-1} \tfrac{1}{2}$. Then θ is the angle shown in Figure 8.10, from which it is easy to see that

$$\sin (\text{Cos}^{-1} \tfrac{1}{2}) = \sin \theta = \frac{\sqrt{3}}{2}$$

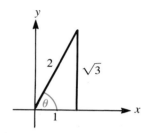

Figure 8.10

Example 6. Find $\cos (\text{Sin}^{-1} x)$.

Solution. We want to find $\cos \theta$, where $\theta = \text{Sin}^{-1} x$. Since the range values of $\text{Sin}^{-1} x$ are $[-\tfrac{1}{2}\pi, \tfrac{1}{2}\pi]$, the angle θ will be one of the angles shown in Figure 8.11. In either case, $\cos \theta = \sqrt{1 - x^2}$.

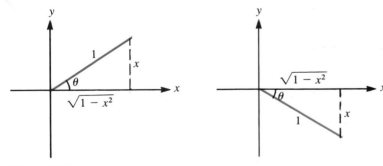

Figure 8.11

Example 7. Find $\sin [\text{Arccos} \tfrac{1}{3} + \text{Arctan} (-2)]$.

We let $\theta = \text{Arccos} \tfrac{1}{3}$ and $\phi = \text{Arctan} (-2)$. See Figure 8.12. Can you tell why θ is drawn as a first-quadrant angle and ϕ as a fourth-quadrant angle? Since $\sin (\theta + \phi) = \sin \theta \cos \phi + \sin \phi \cos \theta$,

$$\sin\left(\operatorname{Arccos}\frac{1}{3} + \operatorname{Arctan}(-2)\right) = \frac{\sqrt{8}}{3}\left(\frac{1}{\sqrt{5}}\right) + \frac{-2}{\sqrt{5}}\left(\frac{1}{3}\right)$$

$$= \frac{-2 + \sqrt{8}}{3\sqrt{5}}$$

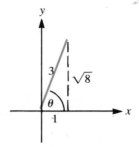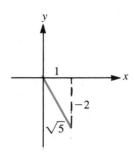

Figure 8.12

Example 8. Verify the identity $\cos(2\operatorname{Arccos} x) = 2x^2 - 1$.

Solution. Letting $\theta = \operatorname{Arccos} x$, we have

$$\cos(2\operatorname{Arccos} x) = \cos 2\theta = \cos^2\theta - \sin^2\theta$$

From Figure 8.13, we read off the values of $\cos\theta$ to be x and the values of $\sin\theta$ to be $\sqrt{1-x^2}$. Hence,

$$\cos(2\operatorname{Arccos} x) = x^2 - (1 - x^2) = 2x^2 - 1$$

Figure 8.13

Example 9. Find x if
(a) $\operatorname{Arccos} x = 0.578$ (b) $\operatorname{Arccos} x = 2\pi$

Solution

(a) If $\operatorname{Arccos} x = 0.578$, then $x = \cos 0.578$, if 0.578 is within the principal value interval, $[0, \pi]$. Since this is so,

$$x = \cos 0.578 = 0.8376$$

(b) Since 2π is not in the principal value interval for the cosine function, the equation $\operatorname{Arccos} x = 2\pi$ has no solution. If you were not on the lookout for the principal value interval, you might conclude that since $\operatorname{Arccos} x = 2\pi$, $x = \cos 2\pi = 1$, which is incorrect.

In passing, we note that

$$\sin(\operatorname{Arcsin} x) = x$$

$$\cos(\operatorname{Arccos} x) = x$$

and so forth, for all six trigonometric functions. This result is a direct consequence of the inverse nature of the functions involved. Also, *if x is limited to the principal value interval* of the function,

$$\text{Arcsin}\,(\sin x) = x$$

$$\text{Arccos}\,(\cos x) = x$$

and so forth. For example, Arcsin $(\sin \frac{1}{4}\pi) = \frac{1}{4}\pi$ since $\frac{1}{4}\pi$ is within the principal value interval of the Arcsin function. But Arcsin $(\sin \frac{5}{6}\pi) = \frac{1}{6}\pi$, since $\frac{5}{6}\pi$ is not within the principal value interval.

Exercises for Section 8.2

In Exercises 1–25, find the exact value of each expression without using tables or a calculator.

1. Arcsin $\frac{1}{2}$

2. Arcsin 1

3. $\text{Tan}^{-1}\,1$

4. $\text{Cos}^{-1}\dfrac{\sqrt{3}}{2}$

5. Arccos $\dfrac{-\sqrt{3}}{2}$

6. $\text{Sin}^{-1}\dfrac{-\sqrt{2}}{2}$

7. Arcsec (-2)

8. Arccot $(-\sqrt{3})$

9. Arcsec 1

10. Arccsc $\sqrt{2}$

11. $\text{Tan}^{-1}\,(-\sqrt{3})$

12. $\sin\,[\text{Arccos}\,(-\frac{3}{5})]$

13. $\cos\,[\text{Sin}^{-1}\,(-\frac{5}{13})]$

14. $\sin\,(\text{Arcsin}\,1)$

15. $\sin\,(\text{Arctan}\,2)$

16. $\sec\,(\text{Arccos}\,\frac{1}{3})$

17. $\cos\,(\text{Arcsin}\,\frac{1}{4})$

18. $\sin\,(2\,\text{Arcsin}\,\frac{1}{3})$

19. $\cos\,(2\,\text{Arcsin}\,\frac{1}{4})$

20. $\cos\,[\text{Cos}^{-1}\,(-\frac{1}{3}) - \text{Sin}^{-1}\,(-\frac{1}{3})]$

21. $\sin\,(\text{Arctan}\,1 - \text{Arctan}\,0.8)$

22. $\cos\,[\text{Arcsin}\,\frac{1}{4} + \text{Arccos}\,(-\frac{1}{3})]$

23. $\tan\,[\text{Arcsin}\,(-\frac{1}{2}) - \text{Arctan}\,(-2)]$

24. $\cos\,(\text{Arctan}\,2 - \text{Arccos}\,\frac{1}{2})$

25. $\tan\,(2\,\text{Tan}^{-1}\,2)$

In Exercises 26–30, simplify the given expression.

26. $\sin\,(\text{Arccos}\,x^2)$

27. $\tan\,(\text{Arcsin}\,x)$

28. $\tan\,(2\,\text{Cos}^{-1}\,x)$

29. $\sin\,(\text{Arcsin}\,y + \text{Arcsin}\,x)$

30. $\cos\,(\text{Arccos}\,x + \text{Arcsin}\,y)$

In Exercises 31–40, solve for x or show that there is no solution.

31. Arccos $x = 0.241$

32. Arcsin $x = -0.314$

33. $\text{Cos}^{-1}\,x = -0.5$

34. Arcsin $x = -\pi$

35. Arctan $x = 1.2$

36. $\text{Tan}^{-1}\,x = 2.43$

37. Arctan $x = -1.34$

38. Arccos $x = \frac{3}{4}\pi$

39. Arcsin $x = 0.8947$

40. $\text{Cos}^{-1}\,x = 2.815$

In Exercises 41–45, verify the given identity.

41. $\cos(2\,\text{Arcsin}\,x) = 1 - 2x^2$

42. $\tan(\text{Arctan}\,x + \text{Arctan}\,1) = \dfrac{1 + x}{1 - x}$

43. $\tan(2\,\text{Arctan}\,x) = \dfrac{2x}{1 - x^2}$

44. $\sin(3\,\text{Sin}^{-1}\,\theta) = 3\theta - 4\theta^3$

45. $\cos(2\,\text{Arccos}\,y) = 2y^2 - 1$

46. Verify that the inverse sine function does not have the linearity property by showing that $\text{Arcsin}\,2x \neq 2\,\text{Arcsin}\,x$.

47. Verify that the inverse cosine function does not have the linearity property by showing that $\text{Arccos}\,x + \text{Arccos}\,y \neq \text{Arccos}\,(x + y)$.

48. Sketch the graph of $\text{Arcsin}\,(\sin x)$.

49. Sketch the graph of $\text{Arcsin}\,(\cos x)$.

50. A picture u ft high is placed on a wall with its base v ft above the level of the observer's eye. If the observer stands x ft from the wall, show that the angle of vision α subtended by the picture is given by

$$\alpha = \text{Arccot}\,\frac{x}{u + v} - \text{Arccot}\,\frac{x}{v}$$

51. Assuming that you can find $\text{Arcsin}\,x$, $\text{Arccos}\,x$, and $\text{Arctan}\,x$ on your calculator, explain how to use it to find the values of $\text{Arccot}\,x$, $\text{Arcsec}\,x$, and $\text{Arccsc}\,x$.

52. The angle θ between a force of magnitude F and its horizontal component F_x is given by $F_x = F\cos\theta$. Solve for θ.

53. The range R of a projectile with muzzle velocity v_0, projected at an angle θ with the horizontal, is given approximately by $R = 16\,v_0^2 \sin 2\theta$. Solve for the angle θ.

54. In using an equal-arm analytical balance, a technician must determine the equilibrium angle θ from the equation

$$Mgl \sin\theta = (m_2 - m_1)gL \cos\theta$$

Solve for the angle θ.

Use your calculator or Table C to determine the values in Exercises 55–61. Give the answer in radians or real numbers.

55. $\text{Arctan}\,2.659$

56. $\text{Arcsin}\,0.7863$

57. $\text{Arcsec}\,5.78$

58. $\text{Arccos}\,0.3547$

59. $\text{Sin}^{-1}\,0.9866$

60. $\text{Tan}^{-1}\,2.76$

61. $\text{Arccot}\,0.8966$

62. $\text{Arccos}\,0.9034$

8.3 Graphs of the Inverse Trigonometric Functions

The graphs of the six inverse trigonometric functions can be determined from their definitions and the graphs of the trigonometric functions. For example, $y = \text{Arcsin } x$ if and only if $x = \sin y$, where $-\frac{1}{2}\pi \leq y \leq \frac{1}{2}\pi$. It follows that $y = \text{Arcsin } x$ has the same shape as a portion of the relation $x = \sin y$. [See Figure 8.14(a).]

The other parts of Figure 8.14 show the graphs of the remaining inverse functions. In each case, you can first think of the graph of the original function wrapped around the y-axis and then consider the portion that corresponds to the principal value interval.

Adding or multiplying by certain constants modifies the graphs of the inverse trigonometric functions in much the same way as it modifies the graphs of the trigonometric functions. Observing the effects of constants on the shape and location of the graphs can facilitate the process of graphing the inverse trigonometric functions. We will explain the modifi-

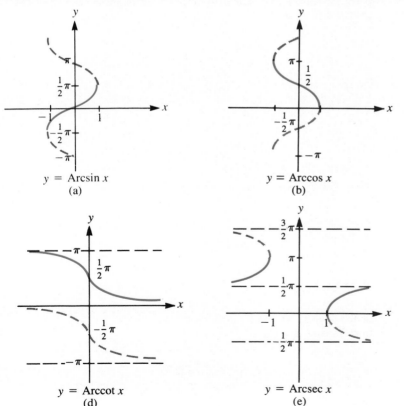

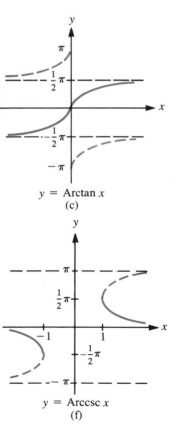

Figure 8.14 The Inverse Trigonometric Functions

cations in terms of Arcsin x, but the results apply to the other inverse trigonometric functions as well.

Multiplication of the Argument by a Constant

The function $y = \text{Arcsin } Ax$ is equivalent to $\sin y = Ax$ or $(1/A) \sin y = x$. We interpret this to mean that the domain of $y = \text{Arcsin } Ax$ is $1/A$ times the domain of $y = \text{Arcsin } x$. Thus, **the constant A expands or contracts the domain of the inverse function**. The principal value interval remains unaltered.

Example 1. Sketch the graph of Arcsin $2x$.

Solution. In this case $A = 2$, so we multiply the domain elements by $\frac{1}{2}$. Thus, the domain of Arcsin $2x$ is $-\frac{1}{2} \leq x \leq \frac{1}{2}$. The graph for $y = \text{Arcsin } 2x$ is shown in Figure 8.15.

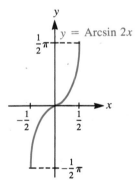

Figure 8.15

Multiplication of the Function by a Constant

Consider the multiplication of Arcsin x by a constant B. In this case, the function $y = B \text{ Arcsin } x$ is equivalent to $\sin y/B = x$. The principal values for $\sin y/B$ can be written $-\frac{1}{2}\pi \leq y/B \leq \frac{1}{2}\pi$, or $-\frac{1}{2}\pi B \leq y \leq \frac{1}{2}\pi B$. We conclude from this that **the constant B alters the principal value interval,** leaving the domain unchanged. To find the principal value interval of Arcsin Bx, simply multiply the endpoints by B.

Example 2. Sketch the graph of 2 Arccos x.

Solution. Let $y = 2$ Arccos x. Since the principal value interval for $y =$ Arccos x is $0 \le y \le \pi$, it follows that the principal value interval for $y = 2$ Arccos x is $0 \le y \le 2\pi$. (See Figure 8.16.)

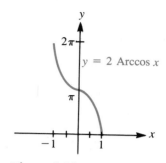

Figure 8.16

Addition of a Constant to the Function

The addition of a constant C to Arcsin x can be represented by $y = C +$ Arcsin x or $y - C =$ Arcsin x. The principal value interval for $C +$ Arcsin x is $-\frac{1}{2} \le y - C \le \frac{1}{2}$, or $-\frac{1}{2} + C \le y \le \frac{1}{2} + C$. Thus, the graph of $y = C +$ Arcsin x is that of $y =$ Arcsin x translated C units up or down. **If C is positive, the graph moves up C units; if it is negative, the graph moves down C units.**

Example 3. Sketch the graph of $y = \frac{1}{4}\pi +$ Arcsin x.

Solution. The graph of this function is just the graph of $y =$ Arcsin x translated up $\frac{1}{4}\pi$ units. Therefore, the principal value interval for $y = \frac{1}{4}\pi +$ Arcsin x is $-\frac{1}{4}\pi \le y \le \frac{3}{4}\pi$. See Figure 8.17.

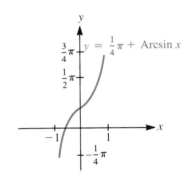

Figure 8.17

Addition of a Constant to the Argument

The function $y =$ Arcsin $(x + D)$ is equivalent to $\sin y = x + D$; thus the graph will be that of Arcsin x moved left or right D units. The translation will be D units to the left if D is positive and D units to the right if D is negative. The graph of $y =$ Arcsin $(Ax + D)$ is translated D/A units left or right.

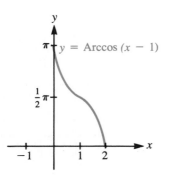

Figure 8.18

Example 4. Sketch the graph of $y = \text{Arccos}\ (x - 1)$.

Solution. Note that the graph is that of $y = \text{Arccos}\ x$ translated 1 unit to the right. The graph is shown in Figure 8.18.

The following example considers the different translations, contractions, and expansions simultaneously.

Example 5. Sketch the graph of $y = \frac{1}{2}\pi + 3\ \text{Arcsin}\ (2x - 1)$.

Solution. The constant $\frac{1}{2}\pi$ causes a vertical upward translation of the function $f(x) = 3\ \text{Arcsin}\ 2x$, and the constant -1 causes a translation $\frac{1}{2}$ unit to the right. The range of $3\ \text{Arcsin}\ 2x$ is

$$3(-\tfrac{1}{2}\pi) \leq y \leq 3(\tfrac{1}{2}\pi)$$

and its domain is

$$-1 \leq 2x \leq 1$$

or $-\frac{1}{2} \leq x \leq \frac{1}{2}$. Figure 8.19 shows the sketch of $f(x)$ and then of the given function as the translated form of $f(x)$. Notice that the graph crosses the x-axis where $y = 0$ [that is, where $\frac{1}{2}\pi + 3\ \text{Arcsin}\ (2x - 1) = 0$]. Solving this equation, we have

$$\text{Arcsin}\ (2x - 1) = -\tfrac{1}{6}\pi$$

which gives

$$2x - 1 = \sin\ (-\tfrac{1}{6}\pi) = -0.5$$
$$x = 0.25$$

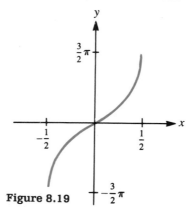

Figure 8.19

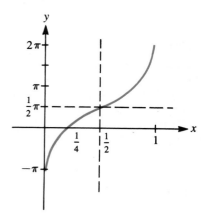

Exercises for Section 8.3

In Exercises 1–22, sketch the graph of each function.

1. $y = \text{Arcsin}\ 2x$

2. $y = \text{Arccos}\ 2x$

3. $y = 2\ \text{Arccos}\ \frac{1}{2}x$

4. $y = \frac{1}{2}\ \text{Arcsin}\ \frac{1}{2}x$

5. $y = \text{Arctan } 3x$ 6. $y = \text{Arctan } x + \frac{1}{2}\pi$

7. $y = 2 \text{ Arcsin } x$ 8. $y = \frac{1}{4} \text{ Arccos } x$

9. $y = \text{Arcsin } x + \pi$ 10. $y = 3 \text{ Arctan } 2x$

11. $y = \frac{1}{2} \text{ Arcsin } x + \frac{1}{2}\pi$ 12. $y = \pi + 2 \text{ Arcsin } \frac{1}{2}x$

13. $y = -\text{Arcsin } x$ 14. $y = -2 \text{ Arccos } x$

15. $y = \text{Arcsin } (x + 1)$ 16. $y = \text{Arccos } (x + 1)$

17. $y = \text{Arctan } (x - 1)$ 18. $y = \text{Arctan } (x + 1)$

19. $y = \text{Arcsin } (x + \frac{1}{2})$ 20. $y = \text{Arccos } (x - \frac{1}{2}\sqrt{3})$

21. $y = \text{Arcsin } (2x - 1)$ 22. $y = \text{Arccos } (2x - \frac{1}{2}\sqrt{3})$

23. The induced emf (electromotive force) in a dynamo is given by the equation $E = E_{\text{max}} \sin 2\pi ft$ where f is the frequency and t is the time in seconds. Graph t as a function of E/E_{max} for $0 \leq E/E_{\text{max}} \leq 1$. Assume f to be 60 Hz.

24. The angle of reflection θ of a ray from air to a medium of index n is given by Snell's Law to be

$$\sin \theta = \frac{1}{n}$$

Solve for θ and plot θ as a function of n for $1 \leq n \leq 3$ in steps of 0.1.

Key Topics for Chapter 8

Define and/or discuss each of the following.

Inverse Functions
Principal Value Intervals of Inverse Trigonometric Functions
Definitions of Inverse Trigonometric Functions
Trigonometric Functions of Inverse Functions
Graphs of Inverse Trigonometric Functions

Review Exercises for Chapter 8

1. If $y = 3x - 2$, for which value of x is $y = 10$?

2. If $y = \dfrac{2x + 1}{x - 1}$, for which value of x is $y = 3$?

3. Determine the function inverse to $y = 2x + 5$.

4. Determine the function inverse to $y = \dfrac{2x + 5}{3x - 1}$.

5. Determine the function inverse to $y = \sin(2x - 5) + 3$.

In Exercises 6–15, find the exact value of the given expression.

6. Arcsin $(-\frac{1}{2})$

7. Arctan $\sqrt{3}$

8. $\sin(\text{Arccos } \frac{1}{3})$

9. $\tan[\text{Sin}^{-1}(-0.4)]$

10. $\cos[2 \text{ Arccos }(0.3)]$

11. $\sin[2 \text{ Arcsin }(-0.6)]$

12. $\sin(\text{Tan}^{-1} 2 - \text{Sin}^{-1} \frac{1}{3})$

13. $\cos[\text{Arccos } \frac{1}{4} + \text{Arccos }(-\frac{1}{3})]$

14. $\tan[\text{Arctan } 1 - \text{Arccot }(-2)]$

15. $\sin[\text{Arccos }(-\frac{1}{3}) + \text{Arcsin }(-\frac{1}{3})]$

In Exercises 16–25, sketch the graph of the given expression.

16. $y = \text{Arcsin } 3x$

17. $y = \frac{1}{2}\pi + \text{Arcsin } x$

18. $y = \text{Arccos } 2x - \frac{1}{4}\pi$

19. $y = \text{Cos}^{-1}(2x - 1)$

20. $y = 2 \text{ Arctan } x$

21. $y = \text{Arctan } 2x + \pi$

22. $y = \text{Arcsin }(x - \frac{1}{2}\sqrt{2})$

23. $y = -\text{Arctan } 2x$

24. $y = 2 \text{ Tan}^{-1}(-x + 1)$

25. $y = -3 \text{ Arccos }(-2x + 1)$

In Exercises 26–27, verify the given equality for x and y in the domain of the Arcsin function.

26. $\sin(2 \text{ Arcsin } x) = 2x(1 - x^2)^{1/2}$

27. $\sin(\text{Arcsin } x + \text{Arcsin } y) = x(1 - y^2)^{1/2} + y(1 - x^2)^{1/2}$

28. To determine the electrostatic distribution in a plate, the equation

$$\frac{y}{x} = 5 \tan(v + \frac{1}{4}\pi) - 10$$

must be solved for v. You do it.

9

Complex Numbers and Polar Equations

9.1 Complex Numbers

We sometimes think of numbers as an invention of the human mind, because they were developed in order to obtain solutions to certain types of equations. For example, the negative integers were invented so that an equation like $x + 7 = 4$ could be solved. Similarly, the set of rational numbers was invented so that linear equations such as $2x = 3$ would have a solution. In order to solve $x^2 = 2$, it was necessary to invent the irrational numbers $\pm\sqrt{2}$. The irrational numbers together with the rational numbers comprise the set of real numbers.

For most applications the set of real numbers is sufficient, but there are instances in which this set is inadequate. For instance, in solving the equation $x^2 + 1 = 0$, we obtain the root $x = \sqrt{-1}$. Since the square of every real number is nonnegative, it is apparent that $\sqrt{-1}$ is not a real number. If we use i to represent $\sqrt{-1}$, with the understanding that i is a number that has the property that $i^2 = -1$, we can write the roots of $x^2 + 1 = 0$ as $x = \pm i$. The number i is called a **pure imaginary*** number and, in general, is the square root of -1. Thus,

$$\sqrt{-4} = 2i$$
$$\sqrt{-7} = i\sqrt{7}$$

Numbers of the form bi, where b is a real number, make up the set of imaginary numbers.

---

* The word "imaginary" is, in a sense, an unfortunate choice of words, since it may lead one to believe that they have a more fictitious character than the so-called real numbers.

Example 1. Solve the equation $x^2 + 9 = 0$.

Solution

$$x^2 + 9 = 0$$
$$x^2 = -9$$
$$x = \pm \sqrt{-9} = \pm 3i$$

In solving the equation $x^2 - 2x + 5 = 0$, we find that $x = 1 \pm \sqrt{-4}$. Using the concept of an imaginary number, we can write this expression as $x = 1 \pm 2i$. Thus, we have a number that is a combination of a real number and an imaginary number; such numbers are called **complex numbers**.

> **Definition 9.1:** A complex number z is any number of the form $z = a + bi$, where a and b are real numbers and $i = \sqrt{-1}$.

The real number a is called the **real part** of z, and the real number b is called the **imaginary part** of z. By convention, if $b = 1$, the number is written $a + i$. Further, if $b = 0$, the imaginary part is customarily omitted and the number is said to be pure real. If $a = 0$ and $b \neq 0$, the real part is omitted and the number is said to be pure imaginary.

Two complex numbers are equal if and only if their real parts are equal and their imaginary parts are equal. Thus, $a + bi$ and $c + di$ are equal if and only if $a = c$ and $b = d$.

Example 2
(a) 3 is a real number.
(b) $2i$ is an imaginary number.
(c) $-1 + 5i$ is a complex number.
(d) If $x + yi = 3 - 2i$, then $x = 3$ and $y = -2$.

> **Comment:** Real numbers are complex numbers in which the imaginary part is 0, and imaginary numbers are complex numbers in which the real part is 0. Thus, in Example 2, $3 = 3 + 0i$ and $2i = 0 + 2i$.

Graphical Representation of Complex Numbers

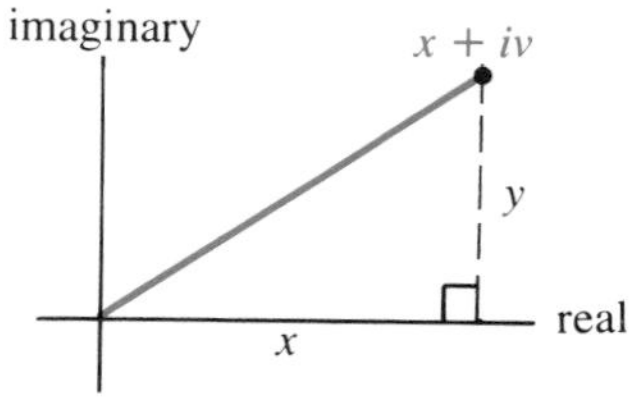

Figure 9.1

Since complex numbers are ordered pairs of real numbers, some two-dimensional configuration is needed to represent them graphically. The Cartesian coordinate system is often used for this purpose, in which case it is called the **complex plane**. The x-axis is used to represent the real part of the complex number, and the y-axis is used to represent the imaginary part. Hence, the two axes are called the real axis and the imaginary axis. The complex number $x + iy$ is represented by the point whose coordinates are (x, y) as shown in Figure 9.1. For this reason, the complex number $z = x + iy$ is said to be written in **rectangular form**.

It is often convenient to think of a complex number $x + iy$ as representing a vector. The complex number $x + iy$ in Figure 9.1 can be represented by the vector drawn from the origin to the point $x + iy$ with coordinates (x, y).

Example 3. Represent $5 + 3i$, $-2 + 4i$, $-1 - 3i$, and $5 - i$ in the complex plane. (See Figure 9.2.)

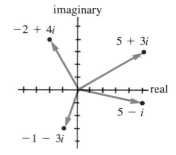

Figure 9.2

Operations on Complex Numbers

Combinations of complex numbers obey the ordinary algebraic rules for real numbers. Thus the sum, difference, product, and quotient of two complex numbers are found in the same manner as are the sum, difference, product, and quotient of two real binomials—with the understanding that $i^2 = -1$.

Example 4. Find the sum and difference of $3 + 5i$ and $-9 + 2i$.

Solution
$$(3 + 5i) + (-9 + 2i) = (3 - 9) + (5 + 2)i$$
$$= -6 + 7i$$
$$(3 + 5i) - (-9 + 2i) = (3 + 9) + (5 - 2)i$$
$$= 12 + 3i$$

It is helpful to consider the graphical representation of the sum of two complex numbers. The numbers $a + bi$, $c + di$, and $(a + c) + (b + d)i$ are represented in Figure 9.3. The result is the same as if we had applied the parallelogram law to the vectors representing $a + bi$ and $c + di$. Note that $c + di$ is subtracted from $a + bi$ by plotting $a + bi$ and $-c - di$ and then using the parallelogram law.

The product of two complex numbers is obtained by expanding the product of two binomials and replacing i^2 with -1.

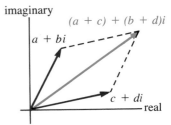

Figure 9.3 Addition of Complex Numbers

Example 5. Find the product $(3 - 2i)(4 + i)$.

Solution
$$(3 - 2i)(4 + i) = 12 + 3i - 8i - 2i^2$$
$$= 12 - 5i + 2$$
$$= 14 - 5i$$

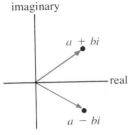

Figure 9.4

The number $2 - 3i$ is called the **conjugate** of $2 + 3i$. The conjugate of a complex number $a + bi$ is denoted $\overline{a + bi}$ and is obtained by changing the sign of the imaginary part. In general, $a - bi$ is the conjugate of $a + bi$. The graphs of $a - bi$ and $a + bi$ are shown in Figure 9.4. Notice that $a - bi$ is the mirror image of $a + bi$ in the x-axis.

The product of a complex number and its conjugate is a real number. To prove this, we note that

$$(a + bi)(a - bi) = a^2 - abi + abi - b^2i^2 = a^2 + b^2$$

The expression $a^2 + b^2$ is a real number.

Example 6

(a) $(-3 + 2i)(-3 - 2i) = (-3)^2 + 2^2 = 13$

(b) The conjugate of $4i$ is $-4i$; the product of $4i$ and its conjugate is 16.

We use the conjugate to find the quotient of two complex numbers. The quotient of two complex numbers is found by multiplying the numerator and the denominator of the quotient by the conjugate of the denominator. This operation makes the denominator a real number. The technique is illustrated in the next example.

Example 7. Find the quotient $\dfrac{2 + 3i}{4 - 5i}$.

Solution

$$\frac{2 + 3i}{4 - 5i} = \frac{(2 + 3i)(4 + 5i)}{(4 - 5i)(4 + 5i)}$$

$$= \frac{8 + (12 + 10)i + 15i^2}{16 - 25i^2}$$

$$= \frac{-7 + 22i}{16 + 25}$$

$$= \frac{-7 + 22i}{41}$$

Exercises for Section 9.1

In Exercises 1–10, plot the number, its negative, and its conjugate on the same coordinate system.

1. $-3 + 2i$ 2. $4 - 3i$ 3. $-2i$

4. $5 + i$ **5.** $-1 - i$ **6.** $3 + 5i$

7. -2 **8.** $3i$ **9.** $1 - i\sqrt{2}$

10. $\sqrt{3} + i\sqrt{5}$

In Exercises 11–34, perform the operations indicated, expressing all answers in the form $a + bi$. Check your answers to Exercises 11–14 graphically.

11. $(3 + 2i) + (4 + 3i)$ **12.** $(6 + 3i) + (5 - i)$

13. $(5 - 2i) + (-7 + 5i)$ **14.** $(-1 + i) + (2 - i)$

15. $(1 + i) + (3 - i)$ **16.** $7 - (5 + 3i)$

17. $(3 + 5i) - 4i$ **18.** $(3 + 2i) + (3 - 2i)$

19. $(2 + 3i)(4 + 5i)$ **20.** $(7 + 2i)(-1 - i)$

21. $(5 - i)(5 + i)$ **22.** $(6 - 3i)(6 + 3i)$

23. $(4 + \sqrt{3}i)^2$ **24.** $(5 - 2i)^2$

25. $6i(4 - 3i)$ **26.** $3i(-2 - i)$

27. $\dfrac{3 + 2i}{1 + i}$ **28.** $\dfrac{4i}{2 + i}$

29. $\dfrac{3}{2 - 3i}$ **30.** $\dfrac{7 - 2i}{6 - 5i}$

31. $\dfrac{1}{5i}$ **32.** $\dfrac{-3 + i}{-2 - i}$

33. $\dfrac{-1 - 3i}{4 - \sqrt{2}i}$ **34.** $\dfrac{i}{2 + \sqrt{5}i}$

35. Show that the sum of a complex number and its conjugate is a real number.

36. Show that the product of a complex number and its conjugate is a real number.

37. In the theory of optics, it is convenient to locate a point in the complex plane, for then its mirror image is its conjugate. Find the mirror image of the point whose complex representation is given by $5 + 2i$.

38. A complex impedance is given by $3 - 5i$. Plot this impedance on the complex plane.

39. The complex expression $x(t) + iy(t)$ is used to represent curves in the plane, where $x = x(t)$ and $y = y(t)$ and t is the parameter. Sketch the curve $\cos t + i(2 \sin t)$.

40. How is the curve $z(t)$ in Exercise 39 related to the curve $\bar{z}(t)$? Sketch the curve $\cos t - i(2 \sin t)$. [Note that $\bar{z}(t)$ is the conjugate of $z(t)$.]

9.2 Polar Representation of Complex Numbers

The rectangular coordinate system was used exclusively in the first eight chapters of this book. Another coordinate system widely used in science and mathematics is the **polar** coordinate system. In this system, the position of a point is determined by specifying a distance from a given point and the direction from a given line. Actually, this concept is not new; we frequently use this system to describe the relative locations of geographic points. When we say that Cincinnati is about 300 miles southeast of Chicago, we are using polar coordinates.

To establish a frame of reference for the polar coordinate system, we begin by choosing a point O and extending a line from this point. The point O is called the **pole**, and the extended line is called the **polar axis**. The position of any point P in the plane can then be determined if we know the distance OP and the angle AOP, as indicated in Figure 9.5. The directed distance OP is called the **radius vector** of P and is denoted by r. The angle AOP is called the **vectorial angle** and is denoted by θ. The coordinates of a point P are then written as the ordered pair (r, θ). Notice that the radius vector is the first element and the vectorial angle is the second.

Polar coordinates, like rectangular coordinates, are regarded as signed quantities. When stating the polar coordinates of a point, it is customary to use the following sign conventions.

(1) The radius vector is positive when measured on the terminal side of the vectorial angle and is negative when measured in the opposite direction.

(2) The vectorial angle is positive when generated by a counterclockwise rotation from the polar axis and negative when generated by a clockwise rotation.

The polar coordinates of a point uniquely determine the location of the point. However, the converse is not true, as Figure 9.6 shows. Ignoring vectorial angles that are numerically greater than 360°, we have four pairs

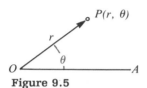

Figure 9.5

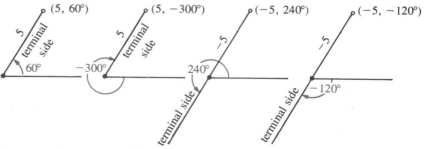

Figure 9.6

of coordinates that yield the same point: $(5, 60°)$, $(5, -300°)$, $(-5, 240°)$, and $(-5, -120°)$.

The graphical representation of a complex number as a point in the complex plane is usually conceptualized as a vector drawn from the origin to the point. We may thus use polar coordinates to describe a complex number. Figure 9.7 shows that a complex number $a + bi$ can be located with the polar coordinates (r, θ), where

$$r = \sqrt{a^2 + b^2} \qquad \text{and} \qquad \tan \theta = \frac{b}{a} \qquad (9.1)$$

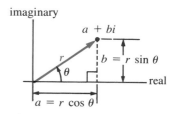

Figure 9.7   Polar Form

Example 1. Determine the polar coordinates for the complex number $1 - i\sqrt{3}$.

Solution. Using Equation (9.1) with $a = 1$ and $b = -\sqrt{3}$, we get

$$r = \sqrt{(1)^2 + (-\sqrt{3})^2} = 2 \qquad \text{and} \qquad \tan \theta = \frac{-\sqrt{3}}{1} = -\sqrt{3}$$

The vectorial angle θ is found by observing that $1 - i\sqrt{3}$ is in the fourth quadrant. (See Figure 9.8.) The reference angle for θ is $\alpha = \text{Arctan }\sqrt{3} = 60°$, so

$$\theta = 360° - 60° = 300°$$

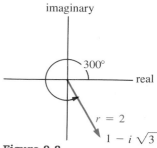

Figure 9.8

Complex numbers are frequently expressed in terms of r and θ. Referring again to Figure 9.7, we see that

$$a = r \cos \theta$$

and

$$b = r \sin \theta$$

Therefore, the complex number $a + bi$ can be written $r \cos \theta + ir \sin \theta$, or in factored form,

$$z = a + bi = r(\cos \theta + i \sin \theta) \qquad (9.2)$$

The right-hand side of Equation (9.2) is called the **polar form** of the complex number $a + bi$. The quantity $\cos \theta + i \sin \theta$ is sometimes written cis θ, in which case the polar form of the complex number is written

$$z = r \text{ cis } \theta \qquad (9.3)$$

The number r is called the **modulus**, or magnitude, of z, and θ is called the **argument**. Note that a given complex number has many arguments, all differing by multiples of 2π. Sometimes we limit the argument to some interval of length 2π and thus obtain the **principal value**. In this book, unless we state otherwise, the principal values will be between $-\pi$ and π; that is, between $-180°$ and $180°$.

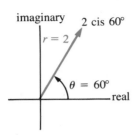

Figure 9.9

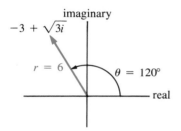

Figure 9.10

Example 2. Represent $z = 1 + \sqrt{3}i$ in polar form. (See Figure 9.9.)

Solution. Since

$$r = \sqrt{1^2 + (\sqrt{3})^2} = \sqrt{4} = 2$$

and

$$\theta = \text{Arctan } \sqrt{3} = 60°$$

we have

$$1 + \sqrt{3}i = 2(\cos 60° + i \sin 60°) = 2 \text{ cis } 60°$$

Example 3. Express $z = 6(\cos 120° + i \sin 120°)$ in rectangular form. (See Figure 9.10.)

Solution. Using the fact that $a = r \cos \theta$ and $b = r \sin \theta$, we have

$$a = 6 \cos 120° = 6\left(-\frac{1}{2}\right) = -3$$

$$b = 6 \sin 120° = 6\left(\frac{\sqrt{3}}{2}\right) = 3\sqrt{3}$$

Therefore,

$$z = a + bi = -3 + 3\sqrt{3}i$$

It is easy to give a geometric interpretation to the product of two complex numbers in polar form. If $z_1 = r_1 \text{ cis } \theta_1$ and $z_2 = r_2 \text{ cis } \theta_2$, the product $z_1 z_2$ may be written

$$\begin{aligned} z_1 z_2 &= r_1(\cos \theta_1 + i \sin \theta_1) \cdot r_2(\cos \theta_2 + i \sin \theta_2) \\ &= r_1 r_2[\cos \theta_1 \cos \theta_2 + i \cos \theta_1 \sin \theta_2 + i \sin \theta_1 \cos \theta_2 \\ &\quad + i^2 \sin \theta_1 \sin \theta_2] \\ &= r_1 r_2[(\cos \theta_1 \cos \theta_2 - \sin \theta_1 \sin \theta_2) \\ &\quad + i(\cos \theta_1 \sin \theta_2 + \sin \theta_1 \cos \theta_2)] \end{aligned}$$

Now, using the identities for the sine and cosine of the sum of two angles, we have

$$z_1 z_2 = r_1 r_2[\cos (\theta_1 + \theta_2) + i \sin (\theta_1 + \theta_2)] = r_1 r_2 \text{ cis } (\theta_1 + \theta_2) \quad (9.4)$$

Therefore, the modulus of the product of two complex numbers is the product of the individual moduli, and the argument of the product is the sum of the individual arguments. Graphically, multiplication of z_1 by z_2 results in a rotation of the vector through z_1 by an angle equal to the argument of z_2, and an expansion or contraction of the modulus depending on whether $|z_2| > 1$ or $|z_2| < 1$. (See Figure 9.11.)

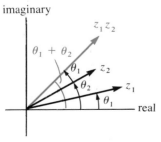

Figure 9.11 Multiplication of
Complex Numbers

Example 4. Multiply $z_1 = -1 + \sqrt{3}i$ and $z_2 = 1 + i$, using the polar form of each.

Solution. Computing the modulus and argument of each complex number yields

$$r_1 = \sqrt{(-1)^2 + (\sqrt{3})^2} = 2 \qquad \tan \theta_1 = \frac{\sqrt{3}}{-1}, \theta_1 = 120°$$

$$r_2 = \sqrt{1^2 + 1^2} = \sqrt{2} \qquad \tan \theta_2 = \frac{1}{1}, \theta_2 = 45°$$

Therefore,

$$z_1 z_2 = (2 \text{ cis } 120°)(\sqrt{2} \text{ cis } 45°) = 2\sqrt{2} \text{ cis } (120° + 45°)$$
$$= 2\sqrt{2} \text{ cis } 165°$$

Using the same procedure we employed to derive Equation (9.4), we can show that if $z_1 = r_1 \text{ cis } \theta_1$ and $z_2 = r_2 \text{ cis } \theta_2$, then

$$\frac{z_1}{z_2} = \frac{r_1}{r_2} [\cos (\theta_1 - \theta_2) + i \sin (\theta_1 - \theta_2)] = \frac{r_1}{r_2} \text{ cis } (\theta_1 - \theta_2) \quad (9.5)$$

In words, the modulus of the quotient of two complex numbers is the quotient of the individual moduli, and the argument is the difference of the individual arguments.

Example 5. Divide $z_1 = 2 \text{ cis } 120°$ by $z_2 = \sqrt{2} \text{ cis } 45°$.

Solution

$$\frac{z_1}{z_2} = \frac{2 \text{ cis } 120°}{\sqrt{2} \text{ cis } 45°} = \frac{2}{\sqrt{2}} \text{ cis } (120° - 45°) = \sqrt{2} \text{ cis } 75°$$

Exercises for Section 9.2

In Exercises 1–10, plot each complex number and then express the number in polar form.

1. $1 - \sqrt{3}i$

2. $3 + 4i$

3. $\sqrt{5} + 2i$.

4. $\sqrt{3} - i$

5. 9

6. $5i$

7. $3 - 4i$

8. $-1 + i$

9. $5 - 6i$

10. $-3 - 4i$

In Exercises 11–20, plot each complex number and then express the number in rectangular form.

11. 2 cis 30°

12. 4 cis 60°

13. 5 cis 135°

14. 10 cis 90°

15. $\sqrt{3}$ cis 210°

16. $\sqrt{5}$ cis 180°

17. 3 cis 300°

18. 7 cis 0°

19. 10 cis 20°

20. 2 cis 100°

In Exercises 21–36, perform the indicated operations. If the complex numbers are not already in polar form, put them in that form before proceeding.

21. (4 cis 30°)(3 cis 60°)

22. (2 cis 120°)($\sqrt{5}$ cis 180°)

23. ($\sqrt{2}$ cis 90°)($\sqrt{2}$ cis 240°)

24. (5 cis 180°)(3 cis 90°)

25. (10 cis 35°)(2 cis 100°)

26. (3 cis 45°)(2 cis 120°)

27. $(3 + 4i)(\sqrt{3} - i)$

28. $3i(2 - i)$

29. $\dfrac{10 \text{ cis } 30°}{2 \text{ cis } 90°}$

30. $\dfrac{5 \text{ cis } 29°}{3 \text{ cis } 4°}$

31. $\dfrac{4 \text{ cis } 26°40'}{2 \text{ cis } 19°10'}$

32. $\dfrac{12 \text{ cis } 100°}{3 \text{ cis } 23°}$

33. $\dfrac{1 - i}{\sqrt{3} + i}$

34. $\dfrac{\sqrt{3} + i}{\sqrt{3} - i}$

35. $\dfrac{4i}{-1 + i}$

36. $\dfrac{5}{1 + i}$

37. Prove *Euler's Identities:*

$$\cos\theta = \frac{1}{2}[\text{cis }\theta + \text{cis}(-\theta)]$$

$$\sin\theta = \frac{1}{2i}[\text{cis }\theta - \text{cis}(-\theta)]$$

38. A force is represented by the complex number 20 cis 2.54. What is the magnitude of the force? What angle, in degrees, does the force make with the positive x-axis? What are the horizontal and vertical components of the force?

39. The power in a circuit is the product of the current, i, and the voltage, v. Write the equation for power, p, if $i = 50$ cis $(-1.7t)$ and $v = 20$ cis $(0.5t)$. What is the power when $t = 1$? Give the horizontal (resistive) and vertical (reactive) components of power.

The square of the complex number $z = r \operatorname{cis} \theta$ is given by

$$z^2 = (r \operatorname{cis} \theta)(r \operatorname{cis} \theta) = r^2 \operatorname{cis} (\theta + \theta)$$
$$= r^2 \operatorname{cis} 2\theta$$

Likewise,

$$z^3 = z^2 \cdot z = (r^2 \operatorname{cis} 2\theta) \cdot (r \operatorname{cis} \theta)$$
$$= r^3 \operatorname{cis} 3\theta$$

We expect the pattern exhibited for z^2 and z^3 to apply as well to z^4, z^5, z^6, and so on. As a matter of fact, if $z = r \operatorname{cis} \theta$, then

$$z^n = r^n \operatorname{cis} n\theta \tag{9.6}$$

This result is known as **DeMoivre's Theorem**. The theorem is true for all real values of n, a fact that we shall accept without proof.

Example 1. Use DeMoivre's Theorem to find $(-2 + 2i)^4$.

Solution. Here we have

$$r = \sqrt{2^2 + (-2)^2} = \sqrt{8} \qquad \text{and} \qquad \theta = 135°$$

Therefore,

$$(-2 + 2i)^4 = (\sqrt{8})^4 (\cos 135° + i \sin 135°)^4$$
$$= (\sqrt{8})^4 [\cos 4(135°) + i \sin 4(135°)]$$
$$= 64[\cos 540° + i \sin 540°]$$
$$= 64[\cos 180° + i \sin 180°]$$
$$= -64$$

In the system of real numbers, there is no square root of -1, no fourth root of -81, and so on. However, if we use complex numbers, we can find the nth root of any number by using DeMoivre's Theorem.

Since DeMoivre's Theorem is valid for all real n, it is possible to evaluate $[r \operatorname{cis} \theta]^{1/n}$ as

$$[r \operatorname{cis} \theta]^{1/n} = r^{1/n} \operatorname{cis} \frac{\theta}{n} = \sqrt[n]{r} \operatorname{cis} \frac{\theta}{n} \tag{9.7}$$

Since $\cos \theta$ and $\sin \theta$ are periodic functions with a period of $360°$, $\cos \theta = \cos (\theta + k \cdot 360°)$ and $\sin \theta = \sin (\theta + k \cdot 360°)$, where k is an integer. Hence,

$$[r \text{ cis } \theta]^{1/n} = \sqrt[n]{r} \text{ cis} \left(\frac{\theta + k \cdot 360°}{n} \right) \qquad (9.8)$$

For a given number n, the right-hand side of Equation (9.8) takes on n distinct values corresponding to $k = 0, 1, 2, \ldots, n - 1$. For $k > n - 1$, the result is merely a duplication of the first n values.

Example 2. Find the square roots of $4i$.

Solution. We first express $4i$ in polar form using

$$r = \sqrt{0^2 + 4^2} = 4 \qquad \text{and} \qquad \theta = 90°$$

Thus,

$$4i = 4 \text{ cis } 90°$$

and the square roots of $4i$ are given by

$$2 \text{ cis} \left(\frac{90° + k \cdot 360°}{2} \right)$$

Therefore, we have, for $k = 0$,

$$2 \text{ cis } 45° = \sqrt{2} + \sqrt{2}i$$

and for $k = 1$,

$$2 \text{ cis } 225° = -\sqrt{2} - \sqrt{2}i$$

It is convenient and informative to plot these values in the complex plane, as shown in Figure 9.12. Notice that both roots are located on a circle of radius 2, but they are 180° apart.

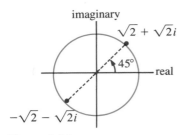

Figure 9.12

Example 3. Find the three cube roots of unity.

Solution. In polar form, the number 1 may be written $1 \text{ cis } 0°$. Thus,

$$\sqrt[3]{1 \text{ cis } 0°} = 1 \text{ cis} \left(\frac{0° + k \cdot 360°}{3} \right)$$

For $k = 0$,

$$1 \text{ cis } 0° = 1$$

For $k = 1$,

$$1 \text{ cis } 120° = \frac{-1 + \sqrt{3}i}{2}$$

For $k = 2$,

$$1 \text{ cis } 240° = \frac{-1 - \sqrt{3}i}{2}$$

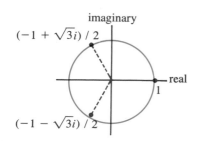

Figure 9.13

These roots are displayed in Figure 9.13. Notice that they are located on a circle of radius 1 at equally spaced intervals of 120°.

Example 4. Find the fourth roots of $-1 + \sqrt{3}i$.

Solution. Writing $-1 + \sqrt{3}i$ in polar form, we have

$$-1 + \sqrt{3}i = 2 \text{ cis } 120°$$

Therefore,

$$[-1 + \sqrt{3}i]^{1/4} = \sqrt[4]{2} \text{ cis } \frac{120° + k \cdot 360°}{4}$$

The four roots corresponding to $k = 0, 1, 2, 3$ are as follows.
For $k = 0$,

$$\sqrt[4]{2} \text{ cis } 30° = \sqrt[4]{2}\left(\frac{\sqrt{3}}{2} + \frac{1}{2}i\right)$$

For $k = 1$,

$$\sqrt[4]{2} \text{ cis } 120° = \sqrt[4]{2}\left(-\frac{1}{2} + \frac{\sqrt{3}}{2}i\right)$$

For $k = 2$,

$$\sqrt[4]{2} \text{ cis } 210° = \sqrt[4]{2}\left(-\frac{\sqrt{3}}{2} - \frac{1}{2}i\right)$$

For $k = 3$,

$$\sqrt[4]{2} \text{ cis } 300° = \sqrt[4]{2}\left(\frac{1}{2} - \frac{\sqrt{3}}{2}i\right)$$

Exercises for Section 9.3

In Exercises 1–10, use DeMoivre's Theorem to evaluate each power. Leave the answer in polar form.

1. $(-1 + \sqrt{3}i)^3$ 2. $(1 + i)^4$

3. $(\sqrt{3} \text{ cis } 60°)^4$ 4. $(\sqrt{3} - i)^6$

5. $(-2 + 2i)^5$ 6. $(-1 + 3i)^3$

7. $(-\sqrt{3} + i)^7$ 8. $(2 \text{ cis } 20°)^5$

9. $(2 + 5i)^4$ 10. $(3 + 2i)^{10}$

In Exercises 11–20, find each root indicated and sketch its location in the complex plane.

11. Fifth roots of 1   12. Cube roots of 64

13. Fourth roots of i 14. Fourth roots of -16

15. Square roots of $1 + i$ 16. Fifth roots of $\sqrt{3} + i$

17. Sixth roots of $-\sqrt{3} + i$ 18. Square roots of $-1 + i$

19. Fourth roots of $-1 + \sqrt{3}i$ 20. Sixth roots of $-i$

21. Obtain an expression for $\cos 2\theta$ and $\sin 2\theta$ in terms of trigonometric functions of θ by making use of DeMoivre's Theorem.

22. Find all roots of the equation $x^4 + 81 = 0$.

23. Find all roots of $x^3 + 64 = 0$.

24. A physicist studying the theory of optics needs to know the $\frac{3}{2}$ power of $2 + i$. This can be found by taking the cube of $2 + i$ and then finding the two square roots of that result. Of these two numbers, the physicist wants the one with the positive real part. Find it for her.

25. In the theory of Laplace transforms, the function $1/(s^4 + 1)$ arises. Determine the values of s for which the denominator is equal to zero.

9.4 Polar Equations and Their Graphs

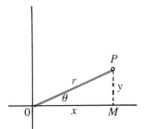

Figure 9.14

The equations $x^2 + y^2 = 25$ and $y^2 = 2x + 1$ are called rectangular equations because x and y represent coordinates in the Cartesian plane. These two equations can be expressed in polar coordinates as $r = 5$ and $r(1 - \cos \theta) = 1$, respectively. The relationship between the rectangular and polar forms of an equation can be found by superimposing the rectangular coordinate system on the polar coordinate system so that the origin corresponds to the pole and the positive x-axis to the polar axis. Under these circumstances, the point P shown in Figure 9.14 has both (x, y) and (r, θ) as coordinates. The desired relationship is then an immediate consequence of triangle OMP. Hence, the equations

$$x = r \cos \theta \tag{9.9}$$

and

$$y = r \sin \theta \tag{9.10}$$

can be used to transform a rectangular equation into a polar equation.

Example 1. Find the polar equation of the circle whose rectangular equation is $x^2 + y^2 = a^2$.

Solution. Substituting Equations (9.9) and (9.10) into the given equation, we have

$$r^2 \cos^2 \theta + r^2 \sin^2 \theta = a^2$$
$$r^2(\cos^2 \theta + \sin^2 \theta) = a^2$$
$$r^2 = a^2$$
$$r = a$$

Hence, $r = a$ is the polar equation of the given circle.

To make the transformation from polar coordinates into rectangular coordinates, we use the following equations.

$$r = \sqrt{x^2 + y^2} \qquad (9.11)$$

$$\sin \theta = \frac{y}{\sqrt{x^2 + y^2}} \qquad (9.12)$$

$$\cos \theta = \frac{x}{\sqrt{x^2 + y^2}} \qquad (9.13)$$

These equations are derived from Figure 9.14.

Example 2. Transform the following polar equation into a rectangular equation:

$$r = 1 - \cos \theta$$

Solution. Substituting Equations (9.11) and (9.13) into the given equation, we have

$$\sqrt{x^2 + y^2} = 1 - \frac{x}{\sqrt{x^2 + y^2}}$$
$$x^2 + y^2 = \sqrt{x^2 + y^2} - x$$

Therefore,

$$x^2 + y^2 + x = \sqrt{x^2 + y^2}$$

is the required equation.

Example 3. Show that

$$r = \frac{1}{1 - \cos \theta}$$

is the polar form of a parabola.

Solution. Here our work is simplified if we multiply both sides of the given equation by $1 - \cos \theta$ before making the substitution. Thus

$$r - r \cos \theta = 1$$

Substituting for r and $\cos \theta$ using Equations (9.11) and (9.13), we get

$$\sqrt{x^2 + y^2} - x = 1$$

Transposing x to the right and squaring both sides of the resulting equation, we get

$$x^2 + y^2 = x^2 + 2x + 1$$
$$y^2 = 2x + 1$$
$$y^2 = 2(x + \tfrac{1}{2})$$

This is the standard form of a parabola having its vertex at $(-\tfrac{1}{2}, 0)$ and a horizontal axis.

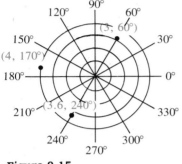

Figure 9.15

A polar equation has a graph in the polar coordinate plane, just as a rectangular equation has a graph in the rectangular coordinate plane. To draw the graph of a polar equation, we start by assigning values to θ and finding the corresponding values of r. The desired graph is then generated by plotting the ordered pairs (r, θ) and connecting them with a smooth curve. Use of polar coordinate paper greatly simplifies graphing a polar equation. Polar coordinate paper is commercially available. As Figure 9.15 shows, this paper consists of equally spaced concentric circles with radial lines extending at equal angles through the pole. Several points are plotted in Figure 9.15 for illustrative purposes.

Example 4.   Sketch the graph of the equation $r = 1 + \cos \theta$.

Solution.   Using increments of $45°$ for θ, we obtain the following table.

θ	0	45°	90°	135°	180°	225°	270°	315°	360°
r	2.00	1.71	1.00	0.29	0.00	0.29	1.00	1.71	2.00

The curve obtained by connecting these points with a smooth curve is called a *cardioid*. (See Figure 9.16.)

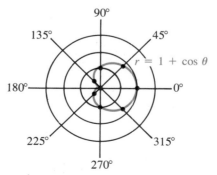

Figure 9.16

Example 5. Sketch the graph of the equation $r = 4 \sin \theta$.

Solution. The following table gives values of r corresponding to the indicated values of θ. Drawing a smooth curve through the plotted points, we obtain the *circle* shown in Figure 9.17. Notice that θ varies only from 0 to π radians. If we allow θ to vary from 0 to 2π radians, the graph will be traced out twice; once for $0 \le \theta \le \pi$ and again for $\pi < \theta \le 2\pi$. You should demonstrate this by plotting points in the interval $\pi < \theta \le 2\pi$.

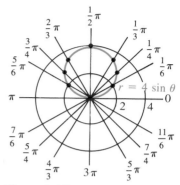

Figure 9.17

θ	0	$\frac{1}{6}\pi$	$\frac{1}{4}\pi$	$\frac{1}{3}\pi$	$\frac{1}{2}\pi$	$\frac{2}{3}\pi$	$\frac{3}{4}\pi$	$\frac{5}{6}\pi$	π
r	0	2	$2\sqrt{2}$	$2\sqrt{3}$	4	$2\sqrt{3}$	$2\sqrt{2}$	2	0

Exercises for Section 9.4

In Exercises 1–10, plot each point on polar coordinate paper.

1. $(5, 30°)$
2. $(3.6, -45°)$
3. $(12, \frac{2}{3}\pi)$
4. $(0.5, 220°)$
5. $(-7.1, 14°)$
6. $(-2, \frac{7}{3}\pi)$
7. $(1.75, -200°)$
8. $(\sqrt{2}, -311°)$
9. $(-5, -30°)$
10. $(5, 150°)$

In Exercises 11–18, convert each rectangular equation into an equation in polar coordinates.

11. $2x + 3y = 6$
12. $y = x$
13. $x^2 + y^2 - 4x = 0$
14. $x^2 - y^2 = 4$
15. $x^2 + 4y^2 = 4$
16. $xy = 1$
17. $x^2 = 4y$
18. $y^2 = 16x$

In Exercises 19–26, convert each polar equation into an equation in rectangular coordinates.

19. $r = 5$
20. $r = \cos \theta$
21. $r = 10 \sin \theta$
22. $r = 2 (\sin \theta - \cos \theta)$
23. $r = 1 + 2 \sin \theta$
24. $r \sin \theta = 10$
25. $r = \dfrac{5}{1 + \cos \theta}$
26. $r(1 - 2 \cos \theta) = 1$

In Exercises 27–40, sketch the graph of the given equation.

27. $r = 5.6$

28. $r = \sqrt{2}$

29. $\theta = \frac{1}{3}\pi$

30. $\theta = 170°$

31. $r = 2 \sin \theta$

32. $r = 0.5 \cos \theta$

33. $r \sin \theta = 1$

34. $r \cos \theta = -10$

35. $r = 1 + \sin \theta$

36. $r = 1 - \cos \theta$

37. $r = \sec \theta$

38. $r = -\sin \theta$

39. $r = 4 \sin 3\theta$

40. $r = \sin 2\theta$

41. The radiation pattern of a particular two-element antenna is a cardioid of the form $r = 100(1 + \cos \theta)$. Sketch the radiation pattern of this antenna.

42. The radiation pattern of a certain antenna is given by

$$r = \frac{1}{2 - \cos \theta}$$

Plot this pattern.

43. By transforming the polar equation in Exercise 42 into rectangular coordinates, show that the indicated radiation pattern is elliptical.

44. The feedback diagram of a certain electronic tachometer can be approximated by the curve $r = \frac{1}{2}\theta$. Sketch the feedback diagram of this tachometer from $\theta = 0$ to $\theta = \frac{7}{6}\pi$.

Key Topics for Chapter 9

Define and/or discuss each of the following.

Complex Numbers
Graphical Representation of a
  Complex Number
Rectangular Form of a Complex
  Number
Complex Conjugate

Polar Coordinates
Polar Form of a Complex Number
DeMoivre's Theorem
Graphical Representation of a Polar
  Equation

Review Exercises for Chapter 9

In Exercises 1–15, perform the indicated operations and express the result in the form $a + bi$. (Notice that $\overline{a + bi} = a - bi$).

1. $(3 - 2i) + (6 - i)$

2. $(3 - 2i) - (i + 2)$

3. $(i + 7) - (i + 2)$

4. $(2 + i)(2 - i)$

5. $(3 - i)\overline{(3 + i)}$

6. $(i - 2)(2 + i)i$

7. $i(i^2 - 1)(i^2 + 1)$

8. $(6 - i)^2\overline{(6 + i)}$

9. $(3 + 2i)(2 - i) + \overline{(3 + 2i)(2 - i)}$

10. $\dfrac{i}{2 + i}$

11. $\dfrac{2i + 1}{i - 1}$

12. $\dfrac{\overline{6 - i}}{i^2}$

13. $\dfrac{i - 1}{\overline{i - 1}}$

14. $\dfrac{(9 - i)(9 - 2i)}{i + 2}$

15. $\dfrac{(4 - i)(3 + 2i)}{i + 1}$

In Exercises 16–20, plot each number and express it in polar form.

16. $1 + i\sqrt{3}$　　　17. i　　　18. $i - 1$　　　19. $1 + i$　　　20. 4

In Exercises 21–25, plot each number and express it in rectangular form.

21. 2 cis 45°

22. -3 cis 75°

23. 4 cis $(-20°)$

24. $(-2$ cis 30°$)^2$

25. $(3$ cis 10°$)^3$

In Exercises 26–30, convert each expression to polar coordinates and sketch.

26. $x^2 + y^2 + y = 0$

27. $y = 2x$

28. $y^2 + x^2 - 3x = 1$

29. $4x^2 + y^2 = 1$

30. $x = 4y^2$

In Exercises 31–40, convert each expression to rectangular coordinates and sketch.

31. $r = 2$

32. $r = 2 + 3 \cos \theta$

33. $r = \dfrac{2}{1 + \sin \theta}$

34. $r \cos \theta = 3$

35. $r = 3 \cos \theta$

36. $r = \theta$

37. $r = \sin 2\theta$

38. $r^2 = \sin \theta$

39. $r^2 - r = 0$

40. $r^2 - 3r + 2 = 0$

In Exercises 41–45, evaluate each expression. Leave the answer in polar form.

41. $(1 + i)^3$

42. $(-2 + 2i)^6$

43. $(-\sqrt{3} + i)^5$

44. $(\sqrt{3}$ cis 60°$)^4$

45. $(3 - i)^{10}$

In Exercises 46–50, find the indicated roots and sketch their location in the complex plane. Leave the answer in polar form.

46. Square roots of i

47. Fifth roots of -1

10

Logarithms

10.1 Definition of the Logarithm

In this chapter you will learn something of the usefulness of logarithms. We begin with the definition of the logarithm and its relation to the exponent.

> **Definition 10.1:** The logarithm of a positive number x to the base $b(\neq 1)$ is the power to which b must be raised to give x. That is,
>
> $$y = \log_b x \quad \text{if and only if} \quad x = b^y$$

The equation $y = \log_b x$ is read, "y is the logarithm of x to the base b." The following examples illustrate this definition.

Example 1
(a) $\log_2 8 = 3$, since $2^3 = 8$
(b) $\log_3 \frac{1}{9} = -2$, since $3^{-2} = \frac{1}{9}$
(c) $\log_{10} 10,000 = 4$, since $10^4 = 10,000$

Example 2. Find the base a if $\log_a 16 = 4$.

Solution. Since $2^4 = 16$, we conclude that the desired base is $a = 2$.

Example 3. Find the number x if $\log_3 x = -3$.

Solution. Since $3^{-3} = \frac{1}{27}$, we conclude that $x = \frac{1}{27}$.

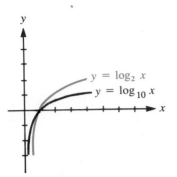

Figure 10.1

Since $y = \log_b x$ means that $x = b^y$, we can sketch the graph of the logarithm function if we first construct a table of values. The tables for $y = \log_2 x$ and $y = \log_{10} x$ are included below, and the corresponding graphs are presented in Figure 10.1.

x	$\frac{1}{16}$	$\frac{1}{8}$	$\frac{1}{4}$	$\frac{1}{2}$	1	2	4	8	16
$\log_2 x$	-4	-3	-2	-1	0	1	2	3	4

x	$\frac{1}{1000}$	$\frac{1}{100}$	$\frac{1}{10}$	1	10	100
$\log_{10} x$	-3	-2	-1	0	1	2

Each of the curves in Figure 10.1 is characteristic of what is called **logarithmic shape**. This figure clearly demonstrates the following functional characteristics of the logarithm function. Any function that obeys these properties is said to *behave logarithmically*.

(1) $\log_a x$ is not defined for $x \leq 0$.

(2) $\log_a 1 = 0$.

(3) $\log_a a = 1$.

(4) $\log_a x$ is negative for $0 < x < 1$ and positive for $x > 1$.

(5) As x approaches 0, $\log_a x$ decreases without bound.

(6) As x increases without bound, $\log_a x$ increases without bound.

The base of a logarithm may theoretically be any positive number except 1, but in practice we seldom use bases other than 10 and e; the number e is an irrational number that is approximated to three decimal places by 2.718. Logarithms with base 10 are called **common** logarithms and are denoted log x with the base 10 understood. Logarithms with base e are called **natural** logarithms and are denoted ln x with the base e understood.

Most scientific calculators have both `log x` *and* `ln x` *buttons or equivalent capability to make these computations. Consult your user's manual for specific instructions on the use of these functions.*

Exercises for Section 10.1

In Exercises 1–5, write a logarithmic equation equivalent to the given exponential equation.

1. $x = 2^3$

2. $y = 3^8$

3. $M = 5^{-3}$

4. $N = 10^{-2}$ 5. $L = 7^2$

In Exercises 6–10, find the base b of the given logarithmic function.

6. $\log_b 8 = 1$ 7. $\log_b 4 = 2$ 8. $\log_b \frac{1}{4} = -2$

9. $\log_b 100 = 2$ 10. $\log_b \frac{1}{3} = -1$

In Exercises 11–28, solve each equation for the unknown.

11. $\log_{10} x = 4$ 12. $\log_5 N = 2$ 13. $\log_x 10 = 1$

14. $\log_x 25 = 2$ 15. $\log_x 64 = 3$ 16. $\log_{16} x = 2$

17. $\log_{27} x = \frac{2}{3}$ 18. $\log_2 \frac{1}{8} = x$ 19. $\log_3 9 = x$

20. $\log_{10} 10^7 = x$ 21. $\log_b b^a = x$ 22. $\log_b x = b$

23. $\log_x 2 = \frac{1}{3}$ 24. $\log_x 0.0001 = -2$ 25. $\log_x 6 = \frac{1}{2}$

26. $\log_6 x = 0$ 27. $6^{\log_6 x} = 6$ 28. $x^{\log_x x} = 3$

29. Let $f(x) = \log_3 x$. Find (a) $f(9)$, (b) $f(\frac{1}{27})$, and (c) $f(81)$.

30. Let $f(x) = \log_2 x$. By example, show that
    (a) $f(x + y) \neq f(x) + f(y)$ (b) $f(ax) \neq af(x)$

31. A power supply has a power output in watts approximated by the equation $P = 64(2)^{-3t}$, where t is in days. How many days does it take for the power supply to reduce to a power output of 1 watt?

32. A certain radioactive material decays exponentially by the equation $A(t) = A_0 2^{-t/5}$. Find the half-life of the material.

10.2 Basic Properties of the Logarithm

Logarithmic expressions must often be rearranged or simplified. These simplifications are accomplished by three basic Rules of Logarithms which correspond precisely to the three fundamental rules for exponents and are necessary consequences of them.

Rule 1:

$$\log_a MN = \log_a M + \log_a N$$

Proof.   Let $u = \log_a M$ and $v = \log_a N$. Then

$$a^u = M \quad \text{and} \quad a^v = N$$

and thus

$$MN = a^u a^v = a^{u+v}$$

Expressed in terms of logarithms, the equation becomes

$$\log_a MN = u + v = \log_a M + \log_a N$$

Rule 2:

$$\log_a M^c = c \log_a M, \quad \text{where } c \text{ is any real number}$$

Proof. Let $u = \log_a M$. Then

$$a^u = M \qquad \text{and} \qquad (a^u)^c = a^{uc} = M^c$$

In terms of logarithms, this may be expressed as

$$\log_a M^c = uc = c \log_a M$$

Rule 3:

$$\log_a \frac{M}{N} = \log_a M - \log_a N$$

Proof. $M/N = M(N)^{-1}$. Now apply the previous two rules.

In words, Rule 1 states that the logarithm of a product is equal to the sum of the logarithms of the individual terms; Rule 2 states that the logarithm of a number to a power is the power times the logarithm of the number; and Rule 3 states that the logarithm of a quotient is the difference of the logarithms of the individual terms. Examine these rules carefully and notice where they apply as well as where they do *not* apply. For example, there is *no* rule for simplifying expressions of the form $\log_a (x + y)$ or $\log_a (x - y)$.

Example 1
(a) $\log_2 (8)(64) = \log_2 8 + \log_2 64 = \log_2 2^3 + \log_2 2^6 = 3 + 6 = 9$
(b) $\log_3 \sqrt{243} = \log_3 243^{1/2} = \frac{1}{2} \log_3 243 = \frac{1}{2}(5) = 2.5$
(c) $\log_2 (\frac{3}{5}) = \log_2 3 - \log_2 5$
(d) $\log (4 \cdot 29/5) = \log 4 + \log 29 - \log 5$. (Note that $\log x$ means $\log_{10} x$.)

Example 2. Write $\log x - 2 \log x + 3 \log (x + 1) - \log (x^2 - 1)$ as a single term.

Solution. Proceed as follows:

$$\log x - 2 \log x + 3 \log (x + 1) - \log (x^2 - 1)$$
$$= \log x - \log x^2 + \log (x + 1)^3 - \log (x^2 - 1)$$
$$= \log \frac{x(x + 1)^3}{x^2(x^2 - 1)} = \log \frac{(x + 1)^2}{x(x - 1)}$$

Example 3. Given $\log_a x = 3$, find $\log_a \dfrac{1}{x}$.

Solution. By definition, $\log_a x = 3$ means

$$x = a^3$$

Letting $\log_a \dfrac{1}{x} = y$, we can write

$$\frac{1}{x} = a^y \qquad \text{or} \qquad x = a^{-y}$$

Since a^3 and a^{-y} are both equal to x, we have

$$a^3 = a^{-y}$$

Therefore, $y = \log_a \dfrac{1}{x} = -3$.

Exercises for Section 10.2

In Exercises 1–8, evaluate the given logarithm.

1. $\log_2 32 \cdot 16$
2. $\log_2 16^5$
3. $\log_5 25^{1/4}$

4. $\log_3 27$
5. $\log_3 27 \cdot 9 \cdot 3$
6. $\log_2 64 \cdot 32 \cdot 8$

7. $\log_2 (8 \cdot 32)^3$
8. $\log_3 (9 \cdot 81)^8$

In Exercises 9–20, find each logarithm, given that $\log 2 = 0.3010$, $\log 3 = 0.4771$, and $\log 7 = 0.8451$.

9. $\log \frac{3}{2}$
10. $\log 4$
11. $\log 12$

12. $\log 30$
13. $\log 90$
14. $\log \sqrt{2}$

15. $\log \sqrt{5}$
16. $\log 21^{1/3}$
17. $\log 2400$

18. $\log 0.00018$
19. $\log 0.0014$
20. $\log 42000$

In Exercises 21–28, write the given expression as a single logarithmic term.

21. $\log_2 x^2 - \log_2 x$
22. $\log_2 (x^2 - 1) - \log_2 (x - 1)$

23. $\log x + \log \dfrac{1}{x}$

24. $\log 3x + 3 \log (x + 2) - \log (x^2 - 4)$

25. $\log 5t + 2 \log (t^2 - 4) - \frac{1}{2} \log (t + 3)$

26. $\log z - 3 \log 3z - \log (2z - 9)$

27. $3 \log u - 2 \log (u + 1) - 5 \log (u - 1)$

28. $\log t + 7 \log (2t - 8)$

29. Let $\log_e y = x + \log_e c$. Show that $y = ce^x$.

30. If y is directly proportional to x^p, what relation exists between $\log y$ and $\log x$?

31. Compare the functions $f(x) = \log x^2$ and $g(x) = 2 \log x$. In what way are they the same? In what way are they different?

32. If $f(x) = \log_a x$ and $g(x) = \log_{1/a} x$, show that $g(x) = f\left(\dfrac{1}{x}\right)$.

33. If $\log_a x = 2$, find $\log_{1/a} x$ and $\log_a \dfrac{1}{x}$.

34. Compare the graphs of the following functions:

$$f(x) = \log_2 2x \qquad g(x) = \log_2 x \qquad h(x) = \log_2 \sqrt{x} \qquad m(x) = \log_2 x^2$$

35. Given the graph of $y = \log x$, explain a convenient way to obtain the following graphs.
    (a) $\log x^p$ (b) $\log px$
    (c) $\log (x + p)$ (d) $\log \dfrac{x}{p}$

36. If $f(x) = \log_b x$, is $f(x + y) = f(x) + f(y)$?

In Exercises 37–40, use a calculator to verify the given statement.

37. $\log [25 \cdot 34] = \log 25 + \log 34$

38. $\log [16 \cdot \pi] = \log 16 + \log \pi$

39. $\log \frac{125}{73} = \log 125 - \log 73$

40. $\log \frac{17}{35} = \log 17 - \log 35$

10.3 Exponential and Logarithmic Equations

Equations in which the variable occurs as an exponent are called **exponential equations**. To solve these equations, we use the fact that the logarithm is a one-to-one function. Thus, $\log x = \log y$ if and only if $x = y$. Hence, if both sides are positive, taking the logarithm of both sides yields an equivalent equation.

Example 1. Solve the exponential equation $3^x = 2^{2x+1}$.

Solution. Taking the logarithm of both sides, we write

$$\log 3^x = \log 2^{2x+1}$$
$$x \log 3 = (2x + 1) \log 2 \qquad \text{Using Rule 2}$$
$$x(\log 3 - 2 \log 2) = \log 2 \qquad \text{Expanding and collecting like terms}$$

$$x = \frac{\log 2}{\log 3 - 2 \log 2} \qquad \text{Solving for } x$$

$$= \frac{0.3010}{0.4771 - 2(0.3010)} \qquad \text{Using a calculator}$$

$$x \approx -2.41$$

Example 2. The expression

$$\frac{e^x - e^{-x}}{2}$$

is called the hyperbolic sine of x and is denoted by sinh x. Solve the equation sinh $x = 3$.

Solution

$$\frac{e^x - e^{-x}}{2} = 3 \qquad \text{Replacing sinh } x \text{ by its equal}$$

$$e^x - e^{-x} = 6 \qquad \text{Multiplying both sides by 2}$$

$$e^{2x} - 6e^x - 1 = 0 \qquad \text{Adding } -6 \text{ to both sides and multiplying by } e^x$$

$$u^2 - 6u - 1 = 0 \qquad \text{Letting } u = e^x$$

$$u = \frac{6 \pm \sqrt{40}}{2} = 3 \pm \sqrt{10} \qquad \text{Using the quadratic formula}$$

Since $u = e^x$ is always positive, discard the root $3 - \sqrt{10}$, which is negative. Thus,

$$e^x = 3 + \sqrt{10}$$

Then

$$x = \ln(3 + \sqrt{10}) \approx 1.82$$

is the desired solution.

Equations involving logarithms are called **logarithmic equations**. Often the use of one of the rules of logarithms will allow you to simplify the equation enough to solve it.

Example 3. Solve the logarithmic equation $\log(x^2 - 1) - \log(x - 1) = 3$.

Solution. We simplify the left-hand side by combining the two logarithm terms:

$$\log \frac{x^2 - 1}{x - 1} = 3$$

$$\log(x + 1) = 3 \qquad \text{Canceling } x - 1 \text{ in } x^2 - 1$$

$$x + 1 = 10^3 \qquad \log M = N \text{ means } M = 10^N$$

$$x = -1 + 10^3 = 999 \qquad \text{Adding } -1 \text{ to both sides}$$

Example 4. Solve the equation

$$x^{\log x} = \frac{x^3}{100}$$

Solution. We take the logarithm of both sides to obtain

$$\log x^{\log x} = \log \frac{x^3}{100}$$

$$(\log x)(\log x) = \log x^3 - \log 100 \qquad \text{Using Rules 2 and 3}$$

$$(\log x)^2 - \log x^3 + \log 100 = 0 \qquad \text{Collecting terms on the left}$$

$$(\log x)^2 - 3 \log x + 2 = 0 \qquad \begin{array}{l} \log x^3 = 3 \log x \text{ and} \\ \log 100 = 2 \end{array}$$

$$(\log x - 2)(\log x - 1) = 0 \qquad \text{Factoring}$$

$$\log x = 2 \text{ or } \log x = 1 \qquad \text{Solving for } \log x$$

Thus, $x = 100$ or 10.

Example 5. A certain power supply has a power output in watts governed by the equation $P = 50e^{-t/250}$, where t is the time in days. If the equipment aboard a satellite requires 10 watts of power to operate properly, what is the operational life of the satellite?

Solution. Letting $P = 10$, we solve the equation $10 = 50e^{-t/250}$ for t. Dividing both sides by 50, we have

$$e^{-t/250} = 0.2$$

$$\frac{-t}{250} = \ln 0.2 \qquad e^y = x \text{ means } y = \ln x$$

$$t = -250 \ln 0.2 \qquad \text{Multiplying both sides by } -250$$

$$= -250 \, (-1.609) \qquad \text{Using a calculator for } \ln 0.2$$

$$\approx 402$$

Hence, the operational life of the satellite is, 402 days.

Exercises for Section 10.3

In Exercises 1–26, solve for x.

1. $7^{x+1} = 2^x$

2. $3^x 2^{2x+1} = 10$

3. $10^{x^2} = 2^x$

4. $8^x = 10^x$

5. $2^{1+x} = 3$

6. $3^{1-x} = 7^x$

7. $2^{x+5} = 3^{x-2}$

8. $(\frac{1}{2})^x > 3$

9. $2^{x+1} > 5$

10. $5^{-(1+x)} < 8$

11. $2^{\log x} = 2$

12. $\log \log \log x = 1$

13. $(\log x)^1 = \log \sqrt{x}$

14. $(\log x)^2 = \log x^2$

15. $(\log x)^3 = \log x^3$

16. $x^{\log x} = 10$

17. $\log (x + 15) + \log x = 2$

18. $\log 3x - \log 2x = \log 3 - \log x$

19. $\log (x - 2) - \log (2x + 1) = \log \dfrac{1}{x}$

20. $\log (x^2 + 1) - \log (x - 1) - \log (x + 1) = 1$

21. $\log (x + 2) - \log (x - 2) - \log x + \log (x - 3) = 0$

22. $\log x + \log (x - 99) = \log 2$

23. $\log x + \log (x^2 - 4) - \log 2x = 0$

24. $3 \log (x - 1) = \log 3(x - 1)$

25. $\log (x + 1) - \log x < 1$

26. $\log (x + 1) + \log (x - 1) < 2$

27. The expression

$$\frac{e^x + e^{-x}}{2}$$

is called the hyperbolic cosine of x and is denoted by cosh x. Solve the equation cosh $x = 2$. (*Hint*: Let $u = e^x$.)

28. Solve the two equations $y = 50e^{-2x}$ and $y = 2^x$ simultaneously by making a sketch of the two equations on the same coordinate system.

29. Explain why if $\log u(x) = v(x)$, then $u(x) = 10^{v(x)}$.

30. Show that if $y = e^{\ln f(x)}$, then $y = f(x)$.

31. The radioactive chemical element strontium 90 has a half-life of approximately 28 years. The element obeys the radioactive decay formula $A(t) = A_0 e^{-kt}$, where A_0 is the original amount and t is the time in years. Find the value of k.

32. Repeat Exercise 31 for the element iodine, whose decay formula is the same type. Express t in days. The half-life is 8 days.

33. What is the half-life of the power supply of Example 5?

34. The difference in intensity level of two sounds with intensities I and I_0 is defined by

$$10 \log \frac{I}{I_0} \text{ db}$$

Find the intensity level in decibels of the sound produced by an electric motor that is 175.6 times greater than I_0.

35. The population growth curve is given by $P = P_0 e^{kt}$, where P_0 is the initial size, t is time in hours, and k is a constant. If $k = 0.0132$ for a bacteria culture, how long does it take for the culture to double in size?

Key Topics for Chapter 10

Define and/or discuss each of the following.

Definition of a Logarithm
Properties of Logarithms
Logarithmic Equations

Review Exercises for Chapter 10

In Exercises 1–8, solve for x.

1. $\log x = 3$
2. $\log_2 x = 5$
3. $\log_3 81 = x$
4. $\log_5 \frac{1}{25} = x$
5. $\log_x 64 = 6$
6. $\log_x 0.027 = -3$
7. $\log_x 0.01 = -2$
8. $\log x = -8$

In Exercises 9–14, evaluate each logarithm without a table or a calculator.

9. $\log_2 (8 \cdot 2)$
10. $\log_2 \frac{1}{16}$
11. $\log_3 \sqrt{27}$
12. $\log_5 (25)^{1/3}$
13. $\log_5 (625 \cdot 125)^3$
14. $\log_2 (128 \cdot 32)^4$

In Exercises 15–20, write the given expression as a single logarithmic term.

15. $\log x + \log x^2$
16. $\log 2x - \log x$
17. $\log_2 2x + 3 \log_2 x$
18. $3 \log_2 x - \log_2 x$
19. $3 \log_5 x - \log_2 (2x - 3)$
20. $5 \log (x + 2) - 2 \log x$

In Exercises 21–30, solve each of the equations.

21. $2^{x+2} = 10$
22. $5^{3-x} = 8$
23. $3^{x+1} = 4^x$
24. $2^x = 3^{x-1}$
25. $\log (\log x) = 1$
26. $(\log x)^2 = \log x$
27. $\log x + \log (x + 1) = \log 6$
28. $\log (x - 2) + \log x = \log 8$
29. $\log x + \log (x - 3) = 1$
30. $\log 5x + \log 2x = 2 + \log x$

31. Find the half-life of a radioactive material that decays exponentially according to the equation $A(t) = A_0 e^{-t/4}$, where t is the time in years and A_0 is the original amount.

32. A colony of bacteria increases according to the law $N(t) = N(0)e^{kt}$. If the colony doubles in 5 hr, find the time needed for the colony to triple.

33. A bacteria colony population is given by $N(t) = 2000\,e^{1.3t}$. Plot a graph of N as a function of t.

34. A principal of \$5000 is invested at the rate r of 9.2% per year compounded continuously. What will be the amount after 20 years? The law governing continuous compounding is $P = P_0 e^{rt}$, where t is the time in years and P_0 is the initial principal.

35. In the dye-dilution procedure for measuring cardiac output, the amount in mg of dye in the heart at any time, t, is given by $D(t) = D(0)e^{-rt/V}$, where $D(0)$ is the amount in mg of dye injected, r is a constant representing the outflow of blood and dye in liters per minute, and V is the volume of the heart in liters. Find the amount of dye in the heart after 5 sec, given that $V = 450$ mL, $r = 1.4$ L/min, and $D(0) = 2.3$ mg. (*Hint*: Use consistent units.)

Appendix A

Review of Elementary Geometry

In this appendix we provide a review of terminology and results of plane geometry that are important in the study of trigonometry.

Geometry is the study of the properties of points, lines, planes, angles, surfaces, and volumes. Certain geometric concepts, such as those of a point, a line, and a plane, are accepted without definition. Technically, a dot on a paper is not really a point since it has dimension; however, by general agreement we make visible marks (dots) to represent points and denote them with capital letters such as A, B, C, and so forth, as shown in Figure A.1.

Similarly, a straight line drawn with a ruler is technically not a line, since a line has no width. In geometry we think of a **line*** as extending indefinitely in both directions. Two points determine a line; thus, a line passing through the points A and B is denoted by $\overleftrightarrow{AB}$. (See Figure A.1.)

A **ray** consists of a point A on a line and that part of the line that is on one side of the point. A ray is denoted by $\overrightarrow{AB}$, as shown in Figure A.2.

(a) The point A (b) The line $\overleftrightarrow{AB}$
Figure A.1

Figure A.2   The Ray $\overrightarrow{AB}$

A **line segment**, $\overline{AB}$, is the part of a line between the two points A and B. Depending on the application, the line segment may or may not include the points A and B. (See Figure A.3.)

Figure A.3   The Line Segment $\overline{AB}$

---

* We shall use *line* to mean a straight line.

Two rays with a common endpoint form an **angle**. The common point is called the **vertex** of the angle, and the rays are called the **sides**. (See Figure A.4.) We generally refer to an angle by mentioning a point on each of its sides and the vertex, although sometimes only the vertex is mentioned. An angle whose two rays form a straight line is called a **straight angle**. (See Figure A.5.)

Figure A.4 ∠BAC or ∠A

Figure A.5 ∠BOC is a straight angle.

Sometimes we conceive of an angle as being "formed" by rotating one of the sides about the angle's vertex while keeping the other side fixed, as shown in Figure A.6. If OA is fixed and OB is rotated about the vertex, OA is called the **initial side** and OB the **terminal side** of the generated angle. The size of the angle depends on the amount of rotation of the terminal side. Thus ∠AOB is considered to be smaller than ∠AOB'.

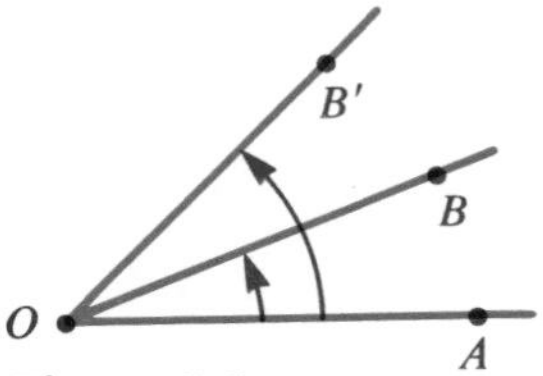

Angles with the same initial and terminal sides are said to be **coterminal**. Thus, even though angles have the same rays forming their sides, they may not be equal. Figure A.7 shows two angles that have the same sides but are not equal because they are formed by different rotations.

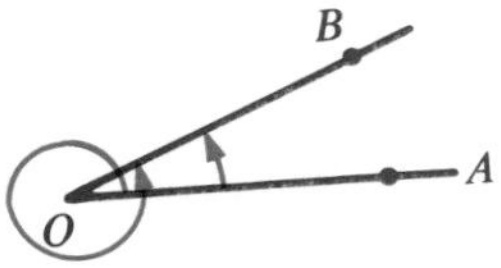

The most common unit of angular measure is the **degree**. The measure of a straight angle is 180°. One half of a straight angle is called a **right angle**; a right angle has a measure of 90°. (See Figure A.8.) An angle is **acute** if its measure is less than 90° and **obtuse** if its measure is more than 90° but less than 180°. (See Figure A.9.)

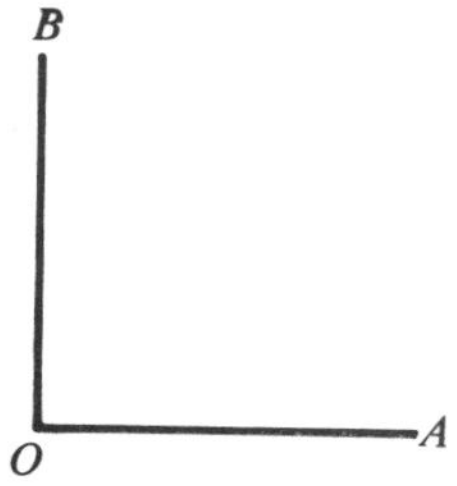

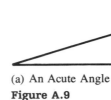

(a) An Acute Angle

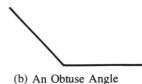

(b) An Obtuse Angle

Figure A.9

Two angles that together form a right angle are **complementary**. If two angles form a straight angle when combined, they are **supplementary** angles. (See Figure A.10.)

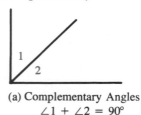

(a) Complementary Angles
∠1 + ∠2 = 90°

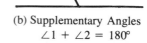

(b) Supplementary Angles
∠1 + ∠2 = 180°

Figure A.10

Figure A.6

Figure A.7

B

O ————— A

Figure A.8   A Right Angle

If two lines have no points in common, they are said to be **parallel**. (See Figure A.11.) Two nonparallel lines intersect at a point to form four angles. In any such intersection, **opposite** angles are equal. Thus, in Figure A.12, $\angle 1$ and $\angle 3$ are equal, as are $\angle 2$ and $\angle 4$.

Figure A.11   Parallel Lines   **Figure A.12**   Intersecting Lines

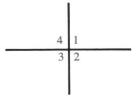

Figure A.13
Perpendicular Lines

If $\angle 1$, $\angle 2$, $\angle 3$, and $\angle 4$ are all right angles, the two lines are perpendicular. (See Figure A.13.)

A line drawn through two parallel lines forms eight angles. The following list summarizes the relationships among these angles. (See Figure A.14.)

- **Corresponding** angles such as $\angle 1$ and $\angle 5$ or $\angle 4$ and $\angle 8$ are equal.
- **Alternate interior** angles such as $\angle 3$ and $\angle 6$ or $\angle 5$ and $\angle 4$ are equal.
- **Alternate exterior** angles such as $\angle 1$ and $\angle 8$ or $\angle 2$ and $\angle 7$ are equal.

The relationships of parallelism and perpendicularity are such that

- Two lines perpendicular to the same line are parallel.
- Two lines parallel to the same line are parallel.

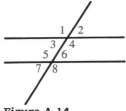

Figure A.14

Polygons

A **polygonal path** is one obtained by connecting line segments as shown in Figure A.15. A polygonal path of more than two segments that begins and ends at the same point, and that has no other points of intersection is called a **polygon**. (See Figure A.16.) A polygon has an interior and an

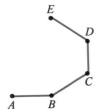

Figure A.15   Polygonal Path

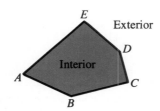

Figure A.16   A Polygon

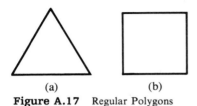

exterior. The interior of a polygon is called its **area**. The **sides** of the polygon are the line segments, and the **angles** of the polygon are those angles interior to the polygon formed by the sides. Polygons are named for the number of sides.

- A **triangle** has three sides.
- A **quadrilateral** has four sides.
- A **pentagon** has five sides.

    A **regular** polygon is one for which all sides have the same length.

- A regular triangle is said to be **equilateral**. [See Figure A.17(a).]
- A regular quadrilateral is a **square**. [See Figure A.17(b).]

Polygons are compared by comparing their line segments and vertices in a specific order. Thus the triangle ABC may be compared with the triangle $A'B'C'$ by any of the six orderings ABC, ACB, BAC, BCA, CAB, or CBA. When a comparison is made for a particular ordering, the sides and angles are said to **correspond**. For example, when $\triangle ABC$ and $\triangle A'C'B'$ are compared, the corresponding angles are $\angle A$ and $\angle A'$, $\angle B$ and $\angle C'$, and $\angle C$ and $\angle B'$; the corresponding sides are $\overline{AB}$ and $\overline{A'C'}$, $\overline{AC}$ and $\overline{A'B'}$, and $\overline{BC}$ and $\overline{C'B'}$. (See Figure A.18.) The notion of corresponding angles and sides is used as a means for comparing the size and shape of two or more polygons.

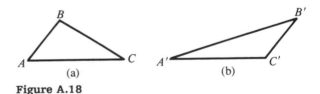

Figure A.18

Two polygons are **similar** if their corresponding angles are equal. In effect, similar polygons have the same shape but not necessarily the same size. The symbol for similarity is $\sim$. In Figure A.19, $\triangle ABC \sim \triangle A'B'C'$.

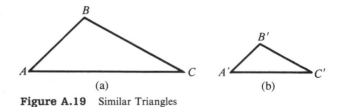

Figure A.19   Similar Triangles

Two polygons are **congruent** if their corresponding angles and sides are equal. Two congruent polygons will coincide exactly in their parts if placed properly one upon the other. The symbol for congruence is $\cong$. In Figure A.20, quadrilateral $ABCD$ is congruent to quadrilateral $A'B'C'D'$.

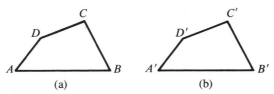

Figure A.20   Congruent Quadrilaterals

Triangles

A triangle is said to be **equiangular** if each of its three angles is exactly the same, and **equilateral** if all three sides have the same length.

- A triangle is equilateral if and only if it is equiangular. [See Figure A.21(a).]
- A triangle is isosceles if two of its angles are equal. [See Figure A.21(b).] In an isosceles triangle, the sides opposite the two equal angles are equal.

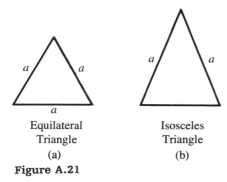

Equilateral
Triangle
(a)

Isosceles
Triangle
(b)

Figure A.21

The perpendicular distance from a vertex to the opposite side of a triangle is called an **altitude**, and the side opposite the vertex is called the **base** for that altitude. Any two triangles with the same base and altitude have the same area. (See Figure A.22.)

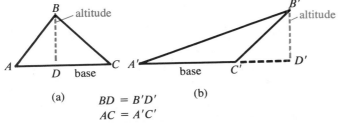

$$BD = B'D'$$
$$AC = A'C'$$

Figure A.22   Since $\overline{BD} = \overline{B'D'}$ and $\overline{AC} = \overline{A'C'}$, these two triangles have equal areas.

A **right** triangle is one in which one of the angles is a right angle. An **oblique** triangle is one without a right angle. In any triangle, the sum of the measures of the angles is 180°. Thus, in an equilateral triangle, each of the angles measures 60°. In a right triangle, each of the non-right angles is acute and the sum of their measures is 90°.

There is a relatively standard method for referencing the sides and the angles of any triangle. For example, consider a triangle with vertices A, B, and C. The sides $\overline{AB}$ and $\overline{AC}$ are called the sides **adjacent** to the angle at vertex A. The side $\overline{BC}$ is called the side **opposite** angle A. The sides opposite and adjacent to angle B and those opposite and adjacent to angle C are referred to similarly. In the special case of a right triangle, the side opposite the right angle is called the **hypotenuse**.

In the right triangle in Figure A.23, side $\overline{AC}$ is called the side adjacent to angle A, side $\overline{BC}$ is called the side opposite angle A, and side $\overline{AB}$ is called the hypotenuse.

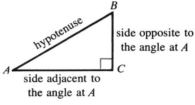

Figure A.23

Similar Triangles

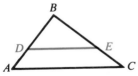

Figure A.24

In a triangle ABC, if $\overline{DE}$ is parallel to $\overline{AC}$, then $\triangle DBE$ is similar to $\triangle ABC$. (See Figure A.24.) The corresponding sides of similar triangles are proportional; that is, the ratios of corresponding sides are equal. Thus, from Figure A.24,

$$\frac{\overline{AB}}{\overline{DB}} = \frac{\overline{BC}}{\overline{BE}} = \frac{\overline{AC}}{\overline{DE}}$$

Congruent Triangles

The following three theorems summarize the three ways in which two triangles can be found to be congruent.

(1) Two triangles are congruent if and only if two sides and the *included* angle of one are equal, respectively, to two sides and the *included* angle

of the other. (See Figure A.25.) For instance, $\triangle ABC$ and $\triangle A'B'C'$ are congruent if $\overline{AB} = \overline{A'B'}$, $\overline{AC} = \overline{A'C'}$, and $\angle A = \angle A'$.

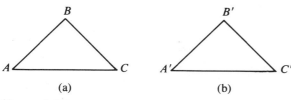

(a)   (b)

Figure A.25

(2) Two triangles are congruent if and only if three sides of one are equal, respectively, to three sides of the other. For instance, in Figure A.25, $\triangle ABC$ and $\triangle A'B'C'$ are congruent if $\overline{AB} = \overline{A'B'}$, $\overline{AC} = \overline{A'C'}$, and $\overline{BC} = \overline{B'C'}$.

(3) Two triangles are congruent if and only if two angles and the *included* side of one are equal, respectively, to two angles and the *included* side of the other. For instance, $\triangle ABC$ and $\triangle A'B'C'$ are congruent if $\angle A = \angle A'$, $\angle B = \angle B'$, and $\overline{AB} = \overline{A'B'}$.

Note that two triangles may NOT be considered congruent merely because two sides of a triangle and the angle opposite one of them are equal, respectively, to two sides of another triangle and the angle opposite one of them. Figure A.26 shows two triangles for which two sides and the angle opposite $\overline{BC}$ are equal. The two triangles are obviously not congruent.

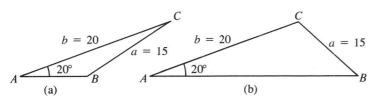

(a)   (b)

Figure A.26

Right Triangles

Right triangles are at the heart of the study of trigonometry. They have the following special properties:

(1) The square of the hypotenuse is equal to the sum of the squares of the other two sides. (See Figure A.27.) This rule is called the **Pythagorean theorem**.

(2) In a 30°–60° right triangle, the length of the side opposite the 30° angle is equal to one-half the length of the hypotenuse. If the hypotenuse is 2 units, the side opposite the 30° angle is 1 unit. Then, by the Pythago-

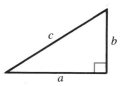

Figure A.27
Pythagorean Theorem
$a^2 + b^2 = c^2$

rean theorem, the side opposite the 60° angle is $\sqrt{3}$. (See Figure A.28.)

(3) In a 45°-45° right triangle, the sides opposite the 45° angles are equal. If the sides opposite the 45° angles have a length of 1 unit, then, by the Pythagorean theorem, the length of the hypotenuse is $\sqrt{2}$. (See Figure A.29.)

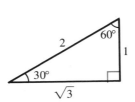

Figure A.28   30°-60° Right Triangle

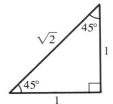

Figure A.29   45°-45° Right Triangle

Quadrilaterals

A **quadrilateral** is a polygon with four sides. (See Figure A.30.) A quadrilateral that has two and only two parallel sides is called a **trapezoid**. If both pairs of opposite sides are parallel line segments, the quadrilateral is called a **parallelogram**. A **rectangle** is a parallelogram with four right angles.

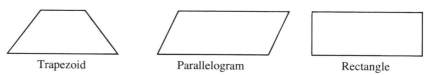

Trapezoid   Parallelogram   Rectangle

Figure A.30

The perpendicular distance between the parallel sides of a trapezoid is called the **altitude** of the trapezoid. The parallel sides are called the **bases**.

Circles

A **circle** is a set of points in a plane that are the same distance from a fixed point. (See Figure A.31.) The fixed point is called the **center** of the circle, and the distance from the center to any point on the circle is called the **radius**. A line segment connecting any two points on a circle is called a **chord**. The **diameter** of the circle is the length of any chord that passes through the center. The total distance around the circle is its **circumference**.

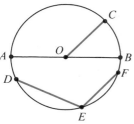

Figure A.31

Figure A.32

An **inscribed angle** is an angle formed by two chords that meet at a common point on the circle. A **central angle** is formed at the center of a circle by two radii. A central angle **subtends** an arc on the circle. A **sector** of a circle is the closed figure formed by a central angle and its subtended arc.

In Figure A.31, $\overline{OA}$, $\overline{AB}$, and $\overline{OC}$ are radii; $\overline{AB}$ is a diameter; $\overline{AB}$, $\overline{DE}$, and $\overline{EF}$ are chords; BOC is a central angle; BOC subtends arc BC; and DEF is an inscribed angle.

Circles have the following properties:

■ An angle inscribed in a semicircle is a right angle. (See Figure A.32.)
■ The ratio of the circumference to the diameter is the same for all circles. This ratio is equal to π.

Formulas from Geometry

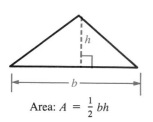

Area: $A = \frac{1}{2}bh$

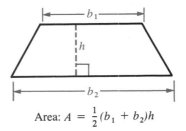

Area: $A = \frac{1}{2}(b_1 + b_2)h$

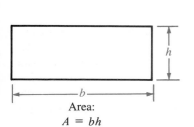

Area:
$A = bh$

Area: $A = \pi r^2 = \frac{1}{4}\pi D^2$

$(D = 2r)$

Circumference: $C = 2\pi r = \pi D$

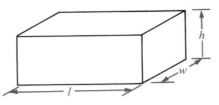

Surface area: $A = 2(lw + lh + wh)$
Volume: $V = hwl$

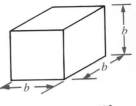

Surface area: $A = 6b^2$
Volume: $V = b^3$

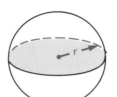

Surface area: $A = 4\pi r^2$
Volume: $V = \frac{4}{3}\pi r^3$

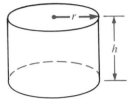

Surface area: $A = 2\pi rh + 2\pi r$
Volume: $V = \pi r^2 h$

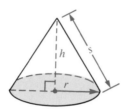
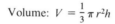

Lateral area: $A = \pi rs$

Volume: $V = \frac{1}{3}\pi r^2 h$

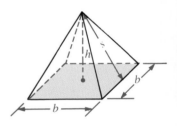

Lateral area: $A = 2\,bs$

Volume: $V = \frac{1}{3}\,b^2 h$

Appendix B

Tables

Table A   Values of the Trigonometric Functions for Degrees, pages 292–299

Table B   Values of Trigonometric Functions—Decimal Subdivisions
(csc θ = 1/sin θ; sec θ = 1/cos θ), pages 300–311

Table C   Values of the Trigonometric Functions for Radians and
Real Numbers, pages 312–315

Table A Values of the Trigonometric Functions for Degrees

x	sin x	cos x	tan x	cot x	sec x	csc x	
0° 0′	.00000	1.0000	.00000	∞	1.0000	∞	90° 0′
10′	.00291	1.0000	.00291	343.77	1.0000	343.78	50′
20′	.00582	1.0000	.00582	171.88	1.0000	171.89	40′
30′	.00873	1.0000	.00873	114.59	1.0000	114.59	30′
40′	.01164	.9999	.01164	85.940	1.0001	85.946	20′
50′	.01454	.9999	.01455	68.750	1.0001	68.757	10′
1° 0′	.01745	.9998	.01746	57.290	1.0002	57.299	89° 0′
10′	.02036	.9998	.02036	49.104	1.0002	49.114	50′
20′	.02327	.9997	.02328	42.964	1.0003	42.976	40′
30′	.02618	.9997	.02619	38.188	1.0003	38.202	30′
40′	.02908	.9996	.02910	34.368	1.0004	34.382	20′
50′	.03199	.9995	.03201	31.242	1.0005	31.258	10′
2° 0′	.03490	.9994	.03492	28.6363	1.0006	28.654	88° 0′
10′	.03781	.9993	.03783	26.4316	1.0007	26.451	50′
20′	.04071	.9992	.04075	24.5418	1.0008	24.562	40′
30′	.04362	.9990	.04366	22.9038	1.0010	22.926	30′
40′	.04653	.9989	.04658	21.4704	1.0011	21.494	20′
50′	.04943	.9988	.04949	20.2056	1.0012	20.230	10′
3° 0′	.05234	.9986	.05241	19.0811	1.0014	19.107	87° 0′
10′	.05524	.9985	.05533	18.0750	1.0015	18.103	50′
20′	.05814	.9983	.05824	17.1693	1.0017	17.198	40′
30′	.06105	.9981	.06116	16.3499	1.0019	16.380	30′
40′	.06395	.9980	.06408	15.6048	1.0021	15.637	20′
50′	.06685	.9978	.06700	14.9244	1.0022	14.958	10′
4° 0′	.06976	.9976	.06993	14.3007	1.0024	14.336	86° 0′
10′	.07266	.9974	.07285	13.7267	1.0027	13.763	50′
20′	.07556	.9971	.07578	13.1969	1.0029	13.235	40′
30′	.07846	.9969	.07870	12.7062	1.0031	12.746	30′
40′	.08136	.9967	.08163	12.2505	1.0033	12.291	20′
50′	.08426	.9964	.08456	11.8262	1.0036	11.868	10′
5° 0′	.08716	.9962	.08749	11.4301	1.0038	11.474	85° 0′
10′	.09005	.9959	.09042	11.0594	1.0041	11.105	50′
20′	.09295	.9957	.09335	10.7119	1.0044	10.758	40′
30′	.09585	.9954	.09629	10.3854	1.0046	10.433	30′
40′	.09874	.9951	.09923	10.0780	1.0049	10.128	20′
50′	.10164	.9948	.10216	9.7882	1.0052	9.839	10′
6° 0′	.10453	.9945	.10510	9.5144	1.0055	9.5668	84° 0′
	cos x	sin x	cot x	tan x	csc x	sec x	x

x	sin x	cos x	tan x	cot x	sec x	csc x	
6° 0′	.1045	.9945	.10510	9.5144	1.0055	9.5668	84° 0′
10′	.1074	.9942	.10805	9.2553	1.0058	9.3092	50′
20′	.1103	.9939	.11099	9.0098	1.0061	9.0652	40′
30′	.1132	.9936	.11394	8.7769	1.0065	8.8337	30′
40′	.1161	.9932	.11688	8.5555	1.0068	8.6138	20′
50′	.1190	.9929	.11983	8.3450	1.0072	8.4647	10′
7° 0′	.1219	.9925	.12278	8.1443	1.0075	8.2055	83° 0′
10′	.1248	.9922	.12574	7.9530	1.0079	8.0157	50′
20′	.1276	.9918	.12869	7.7704	1.0083	7.8344	40′
30′	.1305	.9914	.13165	7.5958	1.0086	7.6613	30′
40′	.1334	.9911	.1346	7.4287	1.0090	7.4957	20′
50′	.1363	.9907	.1376	7.2687	1.0094	7.3372	10′
8° 0′	.1392	.9903	.1405	7.1154	1.0098	7.1853	82° 0′
10′	.1421	.9899	.1435	6.9682	1.0102	7.0396	50′
20′	.1449	.9894	.1465	6.8269	1.0107	6.8998	40′
30′	.1478	.9890	.1495	6.6912	1.0111	6.7655	30′
40′	.1507	.9886	.1524	6.5606	1.0116	6.6363	20′
50′	.1536	.9881	.1554	6.4348	1.0120	6.5121	10′
9° 0′	.1564	.9877	.1584	6.3138	1.0125	6.3925	81° 0′
10′	.1593	.9872	.1614	6.1970	1.0129	6.2772	50′
20′	.1622	.9868	.1644	6.0844	1.0134	6.1661	40′
30′	.1650	.9863	.1673	5.9758	1.0139	6.0589	30′
40′	.1679	.9858	.1703	5.8708	1.0144	5.9554	20′
50′	.1708	.9853	.1733	5.7694	1.0149	5.8554	10′
10° 0′	.1736	.9848	.1763	5.6713	1.0154	5.7588	80° 0′
10′	.1765	.9843	.1793	5.5764	1.0160	5.6653	50′
20′	.1794	.9838	.1823	5.4845	1.0165	5.5749	40′
30′	.1822	.9833	.1853	5.3955	1.0170	5.4874	30′
40′	.1851	.9827	.1883	5.3093	1.0176	5.4026	20′
50′	.1880	.9822	.1914	5.2257	1.0182	5.3205	10′
11° 0′	.1908	.9816	.1944	5.1446	1.0187	5.2408	79° 0′
10′	.1937	.9811	.1974	5.0658	1.0193	5.1636	50′
20′	.1965	.9805	.2004	4.9894	1.0199	5.0886	40′
30′	.1994	.9799	.2035	4.9152	1.0205	5.0159	30′
40′	.2022	.9793	.2065	4.8430	1.0211	4.9452	20′
50′	.2051	.9787	.2095	4.7729	1.0217	4.8765	10′
12° 0′	.2079	.9781	.2126	4.7046	1.0223	4.8097	78° 0′
	cos x	sin x	cot x	tan x	csc x	sec x	x

Table A Values of the Trigonometric Functions for Degrees (continued)

x	sin x	cos x	tan x	cot x	sec x	csc x	
12° 0′	.2079	.9781	.2126	4.7046	1.0223	4.8097	78° 0′
10′	.2108	.9775	.2156	4.6382	1.0230	4.7448	50′
20′	.2136	.9769	.2186	4.5736	1.0236	4.6817	40′
30′	.2164	.9763	.2217	4.5107	1.0243	4.6202	30′
40′	.2193	.9757	.2247	4.4494	1.0249	4.5604	20′
50′	.2221	.9750	.2278	4.3897	1.0256	4.5022	10′
13° 0′	.2250	.9744	.2309	4.3315	1.0263	4.4454	77° 0′
10′	.2278	.9737	.2339	4.2747	1.0270	4.3901	50′
20′	.2306	.9730	.2370	4.2193	1.0277	4.3362	40′
30′	.2334	.9724	.2401	4.1653	1.0284	4.2837	30′
40′	.2363	.9717	.2432	4.1126	1.0291	4.2324	20′
50′	.2391	.9710	.2462	4.0611	1.0299	4.1824	10′
14° 0′	.2419	.9703	.2493	4.0108	1.0306	4.1336	76° 0′
10′	.2447	.9696	.2524	3.9617	1.0314	4.0859	50′
20′	.2476	.9689	.2555	3.9136	1.0321	4.0394	40′
30′	.2504	.9681	.2586	3.8667	1.0329	3.9939	30′
40′	.2532	.9674	.2617	3.8208	1.0337	3.9495	20′
50′	.2560	.9667	.2648	3.7760	1.0345	3.9061	10′
15° 0′	.2588	.9659	.2679	3.7321	1.0353	3.8637	75° 0′
10′	.2616	.9652	.2711	3.6891	1.0361	3.8222	50′
20′	.2644	.9644	.2742	3.6470	1.0369	3.7817	40′
30′	.2672	.9636	.2773	3.6059	1.0377	3.7420	30′
40′	.2700	.9628	.2805	3.5656	1.0386	3.7032	20′
50′	.2728	.9621	.2836	3.5261	1.0394	3.6652	10′
16° 0′	.2756	.9613	.2867	3.4874	1.0403	3.6280	74° 0′
10′	.2784	.9605	.2899	3.4495	1.0412	3.5915	50′
20′	.2812	.9596	.2931	3.4124	1.0421	3.5559	40′
30′	.2840	.9588	.2962	3.3759	1.0430	3.5209	30′
40′	.2868	.9580	.2994	3.3402	1.0439	3.4867	20′
50′	.2896	.9572	.3026	3.3052	1.0448	3.4532	10′
17° 0′	.2924	.9563	.3057	3.2709	1.0457	3.4203	73° 0′
10′	.2952	.9555	.3089	3.2371	1.0466	3.3881	50′
20′	.2979	.9546	.3121	3.2041	1.0476	3.3565	40′
30′	.3007	.9537	.3153	3.1716	1.0485	3.3255	30′
40′	.3035	.9528	.3185	3.1397	1.0495	3.2951	20′
50′	.3062	.9520	.3217	3.1084	1.0505	3.2653	10′
18° 0′	.3090	.9511	.3249	3.0777	1.0515	3.2361	72° 0′
	cos x	sin x	cot x	tan x	csc x	sec x	x

x	sin x	cos x	tan x	cot x	sec x	csc x	
18° 0′	.3090	.9511	.3249	3.0777	1.0515	3.2361	72° 0′
10′	.3118	.9502	.3281	3.0475	1.0525	3.2074	50′
20′	.3145	.9492	.3314	3.0178	1.0535	3.1792	40′
30′	.3173	.9483	.3346	2.9887	1.0545	3.1516	30′
40′	.3201	.9474	.3378	2.9600	1.0555	3.1244	20′
50′	.3228	.9465	.3411	2.9319	1.0566	3.0977	10′
19° 0′	.3256	.9455	.3443	2.9042	1.0576	3.0716	71° 0′
10′	.3283	.9446	.3476	2.8770	1.0587	3.0458	50′
20′	.3311	.9436	.3508	2.8502	1.0598	3.0206	40′
30′	.3338	.9426	.3541	2.8239	1.0609	2.9957	30′
40′	.3365	.9417	.3574	2.7980	1.0620	2.9714	20′
50′	.3393	.9407	.3607	2.7725	1.0631	2.9474	10′
20° 0′	.3420	.9397	.3640	2.7475	1.0642	2.9238	70° 0′
10′	.3448	.9387	.3673	2.7228	1.0653	2.9006	50′
20′	.3475	.9377	.3706	2.6985	1.0665	2.8879	40′
30′	.3502	.9367	.3739	2.6746	1.0676	2.8555	30′
40′	.3529	.9356	.3772	2.6511	1.0688	2.8334	20′
50′	.3557	.9346	.3805	2.6279	1.0700	2.8118	10′
21° 0′	.3584	.9336	.3839	2.6051	1.0712	2.7904	69° 0′
10′	.3611	.9325	.3872	2.5826	1.0724	2.7695	50′
20′	.3638	.9315	.3906	2.5605	1.0736	2.7488	40′
30′	.3665	.9304	.3939	2.5386	1.0748	2.7285	30′
40′	.3692	.9293	.3973	2.5172	1.0760	2.7085	20′
50′	.3719	.9283	.4006	2.4960	1.0773	2.6888	10′
22° 0′	.3746	.9272	.4040	2.4751	1.0785	2.6695	68° 0′
10′	.3773	.9261	.4074	2.4545	1.0798	2.6504	50′
20′	.3800	.9250	.4108	2.4342	1.0811	2.6316	40′
30′	.3827	.9239	.4142	2.4142	1.0824	2.6131	30′
40′	.3854	.9228	.4176	2.3945	1.0837	2.5949	20′
50′	.3881	.9216	.4210	2.3750	1.0850	2.5770	10′
23° 0′	.3907	.9205	.4245	2.3559	1.0864	2.5593	67° 0′
10′	.3934	.9194	.4279	2.3369	1.0877	2.5419	50′
20′	.3961	.9182	.4314	2.3183	1.0891	2.5247	40′
30′	.3987	.9171	.4348	2.2998	1.0904	2.5078	30′
40′	.4014	.9159	.4383	2.2817	1.0918	2.4912	20′
50′	.4041	.9147	.4417	2.2637	1.0932	2.4748	10′
24° 0′	.4067	.9135	.4452	2.2460	1.0946	2.4586	66° 0′
	cos x	sin x	cot x	tan x	csc x	sec x	x

Table A   Values of the Trigonometric Functions for Degrees (continued)

x	sin x	cos x	tan x	cot x	sec x	csc x	
24° 0′	.4067	.9135	.4452	2.2460	1.0946	2.4586	66° 0′
10′	.4094	.9124	.4487	2.2286	1.0961	2.4426	50′
20′	.4120	.9112	.4522	2.2113	1.0975	2.4269	40′
30′	.4147	.9100	.4557	2.1943	1.0990	2.4114	30′
40′	.4173	.9088	.4592	2.1775	1.1004	2.3961	20′
50′	.4200	.9075	.4628	2.1609	1.1019	2.3811	10′
25° 0′	.4226	.9063	.4663	2.1445	1.1034	2.3662	65° 0′
10′	.4253	.9051	.4699	2.1283	1.1049	2.3515	50′
20′	.4279	.9038	.4734	2.1123	1.1064	2.3371	40′
30′	.4305	.9026	.4770	2.0965	1.1079	2.3228	30′
40′	.4331	.9013	.4806	2.0809	1.1095	2.3088	20′
50′	.4358	.9001	.4841	2.0655	1.1110	2.2949	10′
26° 0′	.4384	.8988	.4877	2.0503	1.1126	2.2812	64° 0′
10′	.4410	.8975	.4913	2.0353	1.1142	2.2677	50′
20′	.4436	.8962	.4950	2.0204	1.1158	2.2543	40′
30′	.4462	.8949	.4986	2.0057	1.1174	2.2412	30′
40′	.4488	.8936	.5022	1.9912	1.1190	2.2282	20′
50′	.4514	.8923	.5059	1.9768	1.1207	2.2154	10′
27° 0′	.4540	.8910	.5095	1.9626	1.1223	2.2027	63° 0′
10′	.4566	.8897	.5132	1.9486	1.1240	2.1902	50′
20′	.4592	.8884	.5196	1.9347	1.1257	2.1779	40′
30′	.4617	.8870	.5206	1.9210	1.1274	2.1657	30′
40′	.4643	.8857	.5243	1.9074	1.1291	2.1537	20′
50′	.4669	.8843	.5280	1.8940	1.1308	2.1418	10′
28° 0′	.4695	.8829	.5317	1.8807	1.1326	2.1301	62° 0′
10′	.4720	.8816	.5354	1.8676	1.1343	2.1185	50′
20′	.4746	.8802	.5392	1,8546	1.1361	2.1070	40′
30′	.4772	.8788	.5430	1.8418	1.1379	2.0957	30′
40′	.4797	.8774	.5467	1.8291	1.1397	2.0846	20′
50′	.4823	.8760	.5505	1.8165	1.1415	2.0736	10′
29° 0′	.4848	.8746	.5543	1.8040	1.1434	2.0627	61° 0′
10′	.4874	.8732	.5581	1.7917	1.1452	2.0519	50′
20′	.4899	.8718	.5619	1.7796	1.1471	2.0413	40′
30′	.4924	.8704	.5658	1.7675	1.1490	2.0308	30′
40′	.4950	.8689	.5696	1.7556	1.1509	2.0204	20′
50′	.4975	.8675	.5735	1.7437	1.1528	2.0101	10′
30° 0′	.5000	.8660	.5774	1.7321	1.1547	2.0000	60° 0′
	cos x	sin x	cot x	tan x	csc x	sec x	x

x	sin x	cos x	tan x	cot x	sec x	csc x	
30° 0′	.5000	.8660	.5774	1.7321	1.1547	2.0000	60° 0′
10′	.5025	.8646	.5812	1.7205	1.1567	1.9900	50′
20′	.5050	.8631	.5851	1.7090	1.1586	1.9801	40′
30′	.5075	.8616	.5890	1.6977	1.1606	1.9703	30′
40′	.5100	.8601	.5930	1.6864	1.1626	1.9606	20′
50′	.5125	.8587	.5969	1.6753	1.1646	1.9511	10′
31° 0′	.5150	.8572	.6009	1.6643	1.1666	1.9416	59° 0′
10′	.5175	.8557	.6048	1.6534	1.1687	1.9323	50′
20′	.5200	.8542	.6088	1.6426	1.1708	1.9230	40′
30′	.5225	.8526	.6128	1.6319	1.1728	1.9139	30′
40′	.5250	.8511	.6168	1.6212	1.1749	1.9049	20′
50′	.5275	.8496	.6208	1.6107	1.1770	1.8959	10′
32° 0′	.5299	.8480	.6249	1.6003	1.1792	1.8871	58° 0′
10′	.5324	.8465	.6289	1.5900	1.1813	1.8783	50′
20′	.5348	.8450	.6330	1.5798	1.1835	1.8699	40′
30′	.5373	.8434	.6371	1.5697	1.1857	1.8612	30′
40′	.5398	.8418	.6412	1.5597	1.1879	1.8527	20′
50′	.5422	.8403	.6453	1.5497	1.1901	1.8444	10′
33° 0′	.5446	.8387	.6494	1.5399	1.1924	1.8361	57° 0′
10′	.5471	.8371	.6536	1.5301	1.1946	1.8279	50′
20′	.5495	.8355	.6577	1.5204	1.1969	1.8198	40′
30′	.5519	.8339	.6619	1.5108	1.1992	1.8118	30′
40′	.5544	.8323	.6661	1.5013	1.2015	1.8039	20′
50′	.5568	.8307	.6703	1.4919	1.2039	1.7960	10′
34° 0′	.5592	.8290	.6745	1.4826	1.2062	1.7883	56° 0′
10′	.5616	.8274	.6787	1.4733	1.2086	1.7806	50′
20′	.5640	.8258	.6830	1.4641	1.2110	1.7730	40′
30′	.5664	.8241	.6873	1.4550	1.2134	1.7655	30′
40′	.5688	.8225	.6916	1.4460	1.2158	1.7581	20′
50′	.5712	.8208	.6959	1.4370	1.2183	1.7507	10′
35° 0′	.5736	.8192	.7002	1.4281	1.2208	1.7435	55° 0′
10′	.5760	.8175	.7046	1.4193	1.2233	1.7362	50′
20′	.5783	.8158	.7089	1.4106	1.2258	1.7291	40′
30′	.5807	.8141	.7133	1.4019	1.2283	1.7221	30′
40′	.5831	.8124	.7177	1.3934	1.2309	1.7151	20′
50′	.5854	.8107	.7221	1.3848	1.2335	1.7082	10′
36° 0′	.5878	.8090	.7265	1.3764	1.2361	1.7013	54° 0′
	cos x	sin x	cot x	tan x	csc x	sec x	x

Table A   Values of the Trigonometric Functions for Degrees (continued)

x	sin x	cos x	tan x	cot x	sec x	csc x	
36° 0′	.5878	.8090	.7265	1.3764	1.2361	1.7013	54° 0′
10′	.5901	.8073	.7310	1.3680	1.2387	1.6945	50′
20′	.5925	.8056	.7355	1.3597	1.2413	1.6878	40′
30′	.5948	.8039	.7400	1.3514	1.2440	1.6812	30′
40′	.5972	.8021	.7445	1.3432	1.2467	1.6746	20′
50′	.5995	.8004	.7490	1.3351	1.2494	1.6681	10′
37° 0′	.6018	.7986	.7536	1.3270	1.2521	1.6616	53° 0′
10′	.6041	.7969	.7581	1.3190	1.2549	1.6553	50′
20′	.6065	.7951	.7627	1.3111	1.2577	1.6489	40′
30′	.6088	.7934	.7673	1.3032	1.2605	1.6427	30′
40′	.6111	.7916	.7720	1.2954	1.2633	1.6365	20′
50′	.6134	.7898	.7766	1.2876	1.2662	1.6304	10′
38° 0′	.6157	.7880	.7813	1.2799	1.2690	1.6243	52° 0′
10′	.6180	.7862	.7860	1.2723	1.2719	1.6183	50′
20′	.6202	.7844	.7907	1.2647	1.2748	1.6123	40′
30′	.6225	.7826	.7954	1.2572	1.2779	1.6064	30′
40′	.6248	.7808	.8002	1.2497	1.2808	1.6005	20′
50′	.6271	.7790	.8050	1.2423	1.2837	1.5948	10′
39° 0′	.6293	.7771	.8098	1.2349	1.2868	1.5890	51° 0′
10′	.6316	.7753	.8146	1.2276	1.2898	1.5833	50′
20′	.6338	.7735	.8195	1.2203	1.2929	1.5777	40′
30′	.6361	.7716	.8243	1.2131	1.2960	1.5721	30′
40′	.6383	.7698	.8292	1.2059	1.2991	1.5666	20′
50′	.6406	.7679	.8342	1.1988	1.3022	1.5611	10′
40° 0′	.6428	.7660	.8391	1.1918	1.3054	1.5557	50° 0′
10′	.6450	.7642	.8441	1.1847	1.3086	1.5504	50′
20′	.6472	.7623	.8491	1.1778	1.3118	1.5450	40′
30′	.6494	.7604	.8541	1.1708	1.3151	1.5398	30′
40′	.6517	.7585	.8591	1.1640	1.3184	1.5346	20′
50′	.6539	.7566	.8642	1.1571	1.3217	1.5294	10′
41° 0′	.6561	.7547	.8693	1.1504	1.3250	1.5243	49° 0′
10′	.6583	.7528	.8744	1.1436	1.3284	1.5192	50′
20′	.6604	.7509	.8796	1.1369	1.3318	1.5142	40′
30′	.6626	.7490	.8847	1.1303	1.3352	1.5092	30′
40′	.6648	.7470	.8899	1.1237	1.3386	1.5042	20′
50′	.6670	.7451	.8952	1.1171	1.3421	1.4993	10′
42° 0′	.6691	.7431	.9004	1.1106	1.3456	1.4945	48° 0′
	cos x	sin x	cot x	tan x	csc x	sec x	x

x	sin x	cos x	tan x	cot x	sec x	csc x	
42° 0′	.6691	.7431	.9004	1.1106	1.3456	1.4945	48° 0′
10′	.6713	.7412	.9057	1.1041	1.3492	1.4897	50′
20′	.6734	.7392	.9110	1.0977	1.3527	1.4849	40′
30′	.6756	.7373	.9163	1.0913	1.3563	1.4802	30′
40′	.6777	.7353	.9217	1.0850	1.3600	1.4755	20′
50′	.6799	.7333	.9271	1.0786	1.3636	1.4709	10′
43° 0′	.6820	.7314	.9325	1.0724	1.3673	1.4663	47° 0′
10′	.6841	.7294	.9380	1.0661	1.3711	1.4617	50′
20′	.6862	.7274	.9435	1.0599	1.3748	1.4572	40′
30′	.6884	.7254	.9490	1.0538	1.3786	1.4527	30′
40′	.6905	.7234	.9545	1.0477	1.3824	1.4483	20′
50′	.6926	.7214	.9601	1.0416	1.3863	1.4439	10′
44° 0′	.6947	.7193	.9657	1.0355	1.3902	1.4396	46° 0′
10′	.6967	.7173	.9713	1.0295	1.3941	1.4352	50′
20′	.6988	.7153	.9770	1.0235	1.3980	1.4310	40′
30′	.7009	.7133	.9827	1.0176	1.4020	1.4267	30′
40′	.7030	.7112	.9884	1.0117	1.4061	1.4225	20′
50′	.7050	.7092	.9942	1.0058	1.4101	1.4184	10′
45° 0′	.7071	.7071	1.0000	1.0000	1.4142	1.4142	45° 0′
	cos x	sin x	cot x	tan x	csc x	sec x	x

Table B Values of Trigonometric Functions—Decimal Subdivisions (csc $\theta = 1/\sin \theta$; sec $\theta = 1/\cos \theta$)

θ	$\sin \theta$	$\cos \theta$	$\tan \theta$	$\cot \theta$	
0.0	0.00000	1.0000	0.00000	∞	90.0
.1	.00175	1.0000	.00175	573.0	.9
.2	.00349	1.0000	.00349	286.5	.8
.3	.00524	1.0000	.00524	191.0	.7
.4	.00698	1.0000	.00698	143.24	.6
.5	.00873	1.0000	.00873	114.59	.5
.6	.01047	0.9999	.01047	95.49	.4
.7	.01222	.9999	.01222	81.85	.3
.8	.01396	.9999	.01396	71.62	.2
.9	.01571	.9999	.01571	63.66	.1
1.0	0.01745	0.9998	0.01746	57.29	89.0
.1	.01920	.9998	.01920	52.08	.9
.2	.02094	.9998	.02095	47.74	.8
.3	.02269	.9997	.02269	44.07	.7
.4	.02443	.9997	.02444	40.92	.6
.5	.02618	.9997	.02619	38.19	.5
.6	.02792	.9996	.02793	35.80	.4
.7	.02967	.9996	.02968	33.69	.3
.8	.03141	.9995	.03143	31.82	.2
.9	.03316	.9995	.03317	30.14	.1
2.0	0.03490	0.9994	0.03492	28.64	88.0
.1	.03664	.9993	.03667	27.27	.9
.2	.03839	.9993	.03842	26.03	.8
.3	.04013	.9992	.04016	24.90	.7
.4	.04188	.9991	.04191	23.86	.6
.5	.04362	.9990	.04366	22.90	.5
.6	.04536	.9990	.04541	22.02	.4
.7	.04711	.9989	.04716	21.20	.3
.8	.04885	.9988	.04891	20.45	.2
.9	.05059	.9987	.05066	19.74	.1
3.0	0.05234	0.9986	0.05241	19.081	87.0
.1	.05408	.9985	.05416	18.464	.9
.2	.05582	.9984	.05591	17.886	.8
.3	.05756	.9983	.05766	17.343	.7
.4	.05931	.9982	.05941	16.832	.6
.5	.06105	.9981	.06116	16.350	.5
.6	.06279	.9980	.06291	15.895	.4
.7	.06453	.9979	.06467	15.464	.3
.8	.06627	.9978	.06642	15.056	.2
.9	.06802	.9977	.06817	14.669	.1
4.0	0.06976	0.9976	0.06993	14.301	86.0
	$\cos \theta$	$\sin \theta$	$\cot \theta$	$\tan \theta$	θ

Table B Values of Trigonometric Functions—Decimal Subdivisions (csc θ = 1/sin θ; sec θ = 1/cos θ) (continued)

301

Tables

θ	sin θ	cos θ	tan θ	cot θ	
4.0	0.06976	0.9976	0.06993	14.301	86.0
.1	.07150	.9974	.07168	13.951	.9
.2	.07324	.9973	.07344	13.617	.8
.3	.07498	.9972	.07519	13.300	.7
.4	.07672	.9971	.07695	12.996	.6
.5	.07846	.9969	.07870	12.706	.5
.6	.08020	.9968	.08046	12.429	.4
.7	.08194	.9966	.08221	12.163	.3
.8	.08368	.9965	.08397	11.909	.2
.9	.08542	.9963	.08573	11.664	.1
5.0	0.08716	0.9962	0.08749	11.430	85.0
.1	.08889	.9960	.08925	11.205	.9
.2	.09063	.9959	.09101	10.988	.8
.3	.09237	.9957	.09277	10.780	.7
.4	.09411	.9956	.09453	10.579	.6
.5	.09585	.9954	.09629	10.385	.5
.6	.09758	.9952	.09805	10.199	.4
.7	.09932	.9951	.09981	10.019	.3
.8	.10106	.9949	.10158	9.845	.2
.9	.10279	.9947	.10334	9.677	.1
6.0	0.10453	0.9945	0.10510	9.514	84.0
.1	.10626	.9943	.10687	9.357	.9
.2	.10800	.9942	.10863	9.205	.8
.3	.10973	.9940	.11040	9.058	.7
.4	.11147	.9938	.11217	8.915	.6
.5	.11320	.9936	.11394	8.777	.5
.6	.11494	.9934	.11570	8.643	.4
.7	.11667	.9932	.11747	8.513	.3
.8	.11840	.9930	.11924	8.386	.2
.9	.12014	.9928	.12101	8.264	.1
7.0	0.12187	0.9925	0.12278	8.144	83.0
.1	.12360	.9923	.12456	8.028	.9
.2	.12533	.9921	.12633	7.916	.8
.3	.12706	.9919	.12810	7.806	.7
.4	.12880	.9917	.12988	7.700	.6
.5	.13053	.9914	.13165	7.596	.5
.6	.13226	.9912	.13343	7.495	.4
.7	.13399	.9910	.13521	7.396	.3
.8	.13572	.9907	.13698	7.300	.2
.9	.13744	.9905	.13876	7.207	.1
8.0	0.13917	0.9903	0.14054	7.115	82.0
	cos θ	sin θ	cot θ	tan θ	θ

Table B Values of Trigonometric Functions—Decimal Subdivisions (csc $\theta = 1/\sin \theta$; sec $\theta = 1/\cos \theta$) (continued)

θ	$\sin \theta$	$\cos \theta$	$\tan \theta$	$\cot \theta$	
8.0	0.13917	0.9903	0.14054	7.115	82.0
.1	.14090	.9900	.14232	7.026	.9
.2	.14263	.9898	.14410	6.940	.8
.3	.14436	.9895	.14588	6.855	.7
.4	.14608	.9893	.14767	6.772	.6
.5	.14781	.9890	.14945	6.691	.5
.6	.14954	.9888	.15124	6.612	.4
.7	.15126	.9885	.15302	6.535	.3
.8	.15299	.9882	.15481	6.460	.2
.9	.15471	.9880	.15660	6.386	.1
9.0	0.15643	0.9877	0.15838	6.314	81.0
.1	.15816	.9874	.16017	6.243	.9
.2	.15988	.9871	.16196	6.174	.8
.3	.16160	.9869	.16376	6.107	.7
.4	.16333	.9866	.16555	6.041	.6
.5	.16505	.9863	.16734	5.976	.5
.6	.16677	.9860	.16914	5.912	.4
.7	.16849	.9857	.17093	5.850	.3
.8	.17021	.9854	.17273	5.789	.2
.9	.17193	.9851	.17453	5.730	.1
10.0	0.1736	0.9848	0.1763	5.671	80.0
.1	.1754	.9845	.1781	5.614	.9
.2	.1771	.9842	.1799	5.558	.8
.3	.1788	.9839	.1817	5.503	.7
.4	.1805	.9836	.1835	5.449	.6
.5	.1822	.9833	.1853	5.396	.5
.6	.1840	.9829	.1871	5.343	.4
.7	.1857	.9826	.1890	5.292	.3
.8	.1874	.9823	.1908	5.242	.2
.9	.1891	.9820	.1926	5.193	.1
11.0	0.1908	0.9816	0.1944	5.145	79.0
.1	.1925	.9813	.1962	5.097	.9
.2	.1942	.9810	.1980	5.050	.8
.3	.1959	.9806	.1998	5.005	.7
.4	.1977	.9803	.2016	4.959	.6
.5	.1994	.9799	.2035	4.915	.5
.6	.2011	.9796	.2053	4.872	.4
.7	.2028	.9792	.2071	4.829	.3
.8	.2045	.9789	.2089	4.787	.2
.9	.2062	.9785	.2107	4.745	.1
12.0	0.2079	0.9781	0.2126	4.705	78.0
	$\cos \theta$	$\sin \theta$	$\cot \theta$	$\tan \theta$	θ

Table B   Values of Trigonometric Functions—Decimal Subdivisions (csc θ = 1/sin θ; sec θ = 1/cos θ) (continued)

303

Tables

θ	sin θ	cos θ	tan θ	cot θ	
12.0	0.2079	0.9781	0.2126	4.705	78.0
.1	.2096	.9778	.2144	4.665	.9
.2	.2113	.9774	.2162	4.625	.8
.3	.2130	.9770	.2180	4.586	.7
.4	.2147	.9767	.2199	4.548	.6
.5	.2164	.9763	.2217	4.511	.5
.6	.2181	.9759	.2235	4.474	.4
.7	.2198	.9755	.2254	4.437	.3
.8	.2215	.9751	.2272	4.402	.2
.9	.2233	.9748	.2290	4.366	.1
13.0	0.2250	0.9744	0.2309	4.331	77.0
.1	.2267	.9740	.2327	4.297	.9
.2	.2284	.9736	.2345	4.264	.8
.3	.2300	.9732	.2364	4.230	.7
.4	.2317	.9728	.2382	4.198	.6
.5	.2334	.9724	.2401	4.165	.5
.6	.2351	.9720	.2419	4.134	.4
.7	.2368	.9715	.2438	4.102	.3
.8	.2385	.9711	.2456	4.071	.2
.9	.2402	.9707	.2475	4.041	.1
14.0	0.2419	0.9703	0.2493	4.011	76.0
.1	.2436	.9699	.2512	3.981	.9
.2	.2453	.9694	.2530	3.952	.8
.3	.2470	.9690	.2549	3.923	.7
.4	.2487	.9686	.2568	3.895	.6
.5	.2504	.9681	.2586	3.867	.5
.6	.2521	.9677	.2605	3.839	.4
.7	.2538	.9673	.2623	3.812	.3
.8	.2554	.9668	.2642	3.785	.2
.9	.2571	.9664	.2661	3.758	.1
15.0	0.2588	0.9659	0.2679	3.732	75.0
.1	.2605	.9655	.2698	3.706	.9
.2	.2622	.9650	.2717	3.681	.8
.3	.2639	.9646	.2736	3.655	.7
.4	.2656	.9641	.2754	3.630	.6
.5	.2672	.9636	.2773	3.606	.5
.6	.2689	.9632	.2792	3.582	.4
.7	.2706	.9627	.2811	3.558	.3
.8	.2723	.9622	.2830	3.534	.2
.9	.2740	.9617	.2849	3.511	.1
16.0	0.2756	0.9613	0.2867	3.487	74.0
	cos θ	sin θ	cot θ	tan θ	θ

Table B Values of Trigonometric Functions—Decimal Subdivisions (csc $\theta = 1/\sin \theta$; sec $\theta = 1/\cos \theta$) (continued)

θ	$\sin \theta$	$\cos \theta$	$\tan \theta$	$\cot \theta$	
16.0	0.2756	0.9613	0.2867	3.487	74.0
.1	.2773	.9608	.2886	3.465	.9
.2	.2790	.9603	.2905	3.442	.8
.3	.2807	.9598	.2924	3.420	.7
.4	.2823	.9593	.2943	3.398	.6
.5	.2840	.9588	.2962	3.376	.5
.6	.2857	.9583	.2981	3.354	.4
.7	.2874	.9578	.3000	3.333	.3
.8	.2890	.9573	.3019	3.312	.2
.9	.2907	.9568	.3038	3.291	.1
17.0	0.2924	0.9563	0.3057	3.271	73.0
.1	.2940	.9558	.3076	3.251	.9
.2	.2957	.9553	.3096	3.230	.8
.3	.2974	.9548	.3115	3.211	.7
.4	.2990	.9542	.3134	3.191	.6
.5	.3007	.9537	.3153	3.172	.5
.6	.3024	.9532	.3172	3.152	.4
.7	.3040	.9527	.3191	3.133	.3
.8	.3057	.9521	.3211	3.115	.2
.9	.3074	.9516	.3230	3.096	.1
18.0	0.3090	0.9511	0.3249	3.078	72.0
.1	.3107	.9505	.3269	3.060	.9
.2	.3123	.9500	.3288	3.042	.8
.3	.3140	.9494	.3307	3.024	.7
.4	.3156	.9489	.3327	3.006	.6
.5	.3173	.9483	.3346	2.989	.5
.6	.3190	.9478	.3365	2.971	.4
.7	.3206	.9472	.3385	2.954	.3
.8	.3223	.9466	.3404	2.937	.2
.9	.3239	.9461	.3424	2.921	.1
19.0	0.3256	0.9455	0.3443	2.904	71.0
.1	.3272	.9449	.3463	2.888	.9
.2	.3289	.9444	.3482	2.872	.8
.3	.3305	.9438	.3502	2.856	.7
.4	.3322	.9432	.3522	2.840	.6
.5	.3338	.9426	.3541	2.824	.5
.6	.3355	.9421	.3561	2.808	.4
.7	.3371	.9415	.3581	2.793	.3
.8	.3387	.9409	.3600	2.778	.2
.9	.3404	.9403	.3620	2.762	.1
20.0	0.3420	0.9397	0.3640	2.747	70.0
	$\cos \theta$	$\sin \theta$	$\cot \theta$	$\tan \theta$	θ

Table B Values of Trigonometric Functions—Decimal Subdivisions (csc θ = 1/sin θ; sec θ = 1/cos θ) (continued)

θ	sin θ	cos θ	tan θ	cot θ	
20.0	0.3420	0.9397	0.3640	2.747	70.0
.1	.3437	.9391	.3659	2.733	.9
.2	.3453	.9385	.3679	2.718	.8
.3	.3469	.9379	.3699	2.703	.7
.4	.3486	.9373	.3719	2.689	.6
.5	.3502	.9367	.3739	2.675	.5
.6	.3518	.9361	.3759	2.660	.4
.7	.3535	.9354	.3779	2.646	.3
.8	.3551	.9348	.3799	2.633	.2
.9	.3567	.9342	.3819	2.619	.1
21.0	0.3584	0.9336	0.3839	2.605	69.0
.1	.3600	.9330	.3859	2.592	.9
.2	.3616	.9323	.3879	2.578	.8
.3	.3633	.9317	.3899	2.565	.7
.4	.3649	.9311	.3919	2.552	.6
.5	.3665	.9304	.3939	2.539	.5
.6	.3681	.9298	.3959	2.526	.4
.7	.3697	.9291	.3979	2.513	.3
.8	.3714	.9285	.4000	2.500	.2
.9	.3730	.9278	.4020	2.488	.1
22.0	0.3746	0.9272	0.4040	2.475	68.0
.1	.3762	.9265	.4061	2.463	.9
.2	.3778	.9259	.4081	2.450	.8
.3	.3795	.9252	.4101	2.438	.7
.4	.3811	.9245	.4122	2.426	.6
.5	.3827	.9239	.4142	2.414	.5
.6	.3843	.9232	.4163	2.402	.4
.7	.3859	.9225	.4183	2.391	.3
.8	.3875	.9219	.4204	2.379	.2
.9	.3891	.9212	.4224	2.367	.1
23.0	0.3907	0.9205	0.4245	2.356	67.0
.1	.3923	.9198	.4265	2.344	.9
.2	.3939	.9191	.4286	2.333	.8
.3	.3955	.9184	.4307	2.322	.7
.4	.3971	.9178	.4327	2.311	.6
.5	.3987	.9171	.4348	2.300	.5
.6	.4003	.9164	.4369	2.289	.4
.7	.4019	.9157	.4390	2.278	.3
.8	.4035	.9150	.4411	2.267	.2
.9	.4051	.9143	.4431	2.257	.1
24.0	0.4067	0.9135	0.4452	2.246	66.0
	cos θ	sin θ	cot θ	tan θ	θ

Table B Values of Trigonometric Functions—Decimal Subdivisions (csc $\theta = 1/\sin \theta$; sec $\theta = 1/\cos \theta$) (continued)

θ	$\sin \theta$	$\cos \theta$	$\tan \theta$	$\cot \theta$	
24.0	0.4067	0.9135	0.4452	2.246	66.0
.1	.4083	.9128	.4473	2.236	.9
.2	.4099	.9121	.4494	2.225	.8
.3	.4115	.9114	.4515	2.215	.7
.4	.4131	.9107	.4536	2.204	.6
.5	.4147	.9100	.4557	2.194	.5
.6	.4163	.9092	.4578	2.184	.4
.7	.4179	.9085	.4599	2.174	.3
.8	.4195	.9078	.4621	2.164	.2
.9	.4210	.9070	.4642	2.154	.1
25.0	0.4226	0.9063	0.4663	2.145	65.0
.1	.4242	.9056	.4684	2.135	.9
.2	.4258	.9048	.4706	2.125	.8
.3	.4274	.9041	.4727	2.116	.7
.4	.4289	.9033	.4748	2.106	.6
.5	.4305	.9026	.4770	2.097	.5
.6	.4321	.9018	.4791	2.087	.4
.7	.4337	.9011	.4813	2.078	.3
.8	.4352	.9003	.4834	2.069	.2
.9	.4368	.8996	.4856	2.059	.1
26.0	0.4384	0.8988	0.4877	2.050	64.0
.1	.4399	.8980	.4899	2.041	.9
.2	.4415	.8973	.4921	2.032	.8
.3	.4431	.8965	.4942	2.023	.7
.4	.4446	.8957	.4964	2.014	.6
.5	.4462	.8949	.4986	2.006	.5
.6	.4478	.8942	.5008	1.997	.4
.7	.4493	.8934	.5029	1.988	.3
.8	.4509	.8926	.5051	1.980	.2
.9	.4524	.8918	.5073	1.971	.1
27.0	0.4540	0.8910	0.5095	1.963	63.0
.1	.4555	.8902	.5117	1.954	.9
.2	.4571	.8894	.5139	1.946	.8
.3	.4586	.8886	.5161	1.937	.7
.4	.4602	.8878	.5184	1.929	.6
.5	.4617	.8870	.5206	1.921	.5
.6	.4633	.8862	.5228	1.913	.4
.7	.4648	.8854	.5250	1.905	.3
.8	.4664	.8846	.5272	1.897	.2
.9	.4679	.8838	.5295	1.889	.1
28.0	0.4695	0.8829	0.5317	1.881	62.0
	$\cos \theta$	$\sin \theta$	$\cot \theta$	$\tan \theta$	θ

Table B Values of Trigonometric Functions—Decimal Subdivisions (csc $\theta = 1/\sin \theta$; sec $\theta = 1/\cos \theta$) (continued)

θ	$\sin \theta$	$\cos \theta$	$\tan \theta$	$\cot \theta$	
28.0	0.4695	0.8829	0.5317	1.881	62.0
.1	.4710	.8821	.5340	1.873	.9
.2	.4726	.8813	.5362	1.865	.8
.3	.4741	.8805	.5384	1.857	.7
.4	.4756	.8796	.5407	1.849	.6
.5	.4772	.8788	.5430	1.842	.5
.6	.4787	.8780	.5452	1.834	.4
.7	.4802	.8771	.5475	1.827	.3
.8	.4818	.8763	.5498	1.819	.2
.9	.4833	.8755	.5520	1.811	.1
29.0	0.4848	0.8746	0.5543	1.804	61.0
.1	.4863	.8738	.5566	1.797	.9
.2	.4879	.8729	.5589	1.789	.8
.3	.4894	.8721	.5612	1.782	.7
.4	.4909	.8712	.5635	1.775	.6
.5	.4924	.8704	.5658	1.767	.5
.6	.4939	.8695	.5681	1.760	.4
.7	.4955	.8686	.5704	1.753	.3
.8	.4970	.8678	.5727	1.746	.2
.9	.4985	.8669	.5750	1.739	.1
30.0	0.5000	0.8660	0.5774	1.7321	60.0
.1	.5015	.8652	.5797	1.7251	.9
.2	.5030	.8643	.5820	1.7182	.8
.3	.5045	.8634	.5844	1.7113	.7
.4	.5060	.8625	.5867	1.7045	.6
.5	.5075	.8616	.5890	1.6977	.5
.6	.5090	.8607	.5914	1.6909	.4
.7	.5105	.8599	.5938	1.6842	.3
.8	.5120	.8590	.5961	1.6775	.2
.9	.5135	.8581	.5985	1.6709	.1
31.0	0.5150	0.8572	0.6009	1.6643	59.0
.1	.5165	.8563	.6032	1.6577	.9
.2	.5180	.8554	.6056	1.6512	.8
.3	.5195	.8545	.6080	1.6447	.7
.4	.5210	.8536	.6104	1.6383	.6
.5	.5225	.8526	.6128	1.6319	.5
.6	.5240	.8517	.6152	1.6255	.4
.7	.5255	.8508	.6176	1.6191	.3
.8	.5270	.8499	.6200	1.6128	.2
.9	.5284	.8490	.6224	1.6066	.1
32.0	0.5299	0.8480	0.6249	1.6003	58.0
	$\cos \theta$	$\sin \theta$	$\cot \theta$	$\tan \theta$	θ

Table B Values of Trigonometric Functions—Decimal Subdivisions (csc $\theta = 1/\sin \theta$; sec $\theta = 1/\cos \theta$) (continued)

θ	$\sin \theta$	$\cos \theta$	$\tan \theta$	$\cot \theta$	
32.0	0.5299	0.8480	0.6249	1.6003	58.0
.1	.5314	.8471	.6273	1.5941	.9
.2	.5329	.8462	.6297	1.5880	.8
.3	.5344	.8453	.6322	1.5818	.7
.4	.5358	.8443	.6346	1.5757	.6
.5	.5373	.8434	.6371	1.5697	.5
.6	.5388	.8425	.6395	1.5637	.4
.7	.5402	.8415	.6420	1.5577	.3
.8	.5417	.8406	.6445	1.5517	.2
.9	.5432	.8396	.6469	1.5458	.1
33.0	0.5446	0.8387	0.6494	1.5399	57.0
.1	.5461	.8377	.6519	1.5340	.9
.2	.5476	.8368	.6544	1.5282	.8
.3	.5490	.8358	.6569	1.5224	.7
.4	.5505	.8348	.6594	1.5166	.6
.5	.5519	.8339	.6619	1.5108	.5
.6	.5534	.8329	.6644	1.5051	.4
.7	.5548	.8320	.6669	1.4994	.3
.8	.5563	.8310	.6694	1.4938	.2
.9	.5577	.8300	.6720	1.4882	.1
34.0	0.5592	0.8290	0.6745	1.4826	56.0
.1	.5606	.8281	.6771	1.4770	.9
.2	.5621	.8271	.6796	1.4715	.8
.3	.5635	.8261	.6822	1.4659	.7
.4	.5650	.8251	.6847	1.4605	.6
.5	.5664	.8241	.6873	1.4550	.5
.6	.5678	.8231	.6899	1.4496	.4
.7	.5693	.8221	.6924	1.4442	.3
.8	.5707	.8211	.6950	1.4388	.2
.9	.5721	.8202	.6976	1.4335	.1
35.0	0.5736	0.8192	0.7002	1.4281	55.0
.1	.5750	.8181	.7028	1.4229	.9
.2	.5764	.8171	.7054	1.4176	.8
.3	.5779	.8161	.7080	1.4124	.7
.4	.5793	.8151	.7107	1.4071	.6
.5	.5807	.8141	.7133	1.4019	.5
.6	.5821	.8131	.7159	1.3968	.4
.7	.5835	.8121	.7186	1.3916	.3
.8	.5850	.8111	.7212	1.3865	.2
.9	.5864	.8100	.7239	1.3814	.1
36.0	0.5878	0.8090	0.7265	1.3764	54.0
	$\cos \theta$	$\sin \theta$	$\cot \theta$	$\tan \theta$	θ

Table B Values of Trigonometric Functions—Decimal Subdivisions (csc θ = 1/sin θ; sec θ = 1/cos θ) (continued)

309

Tables

θ	sin θ	cos θ	tan θ	cot θ	
36.0	0.5878	0.8090	0.7265	1.3764	54.0
.1	.5892	.8080	.7292	1.3713	.9
.2	.5906	.8070	.7319	1.3663	.8
.3	.5920	.8059	.7346	1.3613	.7
.4	.5934	.8049	.7373	1.3564	.6
.5	.5948	.8039	.7400	1.3514	.5
.6	.5962	.8028	.7427	1.3465	.4
.7	.5976	.8018	.7454	1.3416	.3
.8	.5990	.8007	.7481	1.3367	.2
.9	.6004	.7997	.7508	1.3319	.1
37.0	0.6018	0.7986	0.7536	1.3270	53.0
.1	.6032	.7976	.7563	1.3222	.9
.2	.6046	.7965	.7590	1.3175	.8
.3	.6060	.7955	.7618	1.3127	.7
.4	.6074	.7944	.7646	1.3079	.6
.5	.6088	.7934	.7673	1.3032	.5
.6	.6101	.7923	.7701	1.2985	.4
.7	.6115	.7912	.7729	1.2938	.3
.8	.6129	.7902	.7757	1.2892	.2
.9	.6143	.7891	.7785	1.2846	.1
38.0	0.6157	0.7880	0.7813	1.2799	52.0
.1	.6170	.7869	.7841	1.2753	.9
.2	.6184	.7859	.7869	1.2708	.8
.3	.6198	.7848	.7898	1.2662	.7
.4	.6211	.7837	.7926	1.2617	.6
.5	.6225	.7826	.7954	1.2572	.5
.6	.6239	.7815	.7983	1.2527	.4
.7	.6252	.7804	.8012	1.2482	.3
.8	.6266	.7793	.8040	1.2437	.2
.9	.6280	.7782	.8069	1.2393	.1
39.0	0.6293	0.7771	0.8098	1.2349	51.0
.1	.6307	.7760	.8127	1.2305	.9
.2	.6320	.7749	.8156	1.2261	.8
.3	.6334	.7738	.8185	1.2218	.7
.4	.6347	.7727	.8214	1.2174	.6
.5	.6361	.7716	.8243	1.2131	.5
.6	.6374	.7705	.8273	1.2088	.4
.7	.6388	.7694	.8302	1.2045	.3
.8	.6401	.7683	.8332	1.2002	.2
.9	.6414	.7672	.8361	1.1960	.1
40.0	0.6428	0.7660	0.8391	1.1918	50.0
	cos θ	sin θ	cot θ	tan θ	θ

Table B Values of Trigonometric Functions—Decimal Subdivisions (csc $\theta = 1/\sin \theta$; sec $\theta = 1/\cos \theta$) (continued)

θ	$\sin \theta$	$\cos \theta$	$\tan \theta$	$\cot \theta$	
40.0	0.6428	0.7660	0.8391	1.1918	50.0
.1	.6441	.7649	.8421	1.1875	.9
.2	.6455	.7638	.8451	1.1833	.8
.3	.6468	.7627	.8481	1.1792	.7
.4	.6481	.7615	.8511	1.1750	.6
.5	.6494	.7604	.8541	1.1708	.5
.6	.6508	.7593	.8571	1.1667	.4
.7	.6521	.7581	.8601	1.1626	.3
.8	.6534	.7570	.8632	1.1585	.2
.9	.6547	.7559	.8662	1.1544	.1
41.0	0.6561	0.7547	0.8693	1.1504	49.0
.1	.6574	.7536	.8724	1.1463	.9
.2	.6587	.7524	.8754	1.1423	.8
.3	.6600	.7513	.8785	1.1383	.7
.4	.6613	.7501	.8816	1.1343	.6
.5	.6626	.7490	.8847	1.1303	.5
.6	.6639	.7478	.8878	1.1263	.4
.7	.6652	.7466	.8910	1.1224	.3
.8	.6665	.7455	.8941	1.1184	.2
.9	.6678	.7443	.8972	1.1145	.1
42.0	0.6691	0.7431	0.9004	1.1106	48.0
.1	.6704	.7420	.9036	1.1067	.9
.2	.6717	.7408	.9067	1.1028	.8
.3	.6730	.7396	.9099	1.0990	.7
.4	.6743	.7385	.9131	1.0951	.6
.5	.6756	.7373	.9163	1.0913	.5
.6	.6769	.7361	.9195	1.0875	.4
.7	.6782	.7349	.9228	1.0837	.3
.8	.6794	.7337	.9260	1.0799	.2
.9	.6807	.7325	.9293	1.0761	.1
43.0	0.6820	0.7314	0.9325	1.0724	47.0
.1	.6833	.7302	.9358	1.0686	.9
.2	.6845	.7290	.9391	1.0649	.8
.3	.6858	.7278	.9424	1.0612	.7
.4	.6871	.7266	.9457	1.0575	.6
.5	.6884	.7254	.9490	1.0538	.5
.6	.6896	.7242	.9523	1.0501	.4
.7	.6909	.7230	.9556	1.0464	.3
.8	.6921	.7218	.9590	1.0428	.2
.9	.6934	.7206	.9623	1.0392	.1
44.0	0.6947	0.7193	0.9657	1.0355	46.0
	$\cos \theta$	$\sin \theta$	$\cot \theta$	$\tan \theta$	θ

Table B Values of Trigonometric Functions—Decimal Subdivisions (csc θ = 1/sin θ; sec θ = 1/cos θ) (continued)

θ	sin θ	cos θ	tan θ	cot θ	
44.0	0.6947	0.7193	0.9657	1.0355	46.0
.1	.6959	.7181	.9691	1.0319	.9
.2	.6972	.7169	.9725	1.0283	.8
.3	.6984	.7157	.9759	1.0247	.7
.4	.6997	.7145	.9793	1.0212	.6
.5	.7009	.7133	.9827	1.0176	.5
.6	.7022	.7120	.9861	1.0141	.4
.7	.7034	.7108	.9896	1.0105	.3
.8	.7046	.7096	.9930	1.0070	.2
.9	.7059	.7083	.9965	1.0035	.1
45.0	0.7071	0.7071	1.0000	1.0000	45.0
	cos θ	sin θ	cot θ	tan θ	θ

Table C Values of the Trigonometric Functions for Radians and Real Numbers

t	sin t	cos t	tan t	cot t	sec t	csc t
.00	.0000	1.0000	.0000	.....	1.000	.....
.01	.0100	1.0000	.0100	99.997	1.000	100.00
.02	.0200	.9998	.0200	49.993	1.000	50.00
.03	.0300	.9996	.0300	33.323	1.000	33.34
.04	.0400	.9992	.0400	24.987	1.001	25.01
.05	.0500	.9988	.0500	19.983	1.001	20.01
.06	.0600	.9982	.0601	16.647	1.002	16.68
.07	.0699	.9976	.0701	14.262	1.002	14.30
.08	.0799	.9968	.0802	12.473	1.003	12.51
.09	.0899	.9960	.0902	11.081	1.004	11.13
.10	.0998	.9950	.1003	9.967	1.005	10.02
.11	.1098	.9940	.1104	9.054	1.006	9.109
.12	.1197	.9928	.1206	8.293	1.007	8.353
.13	.1296	.9916	.1307	7.649	1.009	7.714
.14	.1395	.9902	.1409	7.096	1.010	7.166
.15	.1494	.9888	.1511	6.617	1.011	6.692
.16	.1593	.9872	.1614	6.197	1.013	6.277
.17	.1692	.9856	.1717	5.826	1.015	5.911
.18	.1790	.9838	.1820	5.495	1.016	5.586
.19	.1889	.9820	.1923	5.200	1.018	5.295
.20	.1987	.9801	.2027	4.933	1.020	5.033
.21	.2085	.9780	.2131	4.692	1.022	4.797
.22	.2182	.9759	.2236	4.472	1.025	4.582
.23	.2280	.9737	.2341	4.271	1.027	4.386
.24	.2377	.9713	.2447	4.086	1.030	4.207
.25	.2474	.9689	.2553	3.916	1.032	4.042
.26	.2571	.9664	.2660	3.759	1.035	3.890
.27	.2667	.9638	.2768	3.613	1.038	3.749
.28	.2764	.9611	.2876	3.478	1.041	3.619
.29	.2860	.9582	.2984	3.351	1.044	3.497
.30	.2955	.9553	.3093	3.233	1.047	3.384
.31	.3051	.9523	.3203	3.122	1.050	3.278
.32	.3146	.9492	.3314	3.018	1.053	3.179
.33	.3240	.9460	.3425	2.920	1.057	3.086
.34	.3335	.9428	.3537	2.827	1.061	2.999
.35	.3429	.9394	.3650	2.740	1.065	2.916
.36	.3523	.9359	.3764	2.657	1.068	2.839
.37	.3616	.9323	.3879	2.578	1.073	2.765
.38	.3709	.9287	.3994	2.504	1.077	2.696
.39	.3802	.9249	.4111	2.433	1.081	2.630

t	sin t	cos t	tan t	cot t	sec t	csc t
.40	.3894	.9211	.4228	2.365	1.086	2.568
.41	.3986	.9171	.4346	2.301	1.090	2.509
.42	.4078	.9131	.4466	2.239	1.095	2.452
.43	.4169	.9090	.4586	2.180	1.100	2.399
.44	.4259	.9048	.4708	2.124	1.105	2.348
.45	.4350	.9004	.4831	2.070	1.111	2.299
.46	.4439	.8961	.4954	2.018	1.116	2.253
.47	.4529	.8916	.5080	1.969	1.122	2.208
.48	.4618	.8870	.5206	1.921	1.127	2.166
.49	.4706	.8823	.5334	1.875	1.133	2.125
.50	.4794	.8776	.5463	1.830	1.139	2.086
.51	.4882	.8727	.5594	1.788	1.146	2.048
.52	.4969	.8678	.5726	1.747	1.152	2.013
.53	.5055	.8628	.5859	1.707	1.159	1.978
.54	.5141	.8577	.5994	1.668	1.166	1.945
.55	.5227	.8525	.6131	1.631	1.173	1.913
.56	.5312	.8473	.6269	1.595	1.180	1.883
.57	.5396	.8419	.6410	1.560	1.188	1.853
.58	.5480	.8365	.6552	1.526	1.196	1.825
.59	.5564	.8309	.6696	1.494	1.203	1.797
.60	.5646	.8253	.6841	1.462	1.212	1.771
.61	.5729	.8196	.6989	1.431	1.220	1.746
.62	.5810	.8139	.7139	1.401	1.229	1.721
.63	.5891	.8080	.7291	1.372	1.238	1.697
.64	.5972	.8021	.7445	1.343	1.247	1.674
.65	.6052	.7961	.7602	1.315	1.256	1.652
.66	.6131	.7900	.7761	1.288	1.266	1.631
.67	.6210	.7838	.7923	1.262	1.276	1.610
.68	.6288	.7776	.8087	1.237	1.286	1.590
.69	.6365	.7712	.8253	1.212	1.297	1.571
.70	.6442	.7648	.8423	1.187	1.307	1.552
.71	.6518	.7584	.8595	1.163	1.319	1.534
.72	.6594	.7518	.8771	1.140	1.330	1.517
.73	.6669	.7452	.8949	1.117	1.342	1.500
.74	.6743	.7385	.9131	1.095	1.354	1.483
.75	.6816	.7317	.9316	1.073	1.367	1.467
.76	.6889	.7248	.9505	1.052	1.380	1.452
.77	.6961	.7179	.9697	1.031	1.393	1.437
.78	.7033	.7109	.9893	1.011	1.407	1.422
.79	.7104	.7038	1.009	.9908	1.421	1.408

Table C Values of the Trigonometric Functions for Radians and Real Numbers (continued)

t	$\sin t$	$\cos t$	$\tan t$	$\cot t$	$\sec t$	$\csc t$
.80	.7174	.6967	1.030	.9712	1.435	1.394
.81	.7243	.6895	1.050	.9520	1.450	1.381
.82	.7311	.6822	1.072	.9331	1.466	1.368
.83	.7379	.6749	1.093	.9146	1.482	1.355
.84	.7446	.6675	1.116	.8964	1.498	1.343
.85	.7513	.6600	1.138	.8785	1.515	1.331
.86	.7578	.6524	1.162	.8609	1.533	1.320
.87	.7643	.6448	1.185	.8437	1.551	1.308
.88	.7707	.6372	1.210	.8267	1.569	1.297
.89	.7771	.6294	1.235	.8100	1.589	1.287
.90	.7833	.6216	1.260	.7936	1.609	1.277
.91	.7895	.6137	1.286	.7774	1.629	1.267
.92	.7956	.6058	1.313	.7615	1.651	1.257
.93	.8016	.5978	1.341	.7458	1.673	1.247
.94	.8076	.5898	1.369	.7303	1.696	1.238
.95	.8134	.5817	1.398	.7151	1.719	1.229
.96	.8192	.5735	1.428	.7001	1.744	1.221
.97	.8249	.5653	1.459	.6853	1.769	1.212
.98	.8305	.5570	1.491	.6707	1.795	1.204
.99	.8360	.5487	1.524	.6563	1.823	1.196
1.00	.8415	.5403	1.557	.6421	1.851	1.188
1.01	.8468	.5319	1.592	.6281	1.880	1.181
1.02	.8521	.5234	1.628	.6142	1.911	1.174
1.03	.8573	.5148	1.665	.6005	1.942	1.166
1.04	.8624	.5062	1.704	.5870	1.975	1.160
1.05	.8674	.4976	1.743	.5736	2.010	1.153
1.06	.8724	.4889	1.784	.5604	2.046	1.146
1.07	.8772	.4801	1.827	.5473	2.083	1.140
1.08	.8820	.4713	1.871	.5344	2.122	1.134
1.09	.8866	.4625	1.917	.5216	2.162	1.128
1.10	.8912	.4536	1.965	.5090	2.205	1.122
1.11	.8957	.4447	2.014	.4964	2.249	1.116
1.12	.9001	.4357	2.066	.4840	2.295	1.111
1.13	.9044	.4267	2.120	.4718	2.344	1.106
1.14	.9086	.4176	2.176	.4596	2.395	1.101
1.15	.9128	.4085	2.234	.4475	2.448	1.096
1.16	.9168	.3993	2.296	.4356	2.504	1.091
1.17	.9208	.3902	2.360	.4237	2.563	1.086
1.18	.9246	.3809	2.427	.4120	2.625	1.082
1.19	.9284	.3717	2.498	.4003	2.691	1.077

t	sin t	cos t	tan t	cot t	sec t	csc t
1.20	.9320	.3624	2.572	.3888	2.760	1.073
1.21	.9356	.3530	2.650	.3773	2.833	1.069
1.22	.9391	.3436	2.733	.3659	2.910	1.065
1.23	.9425	.3342	2.820	.3546	2.992	1.061
1.24	.9458	.3248	2.912	.3434	3.079	1.057
1.25	.9490	.3153	3.010	.3323	3.171	1.054
1.26	.9521	.3058	3.113	.3212	3.270	1.050
1.27	.9551	.2963	3.224	.3102	3.375	1.047
1.28	.9580	.2867	3.341	.2993	3.488	1.044
1.29	.9608	.2771	3.467	.2884	3.609	1.041
1.30	.9636	.2675	3.602	.2776	3.738	1.038
1.31	.9662	.2579	3.747	.2669	3.878	1.035
1.32	.9687	.2482	3.903	.2562	4.029	1.032
1.33	.9711	.2385	4.072	.2456	4.193	1.030
1.34	.9735	.2288	4.256	.2350	4.372	1.027
1.35	.9757	.2190	4.455	.2245	4.566	1.025
1.36	.9779	.2092	4.673	.2140	4.779	1.023
1.37	.9799	.1994	4.913	.2035	5.014	1.021
1.38	.9819	.1896	5.177	.1931	5.273	1.018
1.39	.9837	.1798	5.471	.1828	5.561	1.017
1.40	.9854	.1700	5.798	.1725	5.883	1.015
1.41	.9871	.1601	6.165	.1622	6.246	1.013
1.42	.9887	.1502	6.581	.1519	6.657	1.011
1.43	.9901	.1403	7.055	.1417	7.126	1.010
1.44	.9915	.1304	7.602	.1315	7.667	1.009
1.45	.9927	.1205	8.238	.1214	8.299	1.007
1.46	.9939	.1106	8.989	.1113	9.044	1.006
1.47	.9949	.1006	9.887	.1011	9.938	1.005
1.48	.9959	.0907	10.983	.0910	11.029	1.004
1.49	.9967	.0807	12.350	.0810	12.390	1.003
1.50	.9975	.0707	14.101	.0709	14.137	1.003
1.51	.9982	.0608	16.428	.0609	16.458	1.002
1.52	.9987	.0508	19.670	.0508	19.695	1.001
1.53	.9992	.0408	24.498	.0408	24.519	1.001
1.54	.9995	.0308	32.461	.0308	32.476	1.000
1.55	.9998	.0208	48.078	.0208	48.089	1.000
1.56	.9999	.0108	92.620	.0108	92.626	1.000
1.57	1.0000	.0008	1255.8	.0008	1255.8	1.000

Answers to Odd-Numbered Exercises

1, 3, 5.

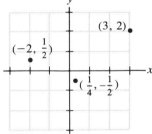

7. I, IV **9.** III **11.** 0

13. 1.414 **15.** $\frac{5}{4}$ **17.** 8.2 **19.** 4.76 **21.** The y-axis

23.

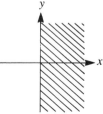

25.

27.

29. $2\sqrt{x + y}$ **31.** $2|y|$ **33.** $\sqrt{5}$

Section 1.2

1. $-6, 6$ **3.** $6, -6$ **5.** $44, -4$ **7.** $A = f(r)$ **9.** $P = f(s)$

11. Domain: all reals   **13.** Domain: all reals ≥ 25
Range: all reals   Range: all reals ≥ 0

15. Domain: all reals $\neq 0$ **17.**
Range: all reals $\neq 0$

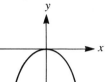

19.

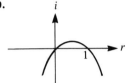

21.

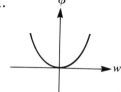

23.

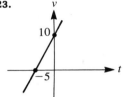

25. Yes   **27.** Yes   **29.** No

31. Yes   **33.** No

Section 1.3

1. Three   **3.** Four   **5.** One   **7.** Two   **9.** Four

11. 9820   **13.** 54.7   **15.** 0.0658   **17.** 39.8   **19.** 1.00

21. 21.50   **23.** 1.92   **25.** 65   **27.** 3.1   **29.** 2280

31. 139.6

Section 1.4

1. 90°　　**3.** 75°15′, 15°5′　　**5.** 109°10′, 8°1′20″
7. 130°55′15″, −10°34′45″　　**9.** 574°10′54″, −92°39′34″
11. −121°1′10″, 39°35′16″　　**13.** 18.4267°　　**15.** 94.2856°
17. 283.6083°　　**19.** 183.2444°　　**21.** 48°15′26″　　**23.** −235°27′0″
25. 45°45′27″　　**27.** 15°15′27″　　**29.** 60°　　**31.** 42.5°　　**33.** 135°
35. 230°　　**37.** 0°　　**39.** 120°　　**41.** 45°　　**43.** 150°
45. 135°　　**47.** −90°　　**49.** 180°　　**51.** 79°28′23″　　**53.** 0.08°
55. 4500°

Section 1.5

1. 74°　　**3.** 127 ft　　**5.** 3.7 m　　**7.** 8.7 cm　　**17.** $a = 8.1, c = 10.3$
19. $a = 9.5, b = 3.5$　　**21.** 33 cm
23. Yes, because angles are all 60°. No, angle size may vary.　　**25.** 80 m
27. $\sqrt{2m}$　　**29.** $\alpha = 7°2′, \phi = 58°46′$

Review for Chapter 1

1. 2, 1, −2　　**3.**　y　　　**5.** $\sqrt{10}$　　**7.** No

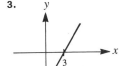

9. $160°$ **11.** $38.7231°$ **13.** $135°$ **15.** $9°47'20''$ **17.** $280°$
19. 2.6 ft **21.** 44.65 **23.** 0.002125 **25.** 353.0 **27.** Three
29. Two **31.** 3.0 km **33.**

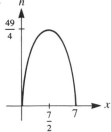

Section 2.1

(The trigonometric functions are listed in order: sine, cosine, tangent, cotangent, secant, cosecant.)

1. $\dfrac{2}{\sqrt{5}}, \dfrac{1}{\sqrt{5}}, 2, \dfrac{1}{2}, \sqrt{5}, \dfrac{\sqrt{5}}{2}$

3. $\dfrac{16}{\sqrt{337}}, -\dfrac{9}{\sqrt{337}}, -\dfrac{16}{9}, -\dfrac{9}{16}, -\dfrac{\sqrt{337}}{9}, \dfrac{\sqrt{337}}{16}$

5. $-\dfrac{7}{\sqrt{53}}, \dfrac{2}{\sqrt{53}}, -\dfrac{7}{2}, -\dfrac{2}{7}, \dfrac{\sqrt{53}}{2}, -\dfrac{\sqrt{53}}{7}$

7. $-\dfrac{1}{\sqrt{10}}, \dfrac{3}{\sqrt{10}}, -\dfrac{1}{3}, -3, \dfrac{\sqrt{10}}{3}, -\sqrt{10}$

9. $-\dfrac{1}{2}, -\dfrac{\sqrt{3}}{2}, \dfrac{\sqrt{3}}{3}, \sqrt{3}, -\dfrac{2\sqrt{3}}{3}, -2$ **11.** (a) I, II (b) I, IV (c) I, III

13. II   **15.** III   **17.** IV   **19.** $\dfrac{3}{5}, \dfrac{4}{5}, \dfrac{3}{4}, \dfrac{4}{3}, \dfrac{5}{4}, \dfrac{5}{3}$

21. $-\dfrac{3}{5}, -\dfrac{4}{5}, \dfrac{3}{4}, \dfrac{4}{3}, -\dfrac{5}{4}, -\dfrac{5}{3}$

23. QI: $\dfrac{1}{2}, \dfrac{\sqrt{3}}{2}, \dfrac{1}{\sqrt{3}}, \sqrt{3}, \dfrac{2}{\sqrt{3}}, 2$; QIV: $-\dfrac{1}{2}, \dfrac{\sqrt{3}}{2}, -\dfrac{1}{\sqrt{3}}, -\sqrt{3}, \dfrac{2}{\sqrt{3}}, -2$

25. $\dfrac{2}{3}, -\dfrac{\sqrt{5}}{3}, -\dfrac{2}{\sqrt{5}}, -\dfrac{\sqrt{5}}{2}, -\dfrac{3}{\sqrt{5}}, \dfrac{3}{2}$

27. QIII: $-\dfrac{1}{2}, -\dfrac{\sqrt{3}}{2}, \dfrac{\sqrt{3}}{3}, \sqrt{3}, -\dfrac{2}{\sqrt{3}}, -2$;

QIV: $-\dfrac{1}{2}, \dfrac{\sqrt{3}}{2}, -\dfrac{\sqrt{3}}{3}, -\sqrt{3}, \dfrac{2}{\sqrt{3}}, -2$

29. $10/\sqrt{101}$, $1/\sqrt{101}$, 10, $1/10$, $\sqrt{101}$, $\sqrt{101}/10$

31. QI: $\frac{5}{13}$, $\frac{12}{13}$, $\frac{5}{12}$, $\frac{12}{5}$, $\frac{13}{12}$, $\frac{13}{5}$; QIV: $-\frac{5}{13}$, $\frac{12}{13}$, $-\frac{5}{12}$, $-\frac{12}{5}$, $\frac{13}{12}$, $-\frac{13}{5}$

33. QII: $\frac{\sqrt{3}}{2}$, $-\frac{1}{2}$, $-\sqrt{3}$, $-\frac{1}{\sqrt{3}}$, -2, $\frac{2}{\sqrt{3}}$;

QIV: $-\frac{\sqrt{3}}{2}$, $\frac{1}{2}$, $-\sqrt{3}$, $-\frac{1}{\sqrt{3}}$, 2, $-\frac{2}{\sqrt{3}}$

35. $\frac{u}{v}$, $\frac{\sqrt{v^2-u^2}}{v}$, $\frac{u}{\sqrt{v^2-u^2}}$, $\frac{\sqrt{v^2-u^2}}{u}$, $\frac{v}{\sqrt{v^2-u^2}}$, $\frac{v}{u}$

37. $\sqrt{1-u^2}$, u, $\frac{\sqrt{1-u^2}}{u}$, $\frac{u}{\sqrt{1-u^2}}$, $\frac{1}{u}$, $\frac{1}{\sqrt{1-u^2}}$ **39.** 1.5 **41.** 0.2925

43. 0.9594

Section 2.2

1. 2 **3.** $\frac{1}{3}$ **5.** $-\frac{1}{2}$ **7.** $\frac{1}{\sqrt{2}}$ **9.** $\frac{\sqrt{3}}{2}$ **11.** $-\frac{5}{12}$

13. $-\frac{1}{\sqrt{5}}$ **15.** $\frac{1}{3}$ **17.** 2 **19.** $-\frac{\sqrt{2}}{\sqrt{7}}$

21. $\frac{1}{3}$, $\frac{\sqrt{8}}{3}$, $\frac{1}{\sqrt{8}}$, $\sqrt{8}$, $\frac{3}{\sqrt{8}}$, 3 **23.** $\frac{1}{\sqrt{5}}$, $-\frac{2}{\sqrt{5}}$, $-\frac{1}{2}$, -2, $-\frac{\sqrt{5}}{2}$, $\sqrt{5}$

25. $-\frac{\sqrt{2}}{\sqrt{3}}$, $-\frac{1}{\sqrt{3}}$, $\sqrt{2}$, $\frac{1}{\sqrt{2}}$, $-\sqrt{3}$, $-\frac{\sqrt{3}}{\sqrt{2}}$ $\bullet$ **27.** 8.658 **29.** 0.5258

31. $\frac{1}{\sqrt{1.6}}$

Section 2.3

5. $30°$ **7.** $60°$ **9.** $45°$ **11.** $45°$ **13.** $30°$ **15.** $\frac{\sqrt{2}}{2}$

17. $\frac{1}{2}$ **19.** $-\sqrt{3}$ **21.** -2 **23.** $\sqrt{2}$ **25.** $-\frac{1}{2}$

27. $-\sqrt{3}$ **29.** $\dfrac{\sqrt{3}}{2}$ **31.** $\dfrac{\sqrt{3}}{3}$ **33.** $-\dfrac{\sqrt{2}}{2}$ **35.** $90° + n \cdot 360°$

37. $90° + n \cdot 360°,\ 270° + n \cdot 360°$ **39.** $30° + n \cdot 360°,\ 150° + n \cdot 360°$

41. $0, \dfrac{3}{2}, \dfrac{3\sqrt{3}}{2}, \dfrac{3\sqrt{3}}{2}, \dfrac{3}{2}, 0$

Section 2.4

1. 0.2250 **3.** 0.3115 **5.** -0.1771 **7.** 0.7771 **9.** -5.6713
11. -1.2690 **13.** 1.0778 **15.** -1.0612 **17.** -0.8337
19. 1.2290 **21.** 1.9781 **23.** -0.1772 **25.** $33°50'$
27. $203°49'$ **29.** $334°50'$ **31.** $18°15'$
33. $41°58'$ **35.** $161°20', 341°20'$ **37.** $41°38', 138°22'$
39. $117°11', 297°11'$

Section 2.5

1. 0.2250 **3.** 0.3153 **5.** 1.0041 **7.** 3.1023
9. 0.9681 **11.** $-\cos 55°$ **13.** $-\sin 45°$ **15.** $-\tan 17°$
17. $\sec 80°$ **19.** $-\cos 87°$ **21.** 0.4384 **23.** -9.5144
25. -1.0439 **27.** 0.2728 **29.** -0.9245 **31.** $334°50'$
33. $198°15'$ **35.** $98°20'$ **37.** $61°6', 298°54'$ **39.** $210°17', 329°43'$

Review for Chapter 2

1. $\dfrac{5}{\sqrt{29}}, -\dfrac{2}{\sqrt{29}}, -\dfrac{5}{2}, -\dfrac{2}{5}, -\dfrac{\sqrt{29}}{2}, \dfrac{\sqrt{29}}{5}$ **3.** $-\dfrac{3}{5}, -\dfrac{4}{5}, \dfrac{3}{4}, \dfrac{4}{3}, -\dfrac{5}{4}, -\dfrac{5}{3}$
5. $44°42'$ **7.** $61.9°$ **9.** $87.8°$ **11.** 0.4245
13. -0.3843 **15.** -6.9273 **17.** -0.9659
19. $234°7'$ and $305°53'$ **21.** $292.6°$ **23.** $322.2°$ **25.** 231.0

Section 3.1

1. $A = 29°45', B = 60°15', c = 8.06$ **3.** $A = 36°52', B = 53°8', b = 16$
5. $B = 80°35', b = 30.15, c = 30.56$ **7.** 143 yd **9.** 5.1 m

Section 3.2

1. 2.24, 63.4°   3. 3.00, 61.9°   5. 5.66, −135°   7. 5, −36.9°

9. 6.24, 106.1°   11. (6.43, 7.66)   13. (11.34, 7.69)

15. (−90.63, 129.43)   17. (−33.32, −27.96)   19. (9.43, −4.40)

21. 25, 36.87°   23. 71.5, 75.83°   25. 0.153, 31.6°   27. 10.6, −41.2°

29. 7.3, 254.1°   31. 35.19, 51.2°   33. 13.36, 73.0°

35. 65.58, 131.4°   37. 395.1, −64.0°

39.

41.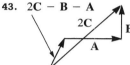

43. 2C − B − A

45.

Section 3.3

1. 76.6 ft/sec, 64.3 ft/sec   3. 36°52′   5. 41.8 ft/sec, 61.4° from vertical

7. 1680 ft/sec, 14.1° below the horizontal   9. 304 mph, 80.5° north of west

11. 197 mph   13. 353.5°, 1 hr 46 min

15. Horizontal 73.3 lb, vertical 15.8 lb   17. 13.46 lb   19. 26.7°

21. 82°49′   23. 33°33′   25. 14°

27. 157.3 lb, 36.8° with the horizontal   29. $X = 52, \theta = 60°$

31. $Z = 224, \theta = 26°34′$ 33. $R = 43.3, Z = 50$

Section 3.4

1. $c = 39.7$ 3. $c = 46.9$ 5. $a = 74.0$ 7. $C = 80°37′$

9. $A = 95°44′$ 11. $B = 118°4′$ 13. $A = 129.7°, B = 19.6°, c = 2.79$

15. $A = 37°48'$, $B = 47°47'$, $c = 195.2$
17. $A = 28°57'$, $B = 46°34'$, $C = 104°29'$ **19.** 3.81 km
21. $30°45'$ **23.** $13°34'$ **25.** $A = 66°35'$, $C = 53°25'$, $\overline{AC} = 3775$ m
27. 346.4 m   **29.** 6.7 mi

Section 3.5

1. $C = 100°$, $b = 14.02$, $c = 18.58$ **3.** $B = 24.5°$, $C = 110.3°$, $c = 11.7$
5. $B = 23.9°$, $C = 120.5°$, $c = 25.9$ **7.** $B = 41°24'$, $C = 17°46'$, $c = 2.35$
9. $B = 28.9°$, $A = 98.1°$, $a = 37.4$ **11.** $A = 44°58'$, $C = 13°2'$, $c = 7.98$
13. $A = 20°$, $a = 17.9$, $c = 49.1$ **15.** $C = 88.6°$, $a = 42.13$, $b = 45.57$
17. 5.72 ft   **19.** 1 hr 31 min   **21.** 4.72 ft
23. Antenna 545 ft, hill 847 ft   **25.** $31.84°$

Section 3.6

1. No solution   **3.** Two solutions   **5.** One solution
7. No solution   **9.** $B = 14°32'$, $C = 15°28'$, $b = 7.53$ **11.** No solution
13. $A = 38°45'$, $B = 113°15'$, $b = 29.4$
    $A = 141°15'$, $B = 10°45'$, $b = 5.96$
15. $B = 41°49'$, $C = 108°11'$, $c = 570$
    $B = 138°11'$, $C = 11°49'$, $c = 122.9$
17. $A = 40°$, $B = 70°$, $a = 68.4$ **19.** $B = 47°48'$, $C = 59°57'$, $c = 0.818$
21. $15 \sin 25° < b < 15$

Section 3.7

1. $C = 45°$, $b = 669.2$, $c = 489.9$ **3.** $A = 60°$, $b = 1.20$, $c = 4.46$
5. No solution   **7.** $A = 32°$, $B = 118°$, $c = 283$ **9.** No solution
11. No solution   **13.** $C = 20°$, $a = 1.759$, $c = 0.695$
15. $C = 50.5°$, $B = 29.5°$, $c = 1.567$ **17.** No solution   **19.** 1.93 mi
21. Ground speed 351 mph, heading $98°42'$

Section 3.8

1. 18.75   **3.** 39.43   **5.** 6   **7.** 0.4479   **9.** 2.239
11. Similar triangles may have unequal areas.   **13.** 85.3 in^2

15. 573.2 ft^2 **17.** 301 m or 499.8 m
19. Other side 5.55 and angles 16.8°, 35.7°, 127.5°; or other side 26.3 and angles
    163.2°, 9.7°, 7.1°   **21.** 122.3 m^2

Review for Chapter 3

1. $B = 58°$, $b = 4.8$, $c = 5.7$ **3.** $A = 45.0°$, $B = 45.0°$, $b = 29$
5. Horizontal 40.1 lb, vertical 29.9 lb   **7.** Magnitude 83.9 lb, angle 59.2°
9. 67.9°   **11.** 472.0 ft   **13.** (13.8, 5.9)
15. $C = 118.9°$, $B = 32.1°$, $a = 15.5$ **17.** $A = 44.3°$, $B = 29.5°$, $C = 106.2°$
19. $A = 148.0°$, $B = 7.8°$, $C = 24.2°$ **21.** $B = 40.2°$, $b = 2175$, $c = 1561$
23. $B = 19.8°$, $C = 135.2°$, $c = 25.0$ **25.** No triangle exists.   **27.** 9.88
29. $A = 24.4°$, $C = 17.6°$, $b = 14.7$, area $= 20.3$ **31.** 90.3 m   **33.** 123 ft
35. 6.5° south of east   **37.** 16.7 ft

Section 4.1

1. A multiple of 2π **3.** $\frac{5}{12}\pi$ **5.** $\frac{8}{3}\pi$ **7.** $-\frac{4}{3}\pi$ **9.** $\frac{19}{36}\pi$
11. $\frac{25}{6}\pi$ **13.** 1.61   **15.** 0.00161   **17.** 4.43   **19.** 0.00785
21. $\frac{1}{4}\pi$ **23.** π **25.** $-\frac{3}{4}\pi$ **27.** $-\frac{1}{2}\pi$ **29.** 229.18°
31. 180°   **33.** π, $-540°$ **35.** -0.53, 5729.58°   **37.** 0, 18,000°
39. 2.8198   **41.** 7.0153   **43.** 1.3476   **45.** -0.4161
47. 2.6937   **49.** 3.6622   **51.** 0.4483   **53.** 5.5224   **55.** $\frac{\sqrt{2}}{2}$
57. $-\sqrt{3}$ **59.** -1 **61.** $\frac{\sqrt{3}}{2}$ **63.** $\sqrt{2}$ **65.** -1
67. $a = 19.3$, $b = 24.1$, $C = 0.97$ rad   **69.** $b = 159$, $c = 273$, $A = 1.28$ radians
73. 4.60

Section 4.2

1. 34.7 cm   **3.** 0.44 rad   **5.** 20.5 ft   **7.** 2.62 ft   **9.** 11.4 in.
11. 4.2 in.   **13.** 1.47 in.   **15.** 57.1°   **17.** 293 in.   **19.** 9.6 in.
21. 107.9 ft^2 **23.** 5.53 in.   **25.** 9.5°, 34.8 ft   **27.** 292°
29. 6.4 ft^3

Section 4.3

1. 12.5 ft/sec   3. 1.72 ft/sec   5. 15.3 ft/sec   7. 0.011 rad/sec
9. 30 rad/sec   11. 628 cm   13. 17.5 ft/sec, 95.2 mi   15. 0.63 in.
17. 16.2 rad/sec   19. $\frac{20}{3}\pi$ ft/sec, $\frac{20}{3}\pi$ rad/sec   21. 858 mph

Review for Chapter 4

1. $\frac{\pi}{5}$ 3. $\frac{11\pi}{18}$ 5. $-\frac{17\pi}{36}$ 7. $10°$ 9. $24923.7°$

11. $-\frac{1}{2}$ 13. $\frac{\sqrt{3}}{3}$ 15. -1 17. -0.9165 19. -0.0715

21. $\frac{5\pi}{3}$ 23. 1.25   25. $\frac{2\pi}{3}$ 27. 4.02   29. 4.25

31. 7.39 or 10.53   33. 6.70 cm   35. 2.62 ft   37. 68.4 cm^2
39. 18.3 rad/sec; 174.3 rpm   41. 44.2 rad/sec

Section 5.1

1. (a) cos 1 = 0.54, sin 1 = 0.84
   (b) cos (−2) = −0.42, sin (−2) = −0.91
   (c) cos 3 = −0.99, sin 3 = 0.14
   (d) cos 10 = −0.84, sin 10 = −.54
   (e) cos 3π = −1, sin 3π = 0
   (f) cos (−4) = −0.65, sin (−4) = 0.76
   (g) cos (−4π) = 1, sin (−4π) = 0
   (h) cos $\frac{1}{3}\pi$ = 0.5, sin $\frac{1}{3}\pi$ = .87
   (i) cos $\frac{1}{3}$ = 0.95, sin $\frac{1}{3}$ = 0.33
   (j) cos $\frac{1}{2}$ = 0.88, sin $\frac{1}{2}$ = 0.48
   (k) cos $\sqrt{7}$ = −0.88, sin $\sqrt{7}$ = 0.48
   (l) cos 5.15 = 0.42, sin 5.15 = −0.91
5. 0   7. 0.5480   9. 0.5   11. 1   13. −1   15. −0.8365

17. 0.2867   19. $\frac{(2n+1)\pi}{2}$ 21. $2n\pi$

23. $-\frac{\pi}{2} < x < \frac{\pi}{2}, \frac{3\pi}{2} < x < \frac{5\pi}{2}$, etc.   25. $x = \frac{(4n+1)\pi}{2}$

27. $x = \frac{(2n+1)\pi}{2}$ 29. 0.985 fps   31. 0.79 sec

1. Domain: all real numbers; range: $[-1, 1]$ 3. Sine function, per. $= \dfrac{2\pi}{3}$

5. Cosine function, per. $= 4\pi$ 7. Even   9. Odd

11.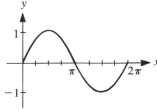

13.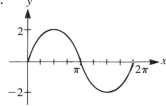

15. per. $= \pi$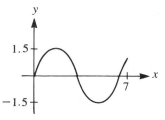

17.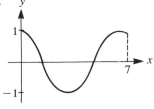

19.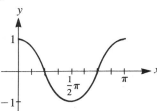

21.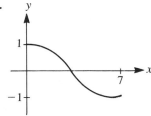

23. per. $= \pi$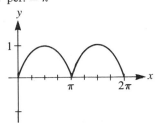

25.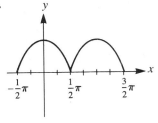

27. Initial vel. $= 1$, $v = 0$ for $t = (2n + 1)(\frac{1}{2}\pi)$. Points where the graph crosses the t-axis are where $v = 0$.

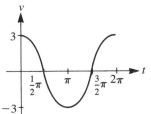

29.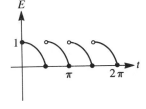

Section 5.3

1. $A = 3$, per. $= 2\pi$

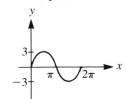

3. $A = 6$, per. $= 2\pi$

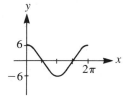

5. $A = 1$, per. $= 3\pi$

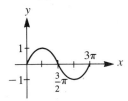

7. $A = 1$, per. $= 2$

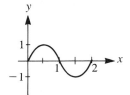

9. $A = \frac{1}{2}$, per. $= \pi$

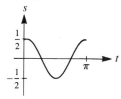

11. $A = 8.2$, per. $= 5\pi$

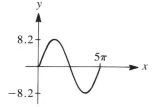

13. $A = 1$, per. $= 5\pi$

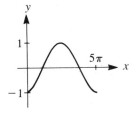

15. $A = \pi$, per. $= \frac{1}{50}\pi$

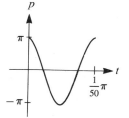

17. $A = 12$, per. $= 10\pi$

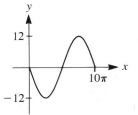

19. $A = \frac{1}{50}$, per. $= \frac{1}{500}$

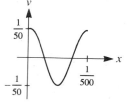

21. $y = \frac{1}{3}\sin\frac{1}{6}\pi x$ **23.** $y = 20\sin\frac{16}{3}\pi x$ **25.** $y = 2.4\sin 6x$

27. $y = \pi \sin \frac{2}{3} x$ **29.** $x = 3.2 \sin \frac{4}{5} \pi t$ **31.**

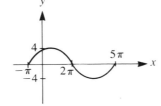

Section 5.4

1. $A = 1$, avg. $= 3$, per. $= \pi$

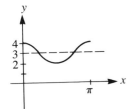

3. $A = 8$, avg. $= 6$, per. $= 2\pi$

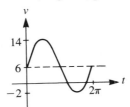

5. $A = 1$, avg. $= -2$, per. $= 4\pi$

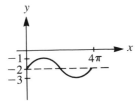

7. $A = 3$, avg. $= 3$, per. $= \frac{2}{3}\pi$

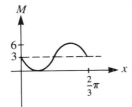

9. $A = 1$, ps $= -\frac{1}{3}\pi$, per. $= 2\pi$

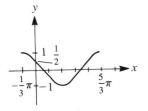

11. $A = 2$, ps $= \pi$, per. $= 4\pi$

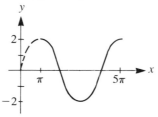

13. $A = 1$, ps $= -\frac{1}{2}\pi$, per. $= \pi$

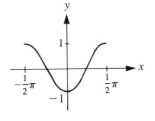

15. $A = 4$, ps $= -\pi$, per. $= 6\pi$

17. $A = 1$, ps $= \frac{1}{4}$, per. $= 2$

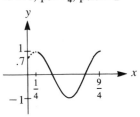

19. $A = 1$, ps $= \frac{1}{4}\pi$, per. $= \frac{1}{2}\pi$, avg. $= 4$

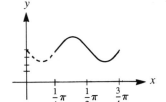

21. $\sin [x + (1 + \pi)]$

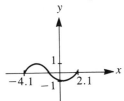

23. $\sin [2\pi x + (\frac{1}{2} + \pi)]$

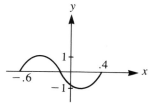

25. $\sin [\pi x + (1 - \frac{1}{2}\pi)]$ **27.** $y = \cos 2(x + \frac{1}{6}\pi)$ **29.** $y = -2 \sin \frac{1}{2}x$

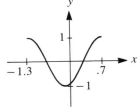

31. They are mirror images in the t-axis.   **33.** They are the same.

35.

37.

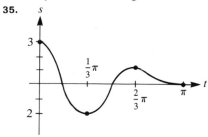

Section 5.5

1. (a) 3.14 sec   (b) -6.24 cm

3. (a) $h = 6 + 7 \cos \frac{1}{5}\pi(t - 2)$

(b)

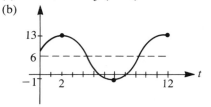

(c) $t = 7.86$ sec

5. (a)

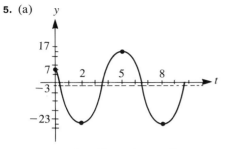

(b) $y = -3 - 20 \cos \frac{1}{3}\pi(t - 2)$

(c) $-16.4,\ 1.2,\ -13$

(d) At $y = 7$

(e) $t = 3.64$ sec

7.

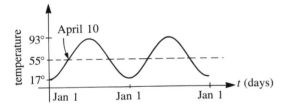

9. Mean temperature $= 55°$

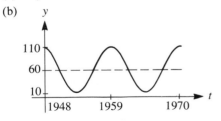

11. (a) 11 years

(b)

(c) $y = 60 + 50 \cos \frac{2}{11}\pi(t - 1948)$

(d) Answers will vary, 52.9

(e) 2003

13. (a) $P = 0.94,\ E = -0.22,\ I = -0.54,$ avg. $= 0.06$

(b) $P = 0.89,\ E = -0.97,\ I = 0.76,$ avg. $= 0.22$

(c) $P = -0.89,\ E = -0.90,\ I = -0.99,$ avg. $= -0.93$

(d) $P = -0.73,\ E = 0.62,\ I = -0.10,$ avg. $= -0.07$

(e) $P = -0.82,\ E = 0.97,\ I = -0.76,$ avg. $= -0.20$

(f) $P = 0.63,\ E = 0.00,\ I = -0.37,$ avg. $= 0.09$

1. per. $= 2\pi$

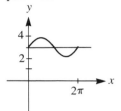

3. per. $= 2\pi$

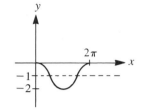

5. per. $= 2\pi$

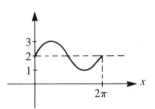

7. Not periodic

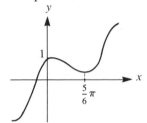

9. Not periodic

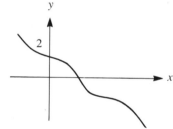

11. Not periodic

13. per. $= 2\pi$

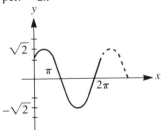

15. per. $= 4\pi$

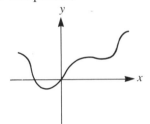

17. per. $= 2\pi$

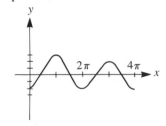
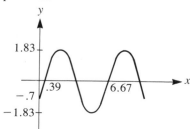

19. Not periodic

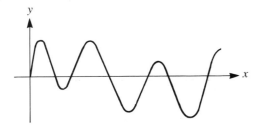

21. They are the same.

23.

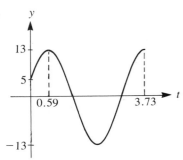

25.

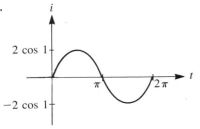

Sections 5.7 and 5.8

1. 2π **3.** 2 **5.** 2 **7.** $\frac{6}{5}\pi$

9. per. $= \frac{1}{2}\pi$, ps $= 0$,
asym. at $x = \frac{1}{4}\pi + \frac{1}{2}n\pi$

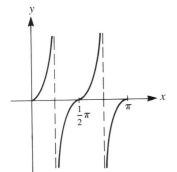

11. per. $= \pi$, ps $= \frac{1}{4}\pi$,
asym. at $x = \frac{1}{4}\pi + n\pi$

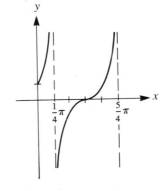

13. per. $= \frac{1}{2}\pi$, ps $= -\frac{1}{6}\pi$,
asym. at $x = \frac{1}{12}\pi + \frac{1}{2}n\pi$

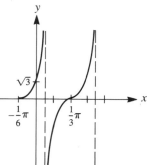

15. per. $= \pi$, ps $= \frac{3}{2}\pi$,
asym. at $x = \frac{1}{2}n\pi$

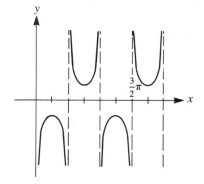

17. per. $= \pi$, ps $= \frac{1}{4}\pi$,
asym. at $x = \frac{3}{4}\pi + n\pi$

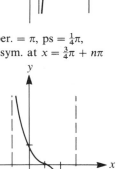

19. $\tan x = -\cot (x + \frac{1}{2}\pi)$

21. They are the same.

23. They are mirror reflections in the y-axis.

25.

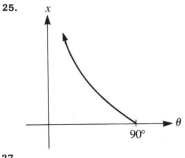

27.

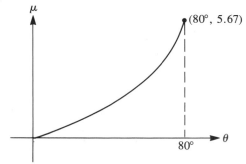

3. Approaches 2

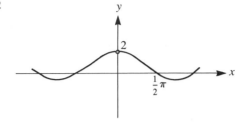

Review for Chapter 5

1. (a) 1.683   (b) 0   (c) 1.683   **3.** $\frac{1}{6}\pi + \frac{1}{3} + \frac{2}{3}n\pi$

5. $A = 1, \text{ps} = \frac{1}{3}\pi, \text{per.} = 2\pi, \text{avg.} = 2$ **7.** $\text{ps} = -\frac{1}{4}\pi, \text{per.} = 2\pi$

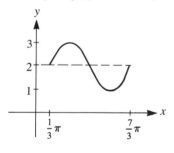

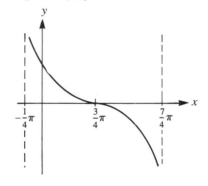

9. $\text{ps} = \pi, \text{per.} = 6\pi, \text{avg.} = 2$ **11.** $A = 9, \text{per.} = \frac{2}{3}\pi$

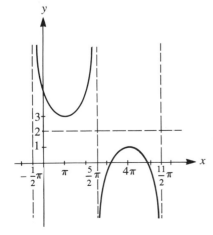

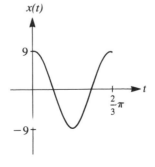

13. $x = -3 + (2n + 1)(\frac{1}{2}\pi)$

15.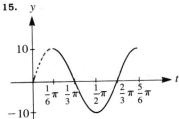

17.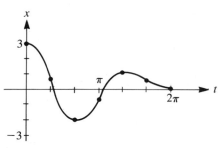

19. -1.13 cm   **21.** $n\pi$, where n is an integer
23. $x = 2 \cos 2\pi t$, $y = 2 \sin 2\pi t$, $(-2, 0)$, $(0, -2)$, $(2, 0)$

Section 6.1

1. $\sec \theta$ **3.** $\sec x$ **5.** $\cot x$ **7.** $\cot x$ **9.** $-\tan^2 x$
11. $\sin x$ **13.** 1 **15.** 1 **17.** $\sec x$ **19.** $\cos x$
21. $a \sec \theta, \dfrac{x}{\sqrt{a^2 + x^2}}$ **23.** $\sin \theta$ **25.** $\sqrt{3} \cos \theta$ **29.** No

Section 6.2

83. Try $t = \pi$.   **85.** Try $x = 0$.   **87.** Try $x = 0$.
89. Try $x = \frac{1}{2}\pi$.   **91.** Try $x = \frac{1}{2}\pi$.   **93.** Try $x = 1$.   **95.** Identity
97. Not an identity. Try $x = 0$.   **99.** Identity
101. Identity   **103.** Identity

Section 6.3

1. $\frac{1}{6}\pi, \frac{5}{6}\pi$

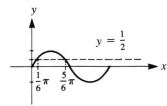

3. $\frac{1}{3}\pi$

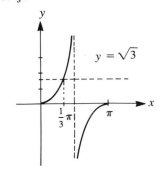

5. $\frac{1}{3}\pi, \frac{2}{3}\pi$

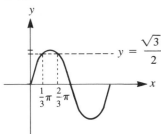

7. $\frac{7}{24}\pi, \frac{11}{24}\pi$

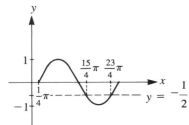

9. $\frac{15}{4}\pi, \frac{23}{4}\pi$

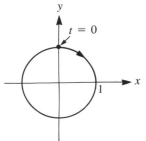

11. $\frac{7}{6}\pi, \frac{11}{6}\pi$

13. 0   **15.** π **17.** $0, \pi, \frac{1}{6}\pi, \frac{5}{6}\pi$ **19.** No solution   **21.** $\frac{1}{4}\pi, \frac{5}{4}\pi$

23. No solution   **25.** $\frac{1}{6}\pi, \frac{5}{6}\pi, \frac{3}{2}\pi$ **27.** $\frac{1}{2}\pi$ **29.** $0, \frac{1}{4}\pi, \pi, \frac{5}{4}\pi$

31. $0, \pi, \frac{1}{4}\pi, \frac{3}{4}\pi, \frac{5}{4}\pi, \frac{7}{4}\pi$ **33.** 0   **35.** $\frac{1}{2}\pi, \pi$ **37.** $\frac{1}{4}\pi, \frac{3}{4}\pi, \frac{5}{4}\pi, \frac{7}{4}\pi$

39. $45°, 135°, 225°, 315°$ **41.** $90°, 270°$ **43.** $90°, 150°, 210°$

45. $0°, 180°, 240°, 300°$ **47.** $0°, 45°, 180°, 225°$ **49.** $30°, 150°$

51. $120°, 300°$ **53.** $0.58, 2.56, 3.72, 5.70$ **55.** $0, 3.14, 3.48, 5.94$

57. $0.875, 2.265, 3.59, 5.83$ **59.** $0.67, 2.47$

61. $\frac{5}{6} + 2n, \frac{11}{6} + 2n$, where n is an integer

63. $-\dfrac{\phi}{\omega_c} + n\pi$, where n is an integer

Section 6.4

1.

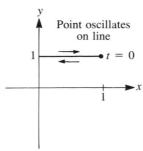

3.

5.

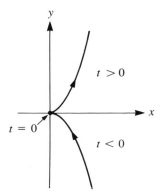

7.

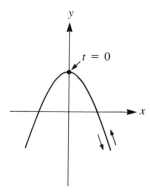

9.

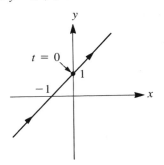

11. $x^2 + y^2 = 1$

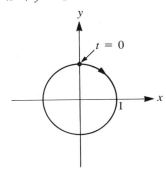

13. $y = x + 1$

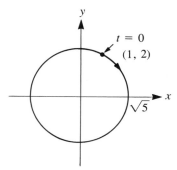

15. $y = x,\ -1 \le x \le 1$

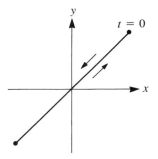

17. $x = \sin^2 y$

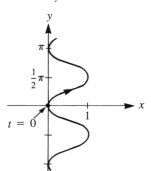

19. $(y - 1) = (x + 3)^2$

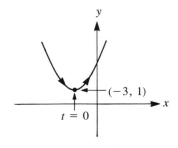

21.

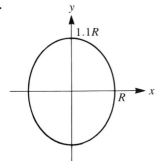

23.

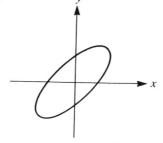

25.

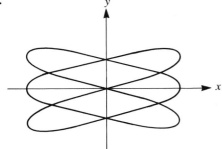

Sections 6.5 and 6.6

1. 0

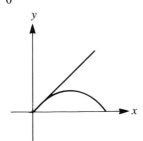

3. 1.2, 3

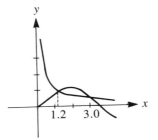

5. 0.8

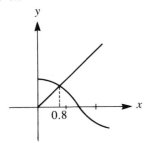

7. 0, 4.67

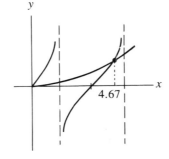

9. 0.7, 2.3

11. $\frac{1}{6}\pi \le x \le \frac{5}{6}\pi$

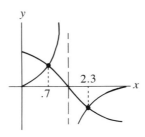

13. $\frac{1}{4}\pi < x \le \frac{3}{4}\pi, \frac{5}{4}\pi \le x \le \frac{7}{4}\pi$

15. $\frac{1}{4}\pi < x < \frac{1}{2}\pi, \pi < x < \frac{5}{4}\pi, \frac{3}{2}\pi < x < 2\pi$

17. $-0.4 < x < 0, x > 0.4$

19. $0 \le x \le \frac{1}{4}\pi, \frac{3}{4}\pi \le x \le \frac{5}{4}\pi, \frac{7}{4}\pi \le x \le 2\pi$

21. $0 < x < 1.16$

Review for Chapter 6

11. $\frac{4}{3}\pi, \frac{5}{3}\pi$

13. $\frac{7}{6}\pi, \frac{11}{6}\pi, \frac{1}{2}\pi$

15. $\frac{1}{4}\pi, \frac{3}{4}\pi, \frac{5}{4}\pi, \frac{7}{4}\pi$

17. $\frac{1}{4}y^2 - x^2 = 1$

19. $y = x$

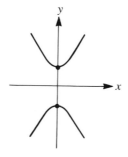

21. 0.88

23. $0 \le x < \frac{1}{4}\pi, \frac{5}{4}\pi < x \le 2\pi$

25.

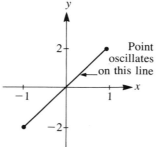

Point oscillates on this line

29. $4 \tan \theta \sec \theta$

31. 0.81

33.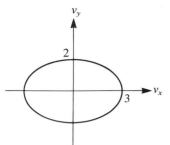

Section 7.1

7. $\dfrac{\sqrt{2}}{4}(1 + \sqrt{3})$ **9.** $\dfrac{-\sqrt{2}}{4}(1 + \sqrt{3})$ **21.** $\cos 8x$ **23.** $\dfrac{\sqrt{3} + \sqrt{8}}{6}$

25. $\frac{3}{5}$ **27.** $\dfrac{5\sqrt{8} + \sqrt{11}}{18}$

29. $A = \sqrt{2}$, ps $= -\frac{1}{4}\pi$ **31.** $A = 2$, ps $= \frac{1}{6}\pi$

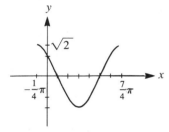   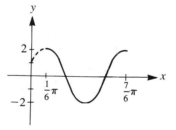

39. $78 \cos 2\pi ft + 78\sqrt{3} \sin 2\pi ft$ **41.** $\frac{12}{13}$ **43.** $y = 3 \cos (2\pi x - ct)$

Section 7.2

1. $\dfrac{\sqrt{2}}{4}(1 + \sqrt{3})$ **3.** $\dfrac{\sqrt{2}}{4}(\sqrt{3} + 1)$ **5.** $\dfrac{\sqrt{3} - 1}{\sqrt{3} + 1}$ **15.** $\tan x$

17. $\tan (x + y + z)$ **19.** $\frac{36}{325}, \frac{36}{323}$

21. $\sqrt{2} \sin (2x + \frac{1}{4}\pi)$ **23.** $2 \sin (\pi x + \frac{1}{6}\pi)$

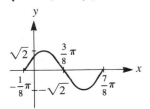   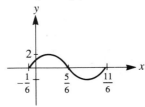

31. $25\sqrt{2}(\sqrt{3}+1)$ **33.** $v(t)=30\sqrt{3}\,\cos 2\pi ft + 30\,\sin 2\pi ft$

35. $\tan\beta = \dfrac{m_2 - m_1}{1 + m_1 m_2}$

Section 7.3

1. $\sin 6x$ **3.** $-\cos 8x$ **5.** $\tan\frac{1}{3}x$

7. max $= \frac{1}{2}$ at $\frac{1}{8}\pi$ **9.** Undefined at $0, \frac{1}{2}\pi, \pi$

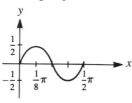

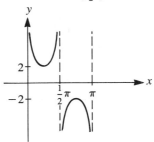

11. Undefined at $x = 0, \frac{1}{2}\pi, \pi$; per. $= \pi$ (same graph as Exercise 9)

13. $\frac{24}{25}, \frac{7}{25}, \frac{24}{7}$ **15.** $\frac{336}{625}, \frac{527}{625}, \frac{336}{527}$ **17.** $\dfrac{\sqrt{2-\sqrt{2}}}{2}$ **19.** $-\dfrac{\sqrt{2}}{2+\sqrt{2}}$

21. $\dfrac{\sqrt{2+\sqrt{3}}}{2}$ **43.** $\frac{1}{5}$ **45.** $\dfrac{7\sqrt{2}}{10}$ **47.** $\frac{1}{4}\pi, \frac{5}{4}\pi$ **49.** $0, \pi, 2\pi$

51. $\frac{3}{8}\pi, \frac{7}{8}\pi, \frac{11}{8}\pi, \frac{15}{8}\pi$ **53.** $\frac{1}{6}\pi, \frac{1}{3}\pi, \frac{2}{3}\pi, \frac{5}{6}\pi, \frac{4}{3}\pi, \frac{7}{6}\pi, \frac{5}{3}\pi, \frac{11}{6}\pi$

65. $\cos\theta = \sqrt{\dfrac{\sqrt{5}+1}{2\sqrt{5}}}$, $\sin\theta = \sqrt{\dfrac{\sqrt{5}-1}{2\sqrt{5}}}$ **67.** $R = 8v_0^2 \sin 2\theta$

Section 7.4

1. $2\sin 2\theta \cos\theta$ **3.** $2\sin 5x \cos 3x$ **5.** $-2\sin 40° \sin 10°$

7. $2\cos\frac{1}{2}\sin\frac{1}{4}$ **9.** $\frac{1}{2}[\sin(\frac{3}{2}x) - \sin(\frac{1}{2}x)]$ **11.** $\frac{1}{2}(\cos 8x + \cos 4x)$

19. $\dfrac{-2\sin\frac{1}{2}(2x+h)\sin\frac{1}{2}h}{h}$ **21.** $0, \frac{1}{6}\pi, \frac{1}{2}\pi, \frac{5}{6}\pi, \pi$ **23.** $0, \frac{2}{5}\pi, \frac{4}{5}\pi$

25. $A = 2\cos 1$, per. $= \frac{2}{3}\pi$ **31.** $2\cos x \cos\frac{1}{2}x$

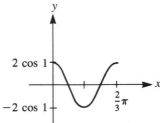

1. $\dfrac{6 - 4\sqrt{5}}{15}$ **3.** $-\dfrac{24}{7}$

21. $A = 13$, per. $= \pi$, ps $= -0.588$

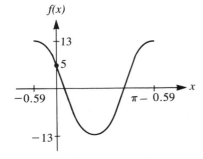

23. $A = \sqrt{2}$, per. $= \frac{2}{3}\pi$, ps $= -\frac{1}{12}\pi$

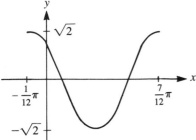

25. $A = 2$, per. $= 2\pi$, ps $= \frac{1}{3}\pi$ **27.** $-\sqrt{\dfrac{3 - \sqrt{8}}{6}}$ **29.** $\frac{24}{25}$

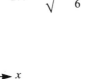

31. $\sin\theta = \dfrac{2\sqrt{5}}{5}$, $\cos\theta = \dfrac{\sqrt{5}}{5}$ **33.** $2 \sin \frac{3}{2}x \cos \frac{1}{2}x$

Section 8.1

1. $\frac{3}{2}$ **3.** $\frac{5}{2}$ **5.** 3; None   **7.** $\{(7, 3), (9, 5), (3, 7), (5, 9)\}$
9. No inverse function   **11.** $\{(3, -2), (4, -1), (0, 0)\}$

13. No inverse function   **15.** $y = x + 3$ **17.** $y = \dfrac{1 - x}{x}$

19. No inverse function   **21.** $y = \dfrac{1 + x}{1 - x}$ **23.** Inverse function

25. Inverse function   **27.** Inverse functions
29. Not inverse functions   **31.** Inverse functions
33. Inverse functions

Section 8.2

1. $\frac{1}{6}\pi$ **3.** $\frac{1}{4}\pi$ **5.** $\frac{5}{6}\pi$ **7.** $\frac{2}{3}\pi$ **9.** 0   **11.** $-\frac{1}{3}\pi$ **13.** $\frac{12}{13}$

15. $\dfrac{2}{\sqrt{5}}$ **17.** $\dfrac{\sqrt{15}}{4}$ **19.** $\frac{7}{8}$ **21.** $\dfrac{1}{\sqrt{82}}$ **23.** $\dfrac{2\sqrt{3} - 1}{2 + \sqrt{3}}$

25. $-\frac{4}{3}$ **27.** $\dfrac{x}{(1 - x^2)^{1/2}}$ **29.** $y\sqrt{1 - x^2} + x\sqrt{1 - y^2}$ **31.** 0.9711

33. No solution   **35.** 2.5722   **37.** -4.2556 **39.** 0.7800

49.

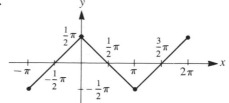

53. $\theta = \frac{1}{2}$ Arcsin $\dfrac{R}{16v_0^2}$

55. 1.211   **57.** 1.397   **59.** 1.407   **61.** 0.840

Section 8.3

1.

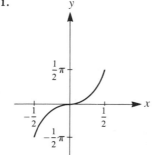

3.

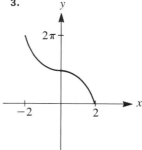

5.

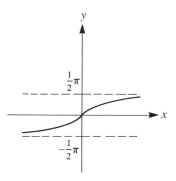

7.

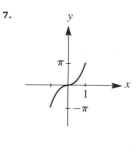

9.

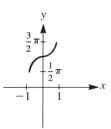

11.

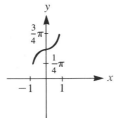

13.

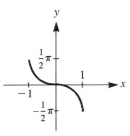

15.

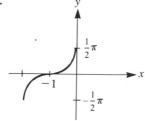

17.

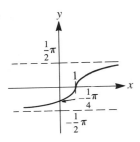

19.

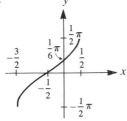

21.

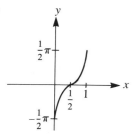

23.

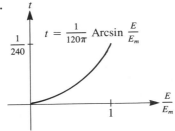

$t = \dfrac{1}{120\pi} \text{ Arcsin } \dfrac{E}{E_m}$

Review for Chapter 8

1. 4   **3.** $y = \dfrac{x-5}{2}$ **5.** $y = \frac{1}{2}[5 + \text{Arcsin } (x - 3)]$ **7.** $\frac{1}{3}\pi$

9. $\dfrac{-2}{\sqrt{21}}$ **11.** $-\frac{24}{25}$ **13.** $\dfrac{-(1 + \sqrt{120})}{12}$ **15.** 1

17.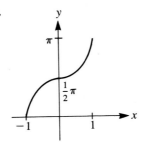

19.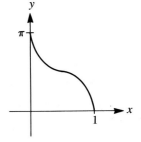

21.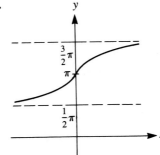

23.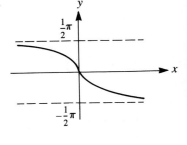

25.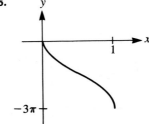

1.

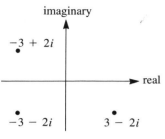

3.

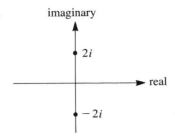

5.

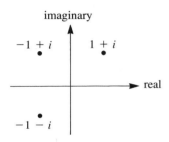

7.

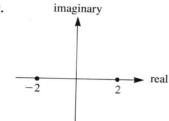

9.

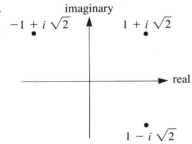

11. $7 + 5i$

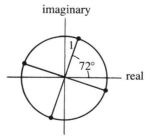

13. $-2 + 3i$

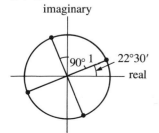

15. 4

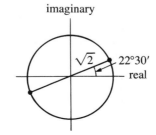

17. $3 + i$

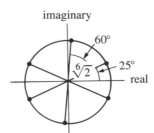

19. $-7 + 22i$

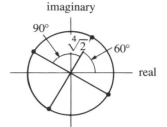

21. 26   **23.** $13 + 8\sqrt{3}i$ **25.** $18 + 24i$ **27.** $\dfrac{5 - i}{2}$ **29.** $\dfrac{6 + 9i}{13}$

31. $\dfrac{-i}{5}$ **33.** $\dfrac{(-4 + 3\sqrt{2}) - (12 + \sqrt{2})i}{18}$ **37.** $5 - 2i$

39.

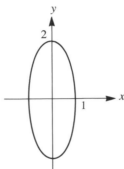

Section 9.2

1. $2 \text{ cis } (-60°)$ **3.** $3 \text{ cis } 41°49'$ **5.** $9 \text{ cis } 0°$ **7.** $5 \text{ cis } (-53°8')$

9. $\sqrt{61} \text{ cis } (-50°12')$ **11.** $\sqrt{3} + i$ **13.** $\dfrac{5\sqrt{2}}{2}(-1 + i)$

15. $\dfrac{-3 - \sqrt{3}i}{2}$ **17.** $\dfrac{3 - 3\sqrt{3}i}{2}$ **19.** $9.397 + 3.420i$ **21.** $12 \text{ cis } 90°$

23. $2 \text{ cis } 330°$ **25.** $20 \text{ cis } 135°$ **27.** $10 \text{ cis } 23°8'$ **29.** $5 \text{ cis } (-60°)$

31. $2 \text{ cis } 7°30'$ **33.** $\dfrac{\sqrt{2}}{2} \text{ cis } (-75°)$ **35.** $\dfrac{4}{\sqrt{2}} \text{ cis } (-45°)$

39. $p = 1000 \text{ cis } (-1.2t)$; $p = 1000 \text{ cis } (-1.2)$; $1000 \cos (-1.2)$, $1000 \sin (-1.2)$

Section 9.3

1. $8 \text{ cis } 0°$ **3.** $9 \text{ cis } 240°$ **5.** $128\sqrt{2} \text{ cis } 315°$

7. $128 \text{ cis } 330°$ **9.** $841 \text{ cis } 272°48'$

11. 1 cis 0° = 1
1 cis 72° = .3090 + .9511i
1 cis 144° = −.8090 + .5878i
1 cis 216° = −.8090 − .5878i
1 cis 288° = .3090 − .9511i

13. 1 cis 22°30′ = .9239 + .3827i
1 cis 112°30′ = −.3827 + .9239i
1 cis 202°30′ = −.9239 − .3827i
1 cis 292°30′ = .3827 − .9239i

15. $\sqrt[4]{2}$ cis 22°30′
$\sqrt[4]{2}$ cis 202°30′

17. $\sqrt[6]{2}$ cis $(25° + M \cdot 60°)$, $M = 0, 1, 2, 3, 4, 5$

19. $\sqrt[4]{2}\left(\dfrac{\sqrt{3}}{2} + \dfrac{i}{2}\right)$

$\sqrt[4]{2}\left(\dfrac{-1}{2} + \dfrac{i\sqrt{3}}{2}\right)$

$\sqrt[4]{2}\left(-\dfrac{\sqrt{3}}{2} - \dfrac{i}{2}\right)$

$\sqrt[4]{2}\left(\dfrac{1}{2} - \dfrac{i\sqrt{3}}{2}\right)$

23. 4 cis 60° = $2 + 2\sqrt{3}\,i$
4 cis 180° = −4
4 cis 300° = $2 - 2\sqrt{3}\,i$

25. $\dfrac{1 \pm i}{\sqrt{2}}, \dfrac{-1 \pm i}{\sqrt{2}}$

Section 9.4

1.

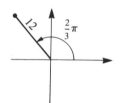

3.

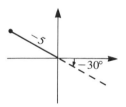

5.

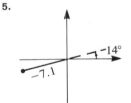

7.

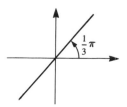

9.

11. $r(2 \cos \theta + 3 \sin \theta) = 6$

13. $r = 4 \cos \theta$

15. $r^2(\cos^2 \theta + 4 \sin^2 \theta) = 4$

17. $r \cos^2 \theta = 4 \sin \theta$

19. $x^2 + y^2 = 25$

21. $x^2 + y^2 = 10y$

23. $x^2 + y^2 = (x^2 + y^2)^{1/2} + 2y$

25. $y^2 = 25 - 10x$

27.

29.

31.

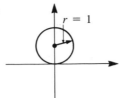

33.

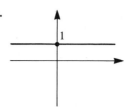

35.

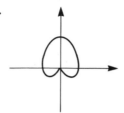

37.

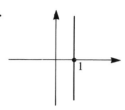

39.

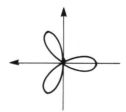

41.

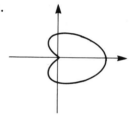

43. $3x^2 + 4y^2 - 2x - 1 = 0$

Review for Chapter 9

1. $9 - 3i$ **3.** 5 **5.** $8 - 6i$ **7.** 0 **9.** 16 **11.** $\dfrac{1 - 3i}{-2}$

13. i **15.** $\frac{19}{2} - \frac{9}{2}i$ **17.** $1 \text{ cis } 90°$ **19.** $\sqrt{2}(\text{cis } 45°)$

21. $\sqrt{2}(1 + i)$ **23.** $4(\cos 20° - i \sin 20°) = 3.8 - 1.4i$ **25.** $\frac{27}{2}(\sqrt{3} + i)$

27. $\tan \theta = 2$ **29.** $r^2 = \dfrac{1}{4 \cos^2 \theta + \sin^2 \theta}$

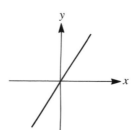

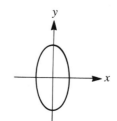

31. $x^2 + y^2 = 4$

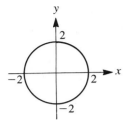

33. $x^2 + 4y = 4$

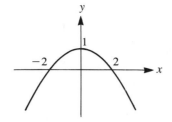

35. $x^2 + y^2 - 3x = 0$

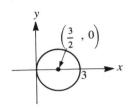

37. $(x^2 + y^2)^{3/2} = 2xy$

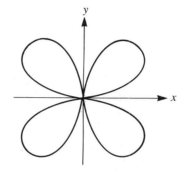

39. $x^2 + y^2 = 1$

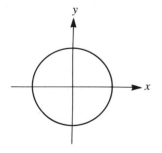

41. $2^{3/2}$ cis $(\frac{3}{4}\pi)$ **43.** 32 cis $(\frac{1}{6}\pi)$

45. 10^5 cis 3.06 **47.** 1 cis $(\frac{1}{5}\pi)$, 1 cis $(\frac{3}{5}\pi)$, 1 cis (π), 1 cis $(\frac{7}{5}\pi)$, 1 cis $(\frac{9}{5}\pi)$

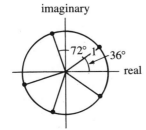

49. $2^{1/6}$ cis $(\frac{1}{12}\pi)$, $2^{1/6}$ cis $(\frac{3}{4}\pi)$, $2^{1/6}$ cis $(\frac{17}{12}\pi)$ **51.** $40 + 25i$ **53.** $\dfrac{19 - 7i}{10}$

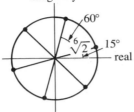

Section 10.1

1. $\log_2 x = 3$ **3.** $\log_5 M = -3$ **5.** $\log_7 L = 2$ **7.** $b = 2$
9. $b = 10$ **11.** $x = 10,000$ **13.** $x = 10$ **15.** $x = 4$
17. $x = 9$ **19.** $x = 2$ **21.** $x = a$ **23.** $x = 8$ **25.** $x = 36$
27. $x = 6$ **29.** (a) 2   (b) -3 (c) 4   **31.** 2 days

Section 10.2

1. 9   **3.** $\frac{1}{2}$ **5.** 6   **7.** 24   **9.** 0.1761   **11.** 1.0791
13. 1.9542   **15.** 0.3495   **17.** 3.3801   **19.** -2.8539

21. $\log_2 x$ **23.** 0   **25.** $\log \dfrac{5t(t^2 - 4)^2}{\sqrt{t + 3}}$ **27.** $\log \dfrac{u^3}{(u + 1)^2(u - 1)^5}$

31. The same for $x > 0$. For $x < 0$, $\log x^2$ is defined, $2 \log x$ is not.
33. Both are -2.
35. (a) Multiply each ordinate value by p.   (b) Move curve up $\log p$ units.
(c) Move curve p units to the left.   (d) Move curve down $\log p$ units.

Section 10.3

1. -1.553 **3.** $x = 0$, $x = \log 2 \approx 0.301$ **5.** 0.5850   **7.** 13.97

9. $x > \dfrac{\ln 2.5}{\ln 2} \approx 1.32$ **11.** $x = 10$ **13.** $x = 1, 10^4$

15. $x = 1$ **17.** $x = 5$ **19.** $x = 2 + \sqrt{5} = 4.236$
$\quad\ x = 10^{\sqrt{3}} = 53.96$
$\quad\ x = 10^{-\sqrt{3}} = 0.0185$

21. $x = 6$ **23.** $x = \sqrt{6}$ **25.** $x > \frac{1}{9}$

27. $x = \ln(2 + \sqrt{3}) = 1.317$
$x = \ln(2 - \sqrt{3}) = -1.317$
29. The logarithm is a one-to-one function. **31.** $k = 0.02476$
33. 173.3 days **35.** 52.5 hr

Review for Chapter 10

1. $x = 1000$ **3.** $x = 4$ **5.** $x = 2$ **7.** $x = 10$ **9.** 4

11. 1.5 **13.** 21 **15.** 3 log x **17.** $\log_2(2x^4)$

19. $\log_5\left(\dfrac{x^3}{2x-3}\right)$ **21.** $x = 1.322$ **23.** $x = 3.819$ **25.** $x = 10^{10}$

27. $x = 2$ **29.** $x = 5$ **31.** 2.77 years

33. 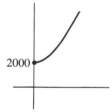 **35.** 1.77 mg

Index

Abscissa, 2
Acute angle, 16
Addition
  of complex numbers, 251
  of ordinates, 164
  of vectors, 74
Adjacent side, 23, 62
Ahmes, 1
Allowable operation, 189
Ambiguous case, 87, 100
Amplitude, 144
Analytical trigonometry, 129
Angle, 15
  acute, 16
  bearing, 79, 99
  complementary, 18, 43
  coterminal, 16, 47
  course, 71
  of depression, 37
  of elevation, 33
  negative, 17
  obtuse, 16
  positive, 17
  quadrantal, 20, 44
  reference, 44
  right, 16
  straight, 16
  supplementary, 18
Angular velocity, 121
Arccos, 238
Arc length, 117
Arcsin, 237
Arctan, 238

Area formulas, 105
Area of a sector, 118
Argument
  of a complex number, 255
  of a function, 8
Asymptotes, vertical, 166
Average angle method, 221
Average value, 149
Axes, 2, 254

Bearing, 79, 99
Beats, 223
Biorhythms, 159
Boundedness, 136

Calculator, 49, 112
Cartesian coordinate system, 2
Circular functions, 131
Cofunctions, 43
Complementary angles, 18
  functions of, 43
Complex numbers
  addition of, 251
  definition of, 249
  graphical representation of, 250
  multiplication of, 251, 256
  polar representation of, 254
  power of, 259
  quotient of, 252
  roots of, 259
Complex plane, 250

Components of a vector, 73
Congruent triangle, 104
Conjugate, 252
Coordinate, 2
Coordinate axes, 2
Coordinate system, 2
Cosecant
    definition of, 32, 62, 131
    graph of, 168
Cosine
    definition of, 32, 62, 131, 146
    of double angle, 214
    graph of, 140
    of half angle, 215
    inverse, 238
    of sum and difference, 204
Cosines, law of, 88
Cotangent
    definition of, 32, 62, 131
    graph of, 166
Coterminal angles, 16, 47
Course, 79
Cycle, 141

Degree, 16, 24, 282
DeMoivre's Theorem, 259
Dependent variable, 7
Differences of sine and cosine, 212
Distance formula, 5
Domain of function, 7, 229
Double-angle formulas, 214

Elevation angle, 33
Equations
    equivalent, 189
    parametric, 194
    trigonometric, 189
Equiangular triangle, 23
Equilateral triangle, 23
Equilibrium, 84
Equivalent equations, 189
Erathosthenes, 26
Euler, 1
Euler's Identities, 258
Even functions, 137
Exponential equations, 274

Force analysis, 82
Frequency, 141
Full-wave rectified sine wave, 143
Function, 6, 229
    argument of, 8

bounded, 136
even, 137
inverse, 232, 238
odd, 137
one-to-one, 231
unbounded, 166
Fundamental identities, 175
Fundamental inequality, 170
Fundamental period, 135
Fundamental relations, 38

Geodesic dome, 92
Graph, 3
    of a function, 9
    of inverse trigonometric
        functions, 243
    of a polar equation, 262
    of secant and cosecant, 168
    of sine and cosine, 140
    of tangent and cotangent, 166
Graphical solutions, 198

Half-angle formulas, 215
Half-wave rectified sine wave, 143
Harmonic motion, 154
Heading, 79
Hipparchus, 1
Historical notes
    Erathosthenes, 26
    Fuller, Buckminster, 92
    sine, 33
Horizontal components of a
    vector, 72
Horizontal translation, 149
Hypotenuse, 23, 62

Identity, 180
Imaginary numbers, 249
Imaginary part of
    complex number, 250
Independent variable, 7
Inequality, trigonometric, 198
Initial side, 15, 282
Interpolation, 54, 114, 262
Inverse, 232
Inverse trigonometric
    functions, 238
    graph of, 243
Isosceles triangle, 23

Law of cosines, 88

Law of sines, 94
Length of arc, 117
Line segment, 281
Linear interpolation, 54, 114
Linear velocity, 122
Lissajous figure, 196
Logarithm
    definition of, 269
    graph of, 270
    properties of, 8, 271
Lunar lander, 84

Magnitude
    of a complex number, 255
    of a vector, 72
Maximum value, 144
Mean value, 149
Minute, 18
Modulus of complex number, 255
Muller, Johann, 1
Multiplication of complex
        numbers, 251, 256

Navigation, 79
Negative angle, 17

Oblique triangles, 23
    solution of, 87
Obtuse angle, 16
Odd functions 137
One-to-one function, 231
Opposite side, 23, 62
Ordered pair, 3
Ordinate, 2
Oriented curve, 195
Origin, 2

Parallelogram rule, 74
Parametric equations, 194
Peak value, 144
Period, 135
Phase shift, 150
Polar coordinates, 254
Polar equation, graph of, 262
Polar representation of a complex
        number, 254
Pole, 254
Positive angle, 17
Predator-prey problem, 157
Principal values, table of, 236
Product formulas, 223

Product of complex numbers, 251, 256
Protractor, 20
Pythagorean relation, 38, 176
Pythagorean Theorem, 4, 23

Quadrant, 2
Quadrantal angle, 20, 44
Quotient of complex numbers, 252
Quotient relations, 176

Radian, 111
Range of a function, 7, 229
Real part of a complex number, 250
Reciprocal relationships, 38, 175
Rectangular coordinate system, 2
Rectangular form, 251
Rectified sine wave, 143
Reference angle, 44
Relation, 7, 230
    inverse, 215
Resultant, 74
Right angle, 16
Right triangle, 23
    solution of, 61
Roots of a number, 259
Rounding off, 12

Scalar, 72
Scalar multiplication, 74
Secant
    definition of, 32, 62, 131
    graph of, 168
Second, 18
Sector, 118
Sides of an angle, 15
    adjacent, 23, 62
    opposite, 23, 62
Significant digits, 11
Similar triangles, 24, 32
Simple harmonic motion, 154
Sine
    definition of, 32, 62, 131
    of double angle, 214
    graph of, 140, 146
    of half angle, 215
    inverse, 238
    of sum and difference, 210
Sines, law of, 94
Sinusoid, 140
Sinusoidal modeling, 154
Solar collector, 68

Solution, 189
  graphical, 198
  of oblique triangles, 87
  of right triangles, 61
Space sciences, 65, 83
Standard position
  of an angle, 20
  of a vector, 72
Straight angle, 16, 282
Sum and difference formulas, 212
Sum and product formulas, 221
Supplementary angle, 18
Symmetry, 138

Tangent
  definition of, 32, 62, 131
  of double angle, 214
  graph of, 166
  of half angle, 216
  inverse, 238
  of sum and difference, 211
Temperature variation, 158
Terminal side, 15, 282
Theorem of Pythagoras, 4, 23
Transit, 21
Translation, 149
Triangle, 23
  oblique, solution of, 87
  right, solution of, 61
  similar, 24, 32
Trigonometric equations, 189
  graphical solution of, 198
Trigonometric expression, 175
Trigonometric functions, 32, 62, 131
  inverse, 235
  table of, 52, 113

at 30°, 45°, 60°, 42
  values of, 49
  variation of, 138
Trigonometric inequality, 198
Trigonometric tables, 52, 113
Trigonometry
  analytical, 129
  right triangle, 62
Truncation, 13

Unboundedness, 166
Unit circle, 130
Unit vector, 74

Variables, 6
  dependent, 7
  independent, 7
Vector, 72
  addition of, 74
  scalar multiplication of, 74
Velocity
  angular, 121
  linear, 122
Vertex, 15, 282
Vertical asymptotes, 166
Vertical components, 73
Vertical translation, 149

x-axis, 2
x-coordinate, 2

y-axis, 2
y-coordinate, 2

Trigonometric Identities

(1) $\sin A = \dfrac{1}{\csc A}$

(2) $\cos A = \dfrac{1}{\sec A}$

(3) $\tan A = \dfrac{1}{\cot A}$

(4) $\tan A = \dfrac{\sin A}{\cos A}$

(5) $\cot A = \dfrac{\cos A}{\sin A}$

(6) $\sin^2 A + \cos^2 A = 1$

(7) $1 + \tan^2 A = \sec^2 A$

(8) $1 + \cot^2 A = \csc^2 A$

(9) $\sin(A + B) = \sin A \cos B + \cos A \sin B$

(10) $\sin(A - B) = \sin A \cos B - \cos A \sin B$

(11) $\cos(A + B) = \cos A \cos B - \sin A \sin B$

(12) $\cos(A - B) = \cos A \cos B + \sin A \sin B$

(13) $\tan(A + B) = \dfrac{\tan A + \tan B}{1 - \tan A \tan B}$

(14) $\tan(A - B) = \dfrac{\tan A - \tan B}{1 + \tan A \tan B}$